Applied Chemistry at Protein Interfaces

Robert E. Baier, *Chairman*

A symposium sponsored by the Division
of Colloid and Surface Chemistry
at the 166th meeting of the American
Chemical Society, Chicago, Ill.,
August 29–31, 1973.

ADVANCES IN CHEMISTRY SERIES **145**

AMERICAN CHEMICAL SOCIETY

WASHINGTON, D. C. **1975**

Library of Congress CIP Data

Applied chemistry at protein interfaces.
(Advances in chemistry series; 145. ISSN 0065-2393)

Includes bibliographical references and index.

1. Surface chemistry—Congresses. 2. Proteins—Congresses.

I. Baier, Robert E., 1939–   II. American Chemical
Society. III. American Chemical Society. Division of
Colloid and Surface Chemistry. IV. Series: Advances in
chemistry series; 145.

QD1.A355 no. 145 (QD506.A1) 540'.8s (541'.3453)
75-25969
ISBN 0-8412-0215-X   ADCSAJ 145 1-399 (1975)

Advances in Chemistry Series

Robert F. Gould, *Editor*

07010

FOREWORD

ADVANCES IN CHEMISTRY SERIES was founded in 1949 by the
American Chemical Society as an outlet for symposia and collections of data in special areas of topical interest that could
not be accommodated in the Society's journals. It provides a
medium for symposia that would otherwise be fragmented,
their papers distributed among several journals or not published at all. Papers are refereed critically according to ACS
editorial standards and receive the careful attention and processing characteristic of ACS publications. Papers published
in ADVANCES IN CHEMISTRY SERIES are original contributions
not published elsewhere in whole or major part and include
reports of research as well as reviews since symposia may
embrace both types of presentation.

CONTENTS

v

PROTEINS AT GAS/LIQUID INTERFACES

PREFACE

This is a missionary book, bringing the message that many seemingly diverse fields are but a single field when attention is focused sharply enough at the various phase boundaries. A common feature often found is that the boundary interactions are dominated by interfacial layers of proteins. Thus, advances in applications of chemistry at protein interfaces will be achieved more rapidly if individual gains from certain isolated disciplines, all of which involve protein interfacial phenomena, can be integrated. For example, this volume demonstrates that most fields of biological adhesion are unified at the fundamental level where one must deal with attachment of single cells or small multicellular aggregates under moist, saline, and biochemically active conditions. The events of maritime fouling, thrombus formation, and dental plaque attachment share marked similarities in the established requirement that all solid surfaces first acquire a conditioning layer of protein which mediates the adhesive interactions with arriving cells. At the electron microscopic level, specimens from the biological milieux of sea water, blood, or saliva can seldom be differentiated. Apparently insurmountable difficulties in some fields have already been overcome in other phases of protein surface science.

This book grew from a perceived need for cross fertilization of these artificially separated disciplines. Chapters were chosen from a symposium entitled "Proteins As and At Substrates," from a concurrent symposium entitled "Biomedical Application of Polymers," and from invited contributions of experts in fields not popularly identified as interfacial chemistry. The authors were asked to provide manuscripts which might stimulate other contributors to this same volume as well as the much broader, interdisciplinary audience whose specific problems are encompassed by one or more of these contributions. Numerous popular aspects of the subject of surface chemistry of proteins have been purposely excluded. The papers gathered here were selected, instead, to exemplify those diverse under-represented fields which also share a common concern for protein interfacial chemistry. Sixteen different institutions are represented, 5 academic, 4 medical, 4 governmental, and 3 industrial. Our single regret is that major industrial concerns which can expect to benefit greatly from the knowledge evident in this volume declined on the grounds of proprietary interest to allow contributions from their outstanding scientific staffs to be included here. Thus, manuscripts on the surface chemistry of hair, of pharmaceutical preparations, and of photographic gelatin are regrettably absent.

ROBERT E. BAIER

Calspan Corp.
Buffalo, N.Y.
January 1975

Applied Chemistry at Protein Interfaces

ROBERT E. BAIER

Calspan Corp., Environmental Systems Department, Buffalo, N. Y. 14221

Some of the areas where interfacial protein layers dominate the boundary chemistry are reviewed, and we introduce some nondestructive analytical methods which can be used simultaneously and/or sequentially to detect and characterize the microscopic amounts of matter at protein or other substrates which spontaneously acquire protein conditioning films. Examples include collagen and gelatin, synthetic polypeptides, nylons, and the biomedically important surfaces of vessel grafts, skin, tissue, and blood. The importance of prerequisite adsorbed films of proteins during thrombus formation, cell adhesion, use of intrauterine contraceptives, development of dental adhesives, and prevention of maritime fouling is discussed. Specifics of protein adsorption at solid/liquid and gas/liquid interfaces are compared.

Numerous surface physicochemical analytical techniques, including internal reflection IR spectrometry, ellipsometry, and determination of critical surface tension and contact potential values, reveal a common interfacial chemistry among seemingly unrelated phenomena. That is, many interfaces are dominated by proteins, either as the original substrates at or within which the key events occur or as the first spontaneously acquired conditioning layers which are prerequisites for subsequent events. Proteins as substrates dominate the surface chemistry of collagenous biomaterials, photographic emulsions, skin, and hair. The interfacial composition and organization of proteins determine the barrier properties of skin and its receptivity to cosmetics and medications. The rapid and often irreversible adsorption of pure protein or glycoprotein constituents from complex media precedes microscopically detectable adhesion of formed (cellular or larval) elements in all known circumstances. All surfaces proposed for blood-compatible implant materials, for example, acquire such a conditioning film as do all structural and engineering materials immersed in the sea, or placed in tissue culture or into the oral or uterine cavities. Protein films are spontaneously formed at and eliminated from most gas/liquid interfaces in nature, and the bubble-stripping of such layers from the sea has been implicated in oceanographic/meteorologic phenomena. Once the common features of interfacial chemistry have been recognized, significant cross-fertilization of the fields mentioned should be stimulated.

Some of the best examples of the dominance of functional interfaces by biological macromolecules occur in the research area of "bioadhesion." This area includes the use of dental restoratives; the use of polymeric surgical adhesives which replace mechanical links such as staples and sutures; the development of prosthetic implants which replace, improve, or supplement almost every part of the human body; the multiplication of extracorporeal circuits for medical treatment in artificial kidney centers, coronary care units, and home dialysis; the study of blood clotting reactions as they are induced by contact with nonphysiologic surfaces ranging from the struts of an implanted heart valve to the transected ends of a normal blood vessel which initiate wound healing; and the various marine fouling events which encrust ships with barnacles and tube worms and which slow oil-carrying supertankers by algal adhesion at the highly stressed water line with large accumulations of seagrass. The common interfacial features of these problems are often overlooked as researchers focus largely on the volume phase effects and the fluid dynamical effects involved.

Analytical Methods and Materials

Figure 1 shows the physicochemical surface methods used extensively in our laboratory to assess the interfacial structure and properties of predominantly protein substrates like skin, collagen, and living cell surfaces and also to assess the initial sequence of events at clean solid substrates upon their exposure to blood, saliva, and sea water.

The molecular structure of the film adsorbed on a substrate such as germanium, silicon, or various common IR-transmitting salts [either before or after their surfaces were modified by standard techniques such as monolayer formation (1, 2)], is readily deduced by the internal reflection technique which has been described (3). When the substrate is a material of high reflectivity and high intrinsic refractive index such as germanium (which is used in most of our experiments), film thickness and refractive index may be determined nondestructively by ellipsometric techniques (4). A third nondestructive and noncontacting technique, which is easily applied to thin film samples on germanium or any conductive substrate, is the determination of the contact potential from vibrating reed electrometer studies using experimental apparatus designed by Bewig and Zisman (5, 6).

A major technique used in all our studies is contact angle data taken according to the methods of Zisman (7) to derive the critical surface tension. All of these techniques can be applied simultaneously or sequentially to the same interfacial area, with a sensitivity for all of the methods great enough to detect layers less than 10 A and representing less than 0.1 μg of material on a substrate surface area similar to a glass microscope slide. For any interfacial film, easily adapted and well-calibrated techniques can determine first the molecular composition of the material, then its degree of organization (from ellipsometric deductions of its thickness and refractive index), its degree of electrical asymmetry (from dipole presence and orientation deduced *via* contact potential data), and finally which of the molecular clusters present within the film (identified by internal reflection IR) must be outermost and in control of interactions with the external environment (using the important contact angle-determined critical surface tension values).

Proteins as Substrates

Contact angle methodology for probing the surface chemical function and architecture of protein-dominated surfaces has been the most revealing technique. The contact angle (θ) of a liquid on a solid surface is defined as that internal angle, generally between 0° and 150°, measured through the liquid drop to the tangent drawn to its peripheral boundary. A plot of the cosine of the contact angle *vs.* the independently determined liquid/vapor surface tension for each diagnostic liquid used forms a linear relationship from which one can infer a critical surface tension value at the cosine $\theta = 1$ axis. The direct correlation which Zisman and coworkers (7) have developed between such critical surface tension intercepts and the true outermost atomic constitution of low energy organic solids is often called a wettability spectrum. For selected low energy surfaces the actual atomic constitution of the surface can be correlated on a 1:1 basis with the critical surface tension experimentally determined.

Collagen and Gelatin

Contact angle techniques used to evaluate thin films of water-soluble collagen, with care to avoid denaturing effects, gave a critical surface tension approaching 40 dynes/cm (8). An anomalous nonwettability by some of the low surface-tension, dispersion-force-only liquids was evident, and this was attributed to organized water adsorbed at the surface of these thin films. There were no special interactions with hydrogen-bonding liquids. On the other hand, it was discovered that by casting thin films of collagen from hot rather than cool water, a marked departure in the wetting properties was exhibited by hydrogen-bonding liquids. Based on previous published work (9, 10), this behavior is explained as randomization of the native protein structure which allows access of the hydrogen-bonding wetting liquids to the hydrogen-bond-susceptible amide links at the interface which serve as a substrate for other interactions (e.g., platelet adhesion to collagen fibers *in vivo* initating thrombosis).

Completely water-swollen collagen differed markedly from the relatively dry [equilibrated at 50% relative humdity (RH)] protein (8) in its wettability by water-immiscible liquids. In the water-swollen case, the apparent critical surface tension diminished to *ca.* 30 dynes/cm, indicating that a protein interface in nature is not correctly modeled by dehydrated specimens.

Protein Analogs

Synthetic polypeptides serve as models for proteins in a number of circumstances, particularly in deducing the influence of backbone chain configurations on the wetting properties of protein-dominated surfaces (9, 11). When, for example, poly(γ-methyl glutamate) was cast as a bulk sheet from solvents which favored the extended chain beta-structure of the polymer (as verified by IR and various diffraction techniques), the interfacial properties were dominated by the organized side chains (methyl ester groups in this case). No evidence of the hydrogen-bonding backbone, which was only a few atomic diameters from the surface, could be found. On the other hand, casting the poly(γ-methyl glutamate) into bulk films from solvents which favored the alpha helical arrangement of the polymer backbone demonstrated a marked increase in the wettability of the surface by those diagnostic wetting liquids characterizing hydrogen-bonding interactions. These films also showed a net increase in the average critical surface tension or apparent surface free energy of the polymer as a result of this backbone reorganization. Similarly, when poly(γ-methyl glutamate) was cast from solvents favoring the random tangle structure of the molecule and again allowing accessibility across the interface to polymeric backbone segments capable of entering into hydrogen-bonding interactions, the wetting results were similar to those for the alpha helical form and completely dissimilar to those for the extended chain intermolecularly hydrogen-bonded beta structure.

An extreme example of the different appearance of a macromolecular surface to an interacting external environment is provided by polyacrylamide (*12*). In this case, the data of all liquids which can enter into hydrogen-bonding interactions fall together on a Zisman-type contact angle plot at a critical surface tension of about 50 dynes/cm; other liquids incapable of this special interaction form a separate straight line which extrapolates to a critical tension some 10 dynes/cm lower (*8, 12*). Thus, protein surfaces can either (a) respond differentially (with the groups present and locked into the surface) to a variety of challenging environments, or (b) through molecular rearrangement, present those molecular groupings most compatible with the liquid adjacent to the solid substrate.

Nylons

Among model materials most relevant to natural proteins, the engineering polymer, nylon 2 (described in biochemical terminology as polyglycine) is important. In polyglycine (*10*) the difference between wetting of the surface by hydrogen-bonding liquids and by nonhydrogen-bonding liquids is striking.

Other nylon polymers showed the same apparent effect (*10*). The frequency distribution of the amide segments in a surface is a significant determinant of the general wettability of the surface as well as of any specific enhanced wettability by hydrogen-bonding liquids. For nylons, because of their lack of masking side chains, the modifications of surface properties resulting from casting of bulk films from various solvents were much less important than in the polypeptides with lengthy side chains which result in significant steric hindrance.

The wettability band for polyamides of the nylon series shifted to show lower slopes and higher critical surface tension intercepts when plotted in the standard Zisman format (*7*) as the amide group density increased (*10*). Unfortunately no theoretical work describes the important factors influencing the slope of such plots which reflect in a general way the strength of solid/liquid interactions.

Biomedically Important Collagenous Substrates

One of the most active areas of surgical research and practice involves collagen-based substitutes for natural blood vessels. What is sought is a blood conduit which is structurally sound and texturally suited for anchoring accumulating blood components without significantly distorting them to cause adverse subsequent reactions (such as thrombus generation, calcification, atherosclerosis). We have presented much surface textural and surface chemical data characterizing the surfaces of modified bovine blood vessels dominated by a collagen matrix and of the natural blood vessels which they can replace (*13*). One of our most

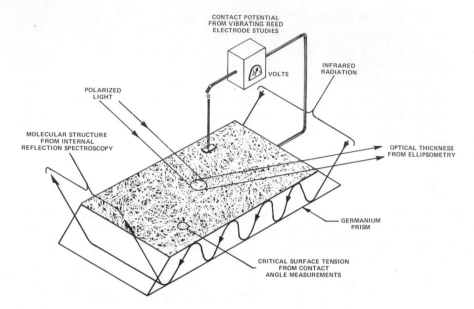

Figure 1.   Simultaneous non-destructive analytical techniques

important findings was that, when the texturally rough collagenous grafts are exposed to fresh flowing blood, they are rapidly overlain with a second protein film of fibrin, deposited spontaneously. We also demonstrated that in certain circumstances cholesterol stearate and fatty acid deposits accumulate along the lumen of collagen vessels. The deposits adversely increased their surface free energy and ultimately became the site of thrombus formation and aneurysmic failure of the collagenous implants.

Skin

Growing from deeper layers is the stratum corneum, a collapsed layer of protein-dominated, keratin-filled membranous sacks which provide the interface with all environments: air, water, and various increasingly hazardous aerosol chemicals. Important contributions to the study of the surface chemistry of this protein substrate were made first in industrial quarters, where the actions of soaps, creams, and other cosmetic applications were sought (*14*). The studies of many laboratories confirm the surface of skin to be a moderately low energy polymer similar to polyethylene in many interactions. Modifications of the method described in Figure 1 have been used for five years to study *in situ* the surface IR characteristics of living skin. Figure 2 shows an internal reflection spectrum obtained simply by touching the forearm to a germanium prism in a horizontal attachment to our internal reflection IR spectrophotometer

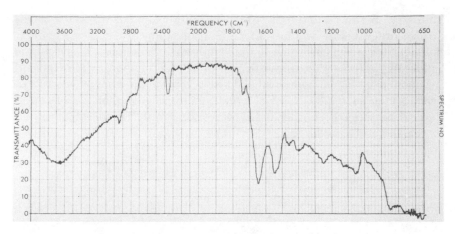

Figure 2.   Skin in situ; spectrum recorded by touching forearm to germanium prism in horizontal attachment

(constructed to our specifications by Harrick Scientific Corp., Ossining, N. Y.). The depth of the skin thus sampled is only a fraction of a micron. This external surface of the body is overwhelmingly dominated by protein components. A similar spectrum characterized old skin *in situ* on the thumb of an experimental subject showing protein domination again with a small contribution of organic ester. Contrasting with this is the situation illustrated spectrally in Figure 3, where the fresh skin of the opposite thumb of the same subject, as generated beneath a blister cap, shows an intermediate skin chemistry wherein the relative abundance of fatty esters

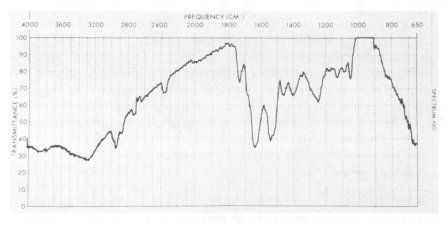

Figure 3.   M.A.I.R. infrared spectrum of fresh skin area beneath peeled away blister cap (1 cm × 3 cm; right thumb in situ, *2 days after first exposure to atmosphere; fingerprint ridges not well developed). Note ratio of hydrocarbon and ester bands to amide bands.*

is greater. Figures 4 and 5 illustrate the difference in skin surface chemistry recorded before and just after the application of a typical commercial moisture cream according to the supplier's instructions. This example highlights the potential diagnostic power of this method, while confirming dominance of the interfacial chemistry by protein components receptive to other materials. Control experiments showed that the observed changes between Figures 4 and 5 were not simply the result of a skin coating.

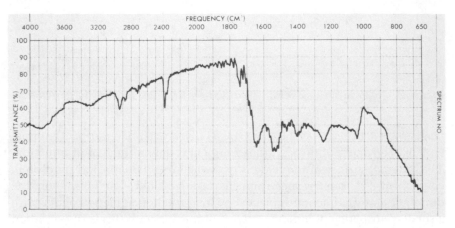

Figure 4.   M.A.I.R. infrared spectrum of virgin skin in situ (female forearm, underside, immediately after soap and water wash, towel dry)

Thus, the outermost layer of living human skin is dominated by proteins; new skin generated under the old is similarly dominated by proteins with the addition of a significant fraction of lipid. To understand the barrier and other protective properties of the skin one must look at the interfacial chemistry of protein layers. Figure 6 shows that even the exudate from a skin wound is dominated by protein components. As discussed later, the proteins in this case are almost exclusively glycoproteins, as indicated in Figure 6 by the relatively strong absorption band at *ca.* 1050 cm^{-1}. In modern protein interfacial chemical literature, glycoproteins dominate the discussion.

Tissue and Blood

One of the most striking differences between protein-dominated substrates (*e.g.*, skin, tissue masses, and blood) and other solid, semi-solid, or liquid surfaces is in their wettability and adhesiveness with other materials. Work on the development of surgical adhesives based upon the poly(α-cyanoacrylates) used successfully in hemostasis for massive

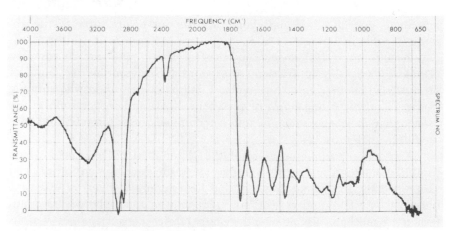

Figure 5.   M.A.I.R. infrared spectrum of modified skin in situ *(female forearm, underside, immediately after application of commercial "moisture cream" according to supplier's instructions)*

wounds, clinical repair of skin incisions, and oral surgery (*15, 16*) provides excellent examples of this differential wettability and adhesion. For example, the homologous series of alkyl cyanoacrylate polymers from methyl through heptyl shows an inverse order of wetting and adhesion on blood and tissue from the order shown on pure water or protein-free fluids. This and ancillary evidence (*16*) shows that interfaces of wet biological masses are dominated by protein molecules which can enter

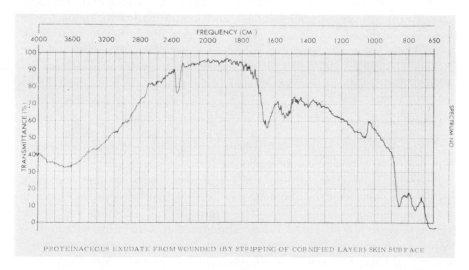

PROTEINACEOUS EXUDATE FROM WOUNDED (BY STRIPPING OF CORNIFIED LAYER) SKIN SURFACE

Figure 6.   Proteinaceous exudate from wounded (by stripping of cornified layer) skin surface

into many special interfacial chemical reactions not possible in the absence of protein.

Protein interfacial layers also control traditional surface physicochemical properties of friction, wear, and lubrication in the joints of articulated bones—e.g., in hip joints. Little is known about the rubbing surfaces and the filling fluids of the natural ball and socket connections except that they are predominately cartilaginous, glycoprotein, and proteoglycan materials.

Another important example of the differential adhesiveness which protein-dominated surfaces can display, is in the development of artificial skin especially for wound dressings and for temporary covers of extensive burns. C. W. Hall and co-workers (17) showed that relative tissue adhesion to mechanically identical velour fabrics constructed of various materials follows the order predicted by the critical surface tensions of construction material.

Mohandas and co-workers (18), confirming previous findings of Weiss and Blumenson (19), have also shown that cells in an environment free of adsorbable proteins (which rapidly modify the surface properties of polymeric or inorganic substrates) will exhibit a similar direct relationship between their adhesion and the critical surface tension of the surface they contacted. Differential adhesion of red blood cells was measured by determining the fraction of cells retained on a surface after the application of well-calibrated shear stresses (18). In protein-free experiments, the red cells (themselves dominated in adhesive interactions by their protein membranes) had greatest adhesion to glass, intermediate adhesion to polyethylene and siliconized glass, and least adhesion to Teflon.

This artificial bioadhesion does not characterize the natural situation, where spontaneous protein adsorption precedes cell-surface contact. Mohandas and co-workers (18) recognized this problem and have extended their studies to protein-coated surfaces as well.

Proteins at Substrates

Thrombus. In over 20,000 substitute heart valves implanted during the mid-1960's, thromboembolism (shedding of small masses of platelet aggregates from their loci on the artificial surfaces of the implanted prostheses) occurred in about one of five patients despite attempts to maintain anticoagulation (20). The events of cell adhesion and breakdown of such adhesion after it has propagated enough so that local shear forces can overcome it is a significant complication of heart valve replacement and similar insertion of nonphysiological material into the cardiovascular system. The first events involve adsorption of proteins, predominantly fibrinogen, as modifying or conditioning films on the implants

(*21, 22, 23*). Then platelets, which had been arriving but not adhering prior to the buildup of a certain thickness of the protein layer, adhere to form a saturated layer (*24, 25, 26*). Depending on the nature of the original substrate as transduced through the nonequilibrium layer of protein present at the time of induction of platelet adhesion, either platelet aggregation into the lumen occurs, or the original platelets are shed. In the former, more common case, the aggregating mass grows downstream in a wake pattern, finally inducing formation of an interaggregate fibrin and red cell mesh, with complete flow block by thrombus and subsequent emboli. Activation of the coagulation factors XII, XIII, etc. is an important independent event which also can begin by surface contact; it enhances the formation of fibrin in the volume phase of the flowing stream. Only two, seldom seen routes to favorable biomedical outcomes have been observed. In the first, the original layer of adherent platelets does not become sticky to arriving siblings. Through a poorly understood secondary adhesion of white cells (predominantly neutrophils), the original platelets are removed to leave a residual protein film in dynamic equilibrium with the blood stream. No further cellular deposition is noted. The second favorable circumstance occurs when, in implants of sufficiently large diameter or in regions of sufficiently high rates of flow, the original layer of platelet thrombus (with or without fibrin strands and trapped erythrocytes) remodels to form a smooth fibrin layer or to support cellular ingrowth (probably of endothelial cells such as those originally lining blood vessels) to provide a passive pseudo intima.

Some general observations on the adhesion of blood platelets can be made based on experiments performed in a variety of carefully designed flow chambers (*24, 27*). A lag of 30–60 sec before platelets deposited, even though arriving in abundance, was observed microscopically and filmed. Ancillary studies by electron microscopy, internal reflection spectroscopy, ellipsometry, staining, antibody, and contact angle techniques provided evidence that no cell adhesion occurs from natural blood without the presence of this intervening layer of protein which is selectively and uniformly deposited on all nonphysiologic substrates (*21, 25, 28, 29*). It has been proposed recently that the interaction of platelets themselves with this deposited protein film is mediated by an extracellular protein layer of contractile protein on the cell surface (*30, 31*). Figure 7 is a highly magnified, electron microscopic view of the edge of a single blood platelet where it contacts a foreign solid (epoxy). This view illustrates both the prerequisite adsorbed film and the platelet surface fuzz which may be involved in this adhesive interaction.

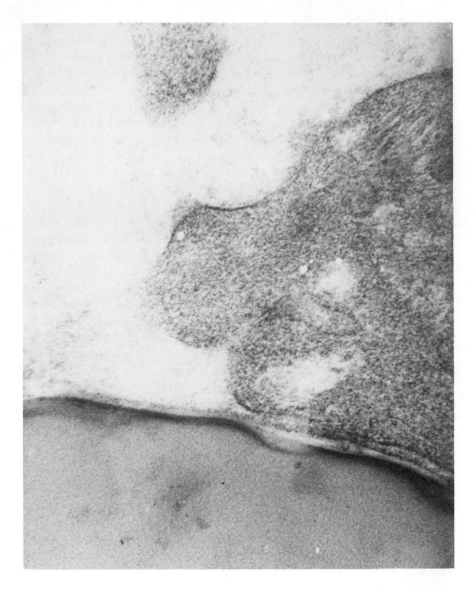

Figure 7.   The edge of a platelet adhering to an epoxy slab, as coated with a protein conditioning film

No explanation exists for the observation routinely confirmed that, although spontaneously adsorbed protein at nonphysiologic interfaces has a uniform initial appearance and chemistry, the arriving blood platelets do not adhere uniformly to that layer. The adherent platelets always leave apparently unoccupied space between themselves and their nearest neighbors, generally giving a saturation population density of between 70 and 90 platelets per 1000 μ^2 (*26*). Actual cell adhesion on preformed protein films can be measured by methods already described (*18*).

The rapidity of the protein adsorption has been demonstrated elsewhere. In as little as 5 sec, for example, the film thickness was already of the order of 50 A (*21*). It is now accepted that adsorbed protein accumulates over the same time period to about the same thickness on all foreign solid surfaces in blood. Electron micrographs of the platelets originally adherent to surfaces having different original characters (*32*) provide some insight on the origin of a continued differential end point existing among thromboresistant or thrombogenic materials. On those surfaces which are thromboresistant after long term implantations, the originally adherent platelets remain more morphologically intact than those which originally adhered to the (empirically found) thrombogenic materials. Those originally adherent platelets which retain their round or disc-like shape and throw out very few pseudopods across the surface are no longer observed on the surface after flow times as short as 2 hrs in many cases (*33*), even though they were present from times of about 1 min (*24*) to 10 min (*26*). Thus, in some unknown way, differences in the adsorbed protein, even the same adsorbed protein, must be marked enough to provide this strikingly different response in adhering blood platelets. Changes in the zeta potential and other electrokinetic properties across adsorbed protein layers are probably not great enough to explain the relative compatibility, or lack of compatibility, of a variety of proposed implant materials (*35, 36*).

Leo Vroman and co-workers, who present additional data in this volume, have made major contributions to our understanding of the fundamentals of this adsorption process. They have shown that events at the substrate/blood interface are not static after the first layer of protein is adsorbed, but that the protein layer is continuously remodeled, reacted with, or converted by other surface-active components in intact plasma (*37*).

The layer which continues to exist in apparent equilibrium with the blood after long term implantations of otherwise inert solid materials remains mysterious; it is not recognized by any of the specific blood component antibodies tried to date. Often, it remains thick enough so that when analyzed by internal reflection spectroscopy (sensitive to, at the most, a few microns of a substrate surface sample), the underlying polymeric substrate cannot be detected at all. An interesting exception to this finding has come from our recent research with inorganic substrates which had been scrupulously cleaned by glow discharge treatment. When borosilicate glass tubing was treated by this process and implanted, it remained thrombus-free for more than a year in a very thrombogenic location (the canine thoracic inferior vena cava). A passivating layer of protein, which was remarkably pure by spectroscopic criteria and abundant after 2 hrs of blood contact, could be barely demonstrated by IR after 480 days (38).

Improved analytical techniques are needed to detect the important configurational and chemical differences among adsorbed films formed spontaneously from complex solutions on various substrates. Internal reflection IR does not reveal significant differences in the adsorbed protein films which accumulate on variously treated Stellite 21 devices (a cobalt-chromium alloy used to make synthetic heart valve struts and seats); in some instances these devices are thrombogenic, and in others they are apparently thromboresistant (39). Yet, scanning electron microscopy reveals that arriving blood platelets can discern differences in the films immediately on contact with them.

Textile Grafts. Flexible knitted and woven tubes (usually of Dacron and Teflon fibers) are often implanted as synthetic blood conduits. It was demonstrated as early as 1958 that these plastic fiber grafts all become coated, on both outer and inner surfaces, with fibrous tissues as soon as 30 days after implantation (40). In addition the nominally higher surface-energy fabrics, constructed from polyamide and polyester, are more prone to accumulate thick layers of such fibrous tissue than are the lower surface-energy grafts of Teflon. Drawing upon such scattered observations, along with our own data on the surface properties of apparently thromboresistant implants (41), we proposed a tentative correlation of the relative surface energies of solids with their biological interactions (13, 24, 42). The most significant feature of our hypothesis is a minimum in relative biological interaction along the scale from very low surface-energy materials (typified by the fluorocarbons) to the higher surface-energy plastics (typified by the various polyesters and polyamides). Evidence (42, 43) suggests that the zone of maximum biocompatibility—as judged by minimum depositions of debris on implants or minimum distortion of cells adherent through an intermediate layer of protein to the surfaces—falls in the critical surface region between 20 and 30 dynes/cm. On the basis of well developed correlations between polymer surface constitution and critical surface tension (7), such a range must be essentially dominated by the $-CH_3$ terminal groups as side chains (as in polydimethylsiloxane) or terminal atomic clusters (as in fatty acids, amides, and long chain aliphatic alcohols) on more complex organic backbones. The apparent critical surface tension of organized water (determined from wetting experiments involving simple liquids which can interact only by dispersion forces and not by polar interactions or hydrogen-bonding interactions) falls also in this zone (45, 46); this suggests that very highly hydrated masses such as hydrogels might also function by this moderate surface-energy mechanism when they remain thromboresistant.

Cell Adhesion. Except for model experiments with red cells in artificial protein-free environments, there is no counter example to the generalization that cell adhesion does not occur to any solid surface without an intervening thin layer of adsorbed or previously deposited protein-dominated matter. An excellent demonstration of the requirement for such adsorbed layers was provided by A. C. Taylor (47) in his studies of adhesion between gingival epithelial cells and hard enamel surfaces at the dental margin and between all living cells which he studied and artificial substrates in culture. More recently, this finding has been confirmed in examples as much in contrast as the adhesion of standard cell lines to unmodified glass surfaces and to completely silicone-masked glass surfaces (48). Such masking was done by a pinhole-free overcoating of a siliconizing material which not only changed the surface free energy of the solid but also reversed its normal electrical charge (from negative to positive); yet protein adsorption preceded cell adhesion both before and after the glass was overcoated.

In this volume Baier and Weiss demonstrate the reality of the spontaneous adsorption of reasonably pure glycoprotein films from common cell culture media (as generally supplemented with calf serum) prior to cell attachment and growth on substrate surfaces. In addition to surface chemical and charge influences on cell adhesion at the solid/solution interface, there is a dependence on the relative sizes and geometries of the cells and their potential substrates (48). In some cases, where the substrates are quite large with respect to the cells, the cells will simply grow over, along, or under such foreign material. When the substrates are very small and fibrous, or otherwise finely particulated, a process akin to phagocytosis occurs—*i.e.*, the cells cluster on and around the foreign solid surfaces. Paul Weiss catagorized these types of reactions as indicative of a tactile chemical response (49). Van Oss and co-workers (50, 51) and Good (52) have recently taken some of these factors into account in their study of mechanisms of phagocytosis as it underlies particle engulfment in health and disease.

An important feature of the incubation of foreign solid surfaces in natural biological media, especially in media containing adsorbable macromolecular components such as serum glycoproteins, is that higher surface free-energy materials (such as glow-discharge cleaned glass) and moderately low surface free-energy materials (such as dichlorodimethylsilane-modified glass) continue to exhibit differential surface properties through an adsorbed protein blanket of equal thickness. These properties are exhibited in both contact angle (wetting and spreading) and cell adhesion experiments which are easily done in the laboratory (47). Well characterized and well behaved glass coating compounds of very low surface free energy are needed to verify that protein adsorption will occur on that substrate in the same way it does on the high and moderate surface free-energy substrates. Cells adhere more tightly and in greater abundance to smooth higher surface free-energy substrates than to smooth moderate (between 20 and 30 dynes/cm) substrates; cells attached to the high surface-free-energy substrates, through an intervening layer of spontaneously adsorbed protein, show much greater frequency of irregular perimeters (*i.e.*, many protrusions, pseudopods, and fibrous projections from the border attaching to the surface). Cells on the biocompatible surfaces remain rounded and poorly adherent. Equally significant is that, in the case of cell contact with protein-coated moderately low-energy surfaces, the cells prefer to aggregate (over a period of time) with one another in tissue-like masses rather than maintain independent and separate adhesion with a substrate (47). We have previously discussed a zone of minimal cell spreading and attachment on such surfaces (41) and reviewed these processes from the point of view of cell motility and migration as necessarily takes place during wound healing (53). Future study must show how, at the leading edge of migrating cells, the adhesion between cell surface structures and the underlying immobilized substrate surface structures can be so rapidly made and broken as to allow the observed rotating-tank tread-like motion of the cell surface and the forward progress of the cell body to occur.

Intrauterine Devices. Leininger (54) has reviewed the utility of various polymers as implants. Objects which are neither a tissue implant nor a cardiovascular implant but which may have features of both, are contraceptive foreign bodies. Since this subject has not been discussed often enough in the surface chemical literature, interactions of devices placed in the uterine cavity remain poorly understood. These devices minimize in some unknown manner, chances for conception and pregnancy. A brief article which discusses the specifics of spontaneous interfacial modification of these contraceptives is included in this volume (57). The major finding is that all such inserts accumulate a remarkably uniform coating of glycoprotein by spontaneous adsorption from the cervical mucous fluid. It has been speculated that this layer of adsorbed interfacial material modestly activates antibody and rejection mechanisms in the surrounding tissue, thereby preventing capacitation of sperm obliged to swim through this subtly chemically modified zone (41).

Dental Adhesion. A problem of increasing importance which is attracting increasing scientific workers, is the establishment in the moist saline, enzymatically active, heat-and-cold-stressed, and mechanically perturbed environment of the mouth good adhesive bonds between the natural surfaces and synthetic prosthetic materials (*50, 57, 58*). This happens spontaneously in most healthy and many diseased persons when dental plaque accumulates and transforms into dental calculus. As we have shown elsewhere (*59*), and as Quintana reviews in this volume (*60*), solid surfaces in the oral cavity even when scrupulously precleaned do not remain free of adsorbed macromolecular components for more than a few seconds. The material which is spontaneously adsorbed is a specific glycoprotein component of saliva and not a heterogeneous random selection from all surface-active components present. In our laboratories, we are determining the surface chemistry of human teeth as they normally rest in a healthy human mouth, by making contact angle measurements (*59*). Once the main features of this interfacial chemistry and the dominance of most interfaces by adsorbed hydrated proteinaceous layers becomes more generally understood, the development of excellent dental adhesives will follow. Such adhesives, supported by special dental treatment, will allow greater adhesion between biological materials and hard substrates, as in implants at the dental margin. In other cases, improved surface chemical knowledge should allow treatments to be developed which will minimize adhesion between cellular elements, such as bacterial flora, and solid surfaces in the mouth so that dental plaque, cavity initiation, and calculus formation will be minimized.

Biological Fouling. Bacterial adhesion is a primary event in the early phases of dental plaque formation. A less popularly understood example of bacterial adhesion is that which occurs on foreign solid surfaces in natural waters. Kevin Marshall in Australia and William Corpe in the United States have led the study of primary bacterial film formation on foreign solid surfaces (*61, 62*). They have demonstrated, independently and in collaboration with this author, the interfacial modification of all foreign substrates, prior to such primary bacterial adhesion, by adsorbed glycoprotein layers. It is not yet certain that these extracellular layers originate from dissolved components in natural seawater or in suspending media in laboratory experiments. These conditioning films of adsorbed protein may result from active participation of the bacteria in extruding such material or from the disintegration of some of the bacteria to provide the adsorbable components in a nonspecific manner. Marshall has also shown that the propagation of these bacteria in chains and clusters proceeds from the surface through the intermingling or adhesion of fibrillar extracellular polymeric material similar to that involved in the original adhesive event (*63*). This observation recalls the involvement of fiber-forming fibrinogen and the mutual aggregation of blood platelets during the initial events of thrombus formation described earlier (*21, 29*).

Even in the absence of microscopically observable bacterial adhesive events, all substrates in natural water, and especially in subtropical areas where maritime fouling abounds, are spontaneously coated with adsorbed protein films (65). In fact, allowing for differences in relative time scales which reflect differences in relative concentrations of adsorbable components, the sequence of events at solid surfaces immersed in common fresh or seawater (containing living organisms) is essentially identical to that in the natural maritime environment. The sequence begins with rapid, nearly monomolecular layer coverage of specifically adsorbed glycoproteins followed by layer thickening and finally adhesion and growth of discrete organisms (65).

With samples prepared by M. Cook in England, we have demonstrated that such adsorbed films on solid substrates are also involved in the adhesion of *mussel byssus* discs to such surfaces (66). A secondary adhesive product, analytically similar to glycoproteins, is that on the underside of the *mussel byssus* discs (67). In the case of the ability of barnacle cyprids to adhere under adverse circumstances to substrates as diverse as Teflon, steel, and highly toxic paint surfaces, the original modification of the solid substrate interface is provided by adsorbed protein-dominated layers. Crisp (68) has shown convincingly that barnacle cyprids in their adherent stage will choose protein-coated substrates for settling rather than freshly inserted or freshly cleaned surfaces. In this case, as with the *mussel byssus* disc and with adherent bacteria, the extruded cement (itself a glycoprotein) must be establishing its great adhesion not directly with the substrate but with the already-present glycoprotein film formed prior to cyprid settling. We suggested earlier that all biological fouling begins by an adsorptive event dominated by accumulation of glycoprotein material and that secondary adhesion of the formed cellular or larval organisms is through a proteoglycan type material (67). We suppose that the secondary cement interacts, carbohydrate-group-to-carbohydrate-group, through its exposed glyco side chains with similar oligosaccharide groups of the originally adherent layer in much the same manner as polysaccharide chains merge in wet cellulosic pulps to give paper products their strength.

Barnacles actually grown on the faces of internal reflection prisms have a cement predominately of the glycoprotein class. In most detachable biological links, such as formed by the *mussel byssus* discs, limpids, snails, and fresh-water removable bacteria, the adhesive is generally of the polysaccharide or proteoglycan class. Based upon immersion studies still in progress, the zone of minimal biological adhesion signaled by the critical surface tension range between 20 and 30 dynes/cm in tissue implantation, cell culture, and blood compatibility experiments will also be the proper functional zone for minimum biological fouling (64).

Specifics of Protein Adsorption

Substrates of differing initial surface properties, when immersed in media containing adsorbable macromolecules such as proteins, attain at equilibrium essentially the same amount of the same adsorbed material (*69*). Although the equilibrium conditions at the surfaces of various substrates are, in the absence of other components and especially of living cells, similar, these surfaces differ remarkably while attaching, adsorbing, or otherwise contacting cellular elements before equilibrium thicknesses or constant chemical conditions have been obtained. To understand bioadhesive events one must know the specific configuration and other details of organization of the films during their conversion from initially attached molecules to the establishment of adsorption–desorption equilibrium. Some aspects of the adsorption process, and especially the differences which continue to exist even under equilibrium conditions, are described in other publications (*69, 70*).

Adsorbed Films

Using the protein beta-lactoglobulin as a model, George Loeb has shown that the ratio of essentially native to configurationally modified protein in spontaneously adsorbed films varies directly with the amount of material adsorbed (*71*). At the lowest amount adsorbed (presumably the first monolayer coverage), the ratio of native to altered molecular configuration is about 0.5. Only as the film achieves adsorption equilibrium at a significantly greater thickness does this ratio become about 0.8 (for films adsorbed on nominally high surface energy materials). Loeb's parameter of structural alteration was the IR-detectable shift in hydrogen-bonding arrangements which differentiate the predominately beta-structured (intermolecularly hydrogen-bonded extended chain) from the alpha-helical configuration (stabilized by intramolecularly hydrogen-bonded amino acid units). Only when the substrate available for adsorption of protein macromolecules from solution was in the moderately low-surface-free-energy class did the ratio of native to denatured material in the adsorbed film approach one. Loeb also showed in a series of experiments in which the concentration of protein in the original solution was systematically modified, that the proportion of native material (configurationally undistorted at the level of secondary structure detected by IR) was lower relative to the lower original concentration of protein in solution. In similar systems studied by ellipsometry, equilibrium film thicknesses are not attained until *ca.* 1500 sec of adsorption; this illustrates the substantial difference in time scales between attainment of equilibrium protein adsorption and the time (30–480 sec) at which cells usually become adhesive to the thickening films. Recent results from the National Bureau of Standards (69) illustrate the further sensitivity of protein adsorption to the specific surface free energy of the solid substrate. Such work has shown that although the mass of material adsorbed on various surfaces might remain essentially constant under equilibrium conditions, the protein extension out into the solution phase can be substantially different. On a variety of surfaces, proteins of special interest in thrombus formation and blood clotting have their greatest extension from those surfaces of lower free energy (69). Such studies have suggested the interpretation, discussed in detail by Nyilas (72), that low interaction energies across solid/protein solution boundaries will ensure short residence times of the adsorbing macromolecules. Proper balance of the interfacial chemistry with the chemistry of the adsorbing species can probably minimize the adverse influences which tenacious, irreversible protein adsorption (and configuration modification) induce (73).

Gas/Liquid Interfaces

Adsorption experiments with protein macromolecules at solid/solution interfaces are often difficult to perform and even more difficult to interpret properly. Fortunately we have reported and reviewed elsewhere (11, 22) that both adsorption and deliberate spreading of proteins and synthetic polypeptides (as model proteins) at air/liquid interfaces provide films whose structures are usually indistinguishable from those formed by adsorption to solids. It has also been shown, as recently reviewed by Malcolm (74) and recapitulated by him in this volume, that gas/liquid interfacial polymeric films can be grossly manipulated, crumpled, and collapsed into fibrous bundles without modification of their fundamental structure (75, 76). Loeb has shown that for predominantly alpha-helical proteins, adsorption at the solid/liquid boundary and more particularly spreading at the air/liquid interface does not significantly degrade their original secondary structure (77). On the other hand, predominantly beta-chain materials, such as beta-lactoglobulin, do during adsorption or air/water interfacial film formation change to a more helical form (71). This was previously shown, by multiple attenuated internal reflection spectroscopy techniques identical to those described here, to occur with the simpler polypeptide poly(γ-methylglutamate) spread from solutions in which its predominant molecular form was beta (extended chain) and in which its air/water interfacial film became about 50% coiled (76). In beta-lactoglobulin, the fraction of native material remaining in a monolayer depended on the surface pressure under which the film was transferred to internal reflection prisms for ultimate analysis. The higher pressures favored more native material in the transferred film. With this exception, it can be generalized that the strong preference is for coiled rather than extended chains of surface-localized macromolecules.

Using direct extensions of film transfer techniques originally described by Langmuir and Blodgett (*1*), we have applied the surface film retrieval method described to films accumulated at natural gas/liquid interfacial boundaries such as those between the sea and the atmosphere above lakes and oceans (*78, 79, 80, 81*). Given prisms which had been scrupulously cleaned in the laboratory prior to packaging, so that their surface properties were predominantly hydrophilic at the time of field immersion, it was demonstrated that even in a reasonably choppy maritime environment where precise immersion and withdrawal was not possible, only a single thickness of the ambient resident film was transferred for analysis (by the combination of techniques described earlier here). It was also shown that, by first conditioning the prism with a hydrophobic coating (such as that provided by siliconization or by a dried monolayer of stearic acid or stearate salt), multiple immersions and withdrawals of a prism through a surface-film-covered liquid would result in the easy transfer of proportionately thicker (multiple ambient layers) films for easier analysis. This technique involves removing a prism in a special small holder from its plastic package, attaching it to a snap hook on a fishing line and lowering it through the natural air/water interface (or outfall of industrial wastes, layer of foam, oil slick) of interest, slowly withdrawing the prism, giving it a brief air drying, and repackaging for later analysis. The first large scale application of the technique was in Chautauqua Lake in New York state, which was repetitively sampled over the entire 1969 recreational season to establish spectroscopic parameters to permit characterization of its surface quality (*78*). More recently, the technique has been extended to the major oceans and seas of the earth. In general, in all nonpolluted locations or in polluted locations which were allowed a few days for natural cleansing to occur at the interface, natural air/water boundaries are dominated by glycoprotein and proteoglycan type films (*80, 81*).

With rare exceptions, such as along the north wall of the Gulf Stream where interfacial films are sometimes dominated by lipid components, contact angle data on the transferred dried films have indicated critical surface tensions between 30 and 40 dynes/cm, thereby confirming the presence of oxygenated and presumably also nitrogenated components as dominant components in such films. The stabilizing film at the gas/liquid boundaries so prolific in long-lasting sea foams is predominantly glycoprotein and proteoglycan material having its origin in sea-surface films contributed primarily by plankton blooms (*81, 82, 83, 84*).

With respect to gas/liquid interfaces created in the bulk of solutions, bubbles rising through solutions containing macromolecules of biological origin, and especially proteins, will not only spontaneously collect these polymers at the new gas/liquid interfaces of the rising bubble but will concentrate them into insoluble fibrous debris which is spun off the disappearing trailing edge of a moving bubble (85). Bubbles traveling to the water surface from the solution also carry, with their gas/liquid boundaries, at least a portion of their adsorbed burden to that surface. During bubble breaking, this material is ejected along with film fragments from the ambient gas/liquid interfacial film already residing there. Blanchard, in this volume, gives an excellent review of the importance of such processes on a worldwide scale and of their implications in oceanography and meteorology. In a more modest but individually more lethal example, the preferred method of blood oxygenation during open heart surgery is to allow a column of gas bubbles to rise through the blood. This creates a vigorous foaming which requires secondary foam breaking and filtering of particulate debris before the blood is reinjected into the patient. In bubbles rising through the ocean, the insoluble protein shed may be contributing to the organic detritis necessary as food for lower-dwelling organisms below the photic zone; in blood oxygenators, the proteins lost irreversibly from the volume phase are usually crucial to the health of the patient. Loss of such crucial material, especially antibodies, could be largely responsible for the frequent deaths from simple infection, pneumonia, and other diseases of patients who have had successful open heart surgery.

Cell-Cell Interactions

Living cells can undergo changes in their surface properties, and these properties dictate the relative adhesiveness of cells to their neighbors (86). This surface chemical interference to adhesion then correlates with decreased strength of cell-to-cell joints and the increased mobility and invasiveness characteristic of malignant cells in tumors and other forms of cancer. In this regard, it is worthwhile to consider mechanisms where the adhesion is between two similar or dissimilar cells; foreign solid substrates, as discussed earlier, are not involved here. It is likely that cell-to-cell adhesions are also mediated by adsorbed macromolecular components, however. Almost certainly, glycoprotein or proteoglycan materials account for the gap of about 100 A shown by electronmicroscopy to exist between closely apposed cell surface membranes. It is a serious task for chemists to decipher the specific constitution, configuration, structure, and function of such glycoproteins and proteoglycans present at interfaces. Since the glycoprotein materials can contain anywhere from 1/10 of 1% to about 15% carbohydrate in side chains, and since the molecular weights of these materials are generally around 1 million, and since proteoglycans (until recently usually called mucopolysaccharides in the biochemical literature) can have every other amino acid along the protein backbone substituted with short sugar chains and also can range in molecular weight from *ca.* 20,000 to > 1 million, it is a formidable challenge to decipher the specifics of interfacial chemistry and organization required for further progress.

Literature Cited

1. Langmuir, I., Blodgett, K. B., *Phys. Rev.* (1937) **51**, 317.
2. Shafrin, E. G., Zisman, W. A., *J. Colloid Sci.* (1949) **4**, 571.
3. Harrick, N. J., "Internal Reflection Spectroscopy," Interscience, New York, 1967.
4. McCrackin, F. L., Passaglia, E., Stromberg, R. R., Steinberg, H. L., *J. Res. Nat. Bur. Stand. Sect. A* (1963) **67**, 363.
5. Bewig, K., "Improvements in the Vibrating Condenser Method of Measuring Contact Potential Differences," Naval Res. Lab. Report (1958) **5096.**
6. Bewig, K. W., Zisman, W. A., *J. Phys. Chem.* (1965) **69**, 4238.
7. Zisman, W. A., ADVAN. CHEM. SER. (1964) **43**, 1.
8. Baier, R. E., Zisman, W. A., ADVAN. CHEM SER. (1975) **145**, 155.
9. Baier, R. E., Zisman, W. A., *Macromolecules* (1970) **3**, 70.
10. *Ibid.* (1970) **3**, 462.
11. Baier, R. E., Loeb, G. I., "Polymer Characterization: Interdisciplinary Approaches," C. D. Craver, Ed., p. 79, Plenum, New York, 1971.
12. Jarvis, N. L., Fox, R. B., Zisman, W. A., ADVAN. CHEM. SER. (1964) **43**, 317.
13. Baier, R. E., DePalma, V. A., "Management of Arterial Occlusive Disease," W. A. Dale, Ed., Chap. 9, Year Book Medical, Chicago, 1971.
14. Ginn, M. E., Noyes, C. M., Jungermann, E., *J. Colloid Interface Sci.* (1968) **26**, 146.
15. Matsumoto, T., "Adhesion in Biological Systems," R. S. Manly, Ed., Chap. 13, Academic, New York, 1970.
16. Leonard, F., Hodge, J., Houston, S., Ousterhout, D., *J. Biomed. Mater. Res.* (1968) **2**, 173.
17. Hall, C. W., Spira, M., Gerow, F., Adams, L., Martin, E., Hardy, B., *Trans. Amer. Soc. Artif. Intern. Organs* (1970) **16**, 12.
18. Mohandas, N., Hochmuth, R. M., Spaeth, E. E., *J. Biomed. Mater. Res.* (1974) **8**, 119.
19. Weiss, L., Blumenson, L. E., *J. Cell. Physiol.* (1967) **70**, 23.
20. Mulder, P., Rosenthal, J., *J. Cardiovasc. Surg.* (1968) **9**, 440.
21. Baier, R. E., Dutton, R. C., *J. Biomed. Mater. Res.* (1969) **3**, 191.
22. Baier, R. E., Loeb, G. I., Wallace, G. T., *Fed. Proc. Fed. Amer. Soc. Exp. Biol.* (1971) **30**, 1523.
23. Baier, R. E., *Bull. N.Y. Acad. Med.* (1972) **48**, 257.
24. Dutton, R. C., Baier, R. E., Dedrick, R. L., Bowman, R. L., *Trans. Amer. Soc. Artif. Intern. Organs* (1968) **14**, 57.
25. Dutton, R. C., Webber, T. J., Johnson, S. A., Baier, R. E., *J. Biomed. Mater. Res.* (1969) **3**, 13.
26. Friedman, L. I., Liem, H., Grabowski, E. F., Leonard, E. F., McCord, C. W., *Trans. Amer. Soc. Artif. Intern. Organs* (1970) **16**, 63.

27. Petschek, H., Adamis, D., Kantrowitz, A. R., *Trans. Amer. Soc. Artif. Intern. Organs* (1968) **14**, 256.
28. Vroman, L., *Thromb. Diath. Haemorrh.* (1964) **10**, 455.
29. Vroman, L., Adams, A. L., *J. Biomed. Mater. Res.* (1969) **3**, 43.
30. Booyse, F. M., Rafelson, Jr., M. E., *Ser. Haematol.* (1971) **4**, 152.
31. Booyse, F. M., Sternberger, L. A., Zschocke, D., Rafelson, Jr., M. E., *J. Histochem. Cytochem.* (1971) **19**, 540.
32. Schoen, F. J., *Fed. Proc. Fed. Amer. Soc. Exp. Biol.* (1971) **30**, 1647.
33. Gott, V. L., Baier, R E., "Materials Compatible with Blood," Vol. 2, *Ann. Rep. Contract PH 43-68-84* (1973) available Nat. Tech. Inf. Serv., Springfield, Va.
34. Moyer, L. S., Gorin, M. H., *J. Biol. Chem.* (1940) **133**, 605.
35. Mirkovitch, V., Beck, R. E., Andrus, P. G., Leininger, R. I., *J. Surg. Res.* (1964) **4**, 395.
36. Vroman, L., Adams, A. L., Klings, M., Fischer, G., ADVAN. CHEM. SER. (1975) **145**, 255.
37. Baier, R. E., DePalma, V. A., Furuse, A., Gott, V. L., Lucas, T., Sawyer, P. N., Srinivasan, S., Stanszweski, B., unpublished data.
38. Baier, R. E., Dutton, R. C., Gott., V. L., *Advan. Exp. Med. Biol.* (1970) **7**, 235.
39. Harrison, J. H., *Amer. J. Surg.* (1958) **95**, 3.
40. Baier, R. E., Gott, V. L., Furuse, A., *Trans. Amer. Soc. Artif. Intern. Organs* (1970) **16**, 50.
41. Baier, R. E., "Adhesion in Biological Systems," R. S. Manly, Ed., Chap. 2, Academic, New York, 1970.
42. Ward, C. A., Stanga, D., Collier, R. P., Zingg, W., *Trans. Amer. Soc. Artif. Intern. Organs* (1974) **20**.
43. Kolobow, T., Stool, E. W., Weathersby, P. K., Pierce, J., Hayano, F., *Trans. Amer. Soc. Artif. Intern. Organs* (1974) **20**.
44. Johnson, Jr., R. E., Dettre, R. H., *J. Colloid Interface Sci.* (1966) **21**, 610.
45. Shafrin, E. G., Zisman, W. A., *J. Phys. Chem.* (1967) **71**, 1309.
46. Taylor, A. C., "Adhesion in Biological Systems," R. S. Manly, Ed., Academic, New York, 1970.
47. Wilkins, J. F., "Adhesion of Ehrich Ascites Tumor Cells to Surface-Modified Glass," Ph.D. Dissertation, State University of New York at Buffalo, 1974.
48. Maroudas, N., *J. Theor. Biol.*, in press.
49. Weiss, P., *Harvey Lect.* (1959) **55**, 13.
50. Neumann, A. W., van Oss, C. J., Szekely, J., *Kolloid Z. Z. Polym.* (1973) **251**, 415.
51. van Oss, C. J., Gillman, C. F., Neumann, A. W., *Immunol. Chem.* (1974) **3**, 77.
52. van Oss, C. J., Good, R. J., Neumann, A. W., *J. Electroanal. Chem.* (1972) **37**, 387.
53. Baier, R. E., "Epidermal Wound Healing," H. I. Maibach and D. T. Rovee, Eds., Chap. 2, Year Book Medical, Chicago, 1972.
54. Leininger, R. I., *Crit. Rev. Bioeng.* (1972) **1**, 333.
55. Baier, R. E., Lippes, J., ADVAN. CHEM. SER. (1975) **145**, 308.
56. Glantz, P., *Odontol. Revy* (1969) Suppl. 17.
57. Glantz, P., Hansson, L., *Odontol. Revy* (1972) **23**, 205.
58. Lasslo, A., Quintana, R. P., Eds., "Surface Chemistry and Dental Integuments," C. C. Thomas, Springfield, 1973.
59. Baier, R. E., "Surface Chemistry and Dental Integuments," A. Lasslo and R. P. Quintana, Eds., Chap. 5, C. C. Thomas, Springfield, 1973.
60. Quintana, R. P., ADVAN. CHEM. SER. (1975) **145**, 290.
61. Marshall, K. C., Stout, R., Mitchell, R., *J. Gen. Microbiol.* (1971) **68**, 337.

62. Corpe, W. A., "Adhesion in Biological Systems," R. S. Manly, Ed., Chap. 4, Academic, New York, 1970.
63. Marshall, K. C., Proc. 3rd Intern. Cong. Marine Corrosion and Fouling, p. 625, Northwestern University Press, Evanston, 1973.
64. Goupil, D. W., DePalma, V. A., Baier, R. E., Proc. Marine Technol. Soc., 9th Ann. Conf., Washington, D.C., 1973, p. 445.
65. Baier, R. E., Proc. 3rd Intern. Cong. Marine Corrosion and Fouling, p. 633, Northwestern University Press, Evanston, 1973.
66. Cook, M., Baier, R. E., Sargent, H., Proc. 21st Mtg. Brit. Div. IADR, 1973, Abst. **116**.
67. Cook, M., "Adhesion in Biological Systems," R. S. Manly, Ed., Chap. 8, Academic, New York, 1970.
68. Crisp, D. J., Ryland, J. S., *Proc. Roy. Soc. London* (1963) **158B**, 364.
69. Fenstermaker, C. A., Grant, W. H., Morrissey, B. W., Smith, L. E., Stromberg, R. R., "Interaction of Plasma Proteins with Surfaces," Nat. Bur. Stand. Rep. (1974) **74-470**.
70. Lee, R. G., Adamson, C., Kim, S. W., *Thromb. Res.* (1974) **4**.
71. Loeb, G. I., *J. Colloid Interface Sci.* (1969) **31**, 572.
72. Nyilas, E., Proc. 23rd Ann. Conf. Eng. Med. Biol. (1970) **12**, 147.
73. Andrade, J. D., *Med. Instrum.* (1973) **7**, 110.
74. Malcolm, B. R., *Progr. Surface Membrane Sci.* (1973) **7**, 183.
75. Loeb, G. I., *J. Colloid Interface Sci.* (1968) **26**, 236.
76. Loeb, G. I., Baier, R. E., *J. Colloid Interface Sci.* (1968) **27**, 38.
77. Loeb, G. I., *Amer. Chem. Soc., Div. Polym. Chem., Prepr.* **11**, 1313 (Chicago, September, 1970).
78. Baier, R. E., Proc. 13th Conf. Great Lakes Res., Ann Arbor, Michigan (1970) p. 114.
79. Baier, R. E., *J. Geophys. Res.* (1972) **77**, 5062.
80. Baier, R. E., Goupil, D. W., *Bull. Union Oceanogr. France* (1973), **A 1**.
81. Baier, R. E., Goupil, D. W., Perlmutter, S., King, R. W., *J. Rech. Atmos.*, in press.
82. Abe, T., *Rec. Oceanogr. Works Jap.* (1954) **1**, 18.
83. Abe, T., *J. Oceanogr. Soc. Jap.* (1962) **20**, 242.
84. Baier, R. E., *Aust. Nat. Hist.* (1972) **17**, 162.
85. Goldacre, R. D., *J. Anim. Ecol.* (1949) **18**, 36.
86. Baier, R. E., Shafrin, E. G., Zisman, W. A., *Science* (1968) **162**, 1360.

RECEIVED June 28, 1974.

2

Collagen Surfaces in Biomedical Applications

KURT H. STENZEL, TERUO MIYATA, ITARU KOHNO,
SUSAN D. SCHLEAR, and ALBERT L. RUBIN

Rogosin Laboratories, The New York Hospital-Cornell Medical Center,
Departments of Surgery and Biochemistry, New York, N.Y. 10021

*Although the biochemical and biophysical properties of
collagen are well known, its surface properties are poorly
understood. Collagen-containing structures, which are exposed to blood when endothelium is injured, initiate clot
formation. Surface properties of restructured collagen have
important biomedical applications. Several types of collagen
films with varying surface morphologies were prepared,
crosslinked with aldehydes, and implanted in rabbits. Results indicate that biologic degradation of collagen can be
controlled and delayed for at least 90 days. Inflammatory
reactions are minimal. Crosslinking decreases swelling ratios
and increases shrinkage temperature and resistance to bacterial collagenase. These studies are a base to develop
collagen for specific biomaterials and to study collagen
surface-blood interactions.*

Collagen has evolved in nature as the primary connective tissue protein
in animals. Electron micrographs of collagen from widely divergent
species reveal few, if any, differences. As a supporting structure and as a
surface for growth of cells, collagen makes an attractive biomaterial.
Methods exist for solubilizing large amounts of collagen and for restructuring it into a variety of forms for biomedical applications (1, 2).

Surface properties of this ubiquitous protein, which are of paramount
importance for many of these applications, are, however, the least studied
and least understood.

The structure and physical properties of collagen are well known.
Several recent reviews are available to interested readers (3, 4, 5, 6, 7, 8).
Some of the work pertinent to the problems of using collagen as a biomaterial along with some of our recent work on reconstituting collagen
surfaces are reviewed here.

Collagen molecules, tropocollagen, consist of three peptide chains wound together as a triple helix. The molecule is markedly asymmetrical with a length of 2800 A and a width of 15 A. It readily polymerizes to form fibrils and fibers, and in nature it exists largely as an insoluble macromolecular complex of crosslinked molecules. Only small amounts of native collagen are soluble, either in dilute acid or salt solutions. Most of the chemical and physical studies on the structure of collagen have been done with acid-soluble preparations. Of prime importance in using reconstituted collagen as a biomaterial was the discovery of non-helical appendages, termed telopeptides, on the basic tropocollagen triple helix.

F. O. Schmitt's group at MIT found that proteolytic enzymes, such as pepsin and trypsin, digested a small portion of tropocollagen but left the triple helix intact (9). Treatment of tropocollagen with proteolytic enzymes, other than collagenase, altered the interaction properties of collagen but did not result in denaturation or degradation. Nishihara and Miyata (1) developed techniques for solubilizing and purifying large quantities of insoluble collagen with controlled proteolysis at a low pH. The solubilized collagen is easily purified by repeated precipitation, washing, and resolubilization. Both physical and chemical methods have been used for restructuring or recrosslinking the enzyme-solubilized collagen as specific biomaterials. Virtually all of our work has utilized this enzyme-solubilized collagen material.

The individual peptide chains assume a random coil configuration when collagen is denatured and can exist in any one of three states. If the three polypeptide strands are not covalently linked, the collagen is known as alpha collagen. If two of the strands are covalently crosslinked, the collagen is known as beta collagen. All three polypeptide chains linked together is known as gamma collagen. The individual chains themselves can exist in either of two species, each with a similar but distinct primary structure. The two most common types of collagen polypeptide chains found in mammalian tissues are known as alpha-1 and alpha-2 chains. Most collagens are made up of two alpha-1 and one alpha-2 chains, although there are significant differences in this makeup which will be noted later. Both of the chains have a fairly typical sequence of amino acids characterized by a repeating unit of three amino acids, glycine appearing at every third position. There are large numbers of imino acids, especially proline and hydroxyproline. These probably account for the typical tertiary structure of the molecule. The polypeptide chains contain about 40 sets of polar and apolar amino acid groups. These groups may be extremely important in the surface properties of collagen materials. The typical repeating sequence with glycine in the third position does not exist for the telopeptide end regions. This part of the molecule is not in a helical configuration.

Collagen contains carbohydrate, predominantly in the form of glycosylated hydroxylysine residues, as O-galactosyl-β-glucosyl side chains. The intermolecular crosslinks in collagen occur when the enzymatic oxidation of lysine residues by lysine oxidase spontaneously crosslink *via* aldimine and aldol bonds (*10*).

Immunologic properties must be considered when any protein material is to be used as an implant or surface for blood flow in a foreign species. Since collagen has changed very little in the course of evolution, there are few antigenic determinants that are recognized in heterologous systems. The strongest of these determinants appears to reside in the protease labile, non-helical end regions. These are known as P-specific antigens and are found in the 300 A region at the C terminus of both types of alpha chains. These are species specific and protease labile. A-specific antigens have a general collagen specificity, and they crossreact among species and with enzyme-treated collagen. S-specific determinants are species specific and crossreact with both native and enzyme-treated collagen (*11, 12*). Enzyme-solubilized collagen, therefore, lacks one of the strongest antigenic determinants of the molecule. Crosslinking with a variety of reagents diminishes the immunologic activity even further. From a practical or clinical point of view, antigenicity has not been a problem even when materials such as treated bovine carotid arteries have been placed in foreign species. Thus, although the antigenic structure of collagen is important in understanding its biochemistry, it is of minor importance in its clinical application.

More pertinent to the problem of surface structure is the effect of collagen on various clotting factors. When the endothelium of blood vessels is injured or damaged, platelets adhere to the exposed subendothelial material, and a complex set of reactions that lead to thrombosis and hemostasis is initiated (*13*). The subendothelium is rich in collagen and other connective tissue proteins, but the collagen, especially, has been implicated as an initiator of the clotting mechanism. When collagen is added to platelet-rich plasma, the platelets agglutinate (*14*). Collagen has also been shown to activate Hageman factor, or Factor XII (*15*). Numerous studies have indicated that the cluster of charged groups along the collagen fibrils are important in these reactions (*14, 15, 16*). More interesting is what takes place at the collagen surface, but few studies direct themselves to this problem.

Recently, Jaffe and Deykin (*16*) presented evidence for a structural requirement for collagen-induced platelet aggregation. They found that particulate, salt-precipitated collagen was inactive in terms of platelet agglutination and that soluble monomeric collagen resulted in platelet agglutination only after a lag of about 3 min. Soluble microfibrillar collagen, however, was as active as native particulate collagen in aggregating platelets. This evidence suggests that tropocollagen is not sufficient for platelet aggregation nor is randomly precipitated (salt) collagen, but that an architectural arrangement of the molecules is required to initiate platelet aggregation. These requirements may also be necessary for platelet adhesion.

Collagen in basement membrane has a different quarternary structure from skin or fibrous collagen (*17, 18*). This type of collagen consists of three α-1 chains, has a high glycosylated hydroxylysine content, and apparently has disulfide bonds connecting it to other protein constituents. These disulfide bonds may be located in the telopeptide, or non-helical, region of the molecule. The surface of basement membrane may provide a clue to the physiological architecture required for initiation of thrombosis. Such knowledge would be of use, not only in biomaterials research, but also in understanding various kidney diseases and possibly even atherosclerosis.

Several natural collagen materials are currently being used in clinical medicine. Bovine carotid heterografts, for instance, function well in man (*19*). These grafts are prepared by cleaning bovine carotid arteries, treating them with a proteolytic enzyme (ficin), and crosslinking them with an aldehyde (dialdehyde starch). Autologous and homologous vein grafts are also used clinically. The endothelium varies from relatively poorly preserved to virtually absent in these vein grafts. The bovine vessels are all placed in areas of relatively high flow and, in the case of vein grafts, endothelialization occurs over the surfaces (*20*).

The crosslinking reagent is important in the case of bovine carotid heterografts. Chrome-tanned grafts had a high incidence of thrombosis whereas aldehyde-treated ones functioned well. Formalin-treated grafts tended to be weak, but glutaraldehyde-treated ones did not rupture (*21*).

A variety of methods for preparing collagen films from enzyme-solubilized, monomeric collagen with different surface structures were evaluated. Our initial interest was to determine the effects of various preparative techniques on the *in vivo* behavior of the films.

Experimental

In method I, the collagen solution was poured onto a methyl methacrylate plate and lowered into a $0.02M$ dibasic Na_2HPO_4 solution (pH 8.5) containing 0.1% glutaraldehyde. Collagen precipitates as native-type fibrils in phosphate buffer at this pH. The films were crosslinked as the fibers were forming before the films were dried. Crosslinking was allowed to continue for 7 or 24 hrs. The films were washed with water, plasticized with 2% glycerine, and air dried.

In method II, the collagen was not precipitated. Acidic solutions of collagen were air dried and the resulting films were crosslinked using

0.5% glutaraldehyde in .02M Na$_2$HPO$_4$ for 10 or 20 mins. These films were also washed with water and plasticized.

In method III, the collagen was first precipitated in phosphate buffer, as in method I, but the film was dried prior to crosslinking. The crosslinking was performed as in method II.

Another series of films was prepared by the same methods but crosslinked with dialdehyde starch rather than with glutaraldehyde. The crosslinking conditions were similar to those used for the glutaraldehyde crosslinked films with the exception of method III. These latter films were prepared by mixing collagen and dialdehyde starch at a ratio of 1 to 1.5. This mixture was then dialyzed against .02M Na$_2$HPO$_4$ to precipitate collagen fibers. After crosslinking, the films were washed, plasticized, and dried.

Control films of each type were also prepared using the same procedures but without the crosslinking reagent.

A third group of films was prepared, using a mixture of 20% enzyme-solubilized collagen and 80% insoluble collagen to increase the initial film strength. These films were crosslinked using glutaraldehyde in the same manner as the enzyme-solubilized collagen films. Since the solutions contained a high initial content of fiber, method III was eliminated in this group. Control films of each were also made by eliminating the glutaraldehyde.

Table I.   Swelling Ratios of Collagen Films

Method	Glutaraldehyde		Dialdehyde Starch		Composite 80% Insoluble Collagen Fiber—20% Soluble Collagen	
	Control	Cross-linked	Control	Cross-linked	Control	Cross-linked
I	19	2	19	2	7	2
II	12	2	12	2	6	2
III	19	2	19	2	—	—

Results

Swelling ratios, shrinkage temperatures, and resistance to collagenase were measured for all the films. Swelling ratios were determined by measuring the change in weight before and after hydration of the films. Crosslinking decreased the swelling ratios as compared with control films (Table I).

Shrinkage temperatures were determined as an indicator of the denaturation temperature of the films. In all cases, crosslinking greatly increased shrinkage temperatures for each preparation. Thus, crosslinking stabilizes collagen molecules and retards denaturation (Table II).

Another important aspect of these films is their resistance to degradation by biologic substances like collagenase. Each of the films was

Table II. Shrinkage Temperatures

Method	Glutaraldehyde		Dialdehyde Starch Collagen Films		Composite 80% Insoluble Collagen Fiber—20% Soluble Collagen	
	Control	Cross-linked	Control	Cross-linked	Control	Cross-linked
I	45.0°C	66.0°C	45.0°C	65.0°C	47.5°C	77.0°C
II	45.8	71.8	45.8	61.5	48.0	71.0
III	45.0	78.1	45.0	67.0	—	—

incubated for 2 hrs at 37°C with bacterial collagenase. This enzyme is less specific and more destructive than mammalian collagenase. The degradation of the films was measured by the micromoles of amino acids released per milligram film. In each case, the crosslinking increased resistance to bacterial collagenase. In some cases the films were completely resistant to it (Table III).

Each type of film was then implanted subcutaneously and intramuscularly in rabbits to assess tissue reaction and rate of resorption. The film was carefully cut into a circle with a diameter of 10 mm. Sham incisions were also made in each rabbit as controls for the amount of inflammation resulting from surgical trauma alone. The animals were killed after 7 to 180 days. The implants were examined grossly for inflammation, changes in size, and appearance of the films. Histologic preparations were also made of each film to examine the cellular reaction. By 7 days, inflammatory cuffs had encircled the implants. The control films showed evidence of being digested; they were swollen, weaker, and usually thinner.

Figure 1 is a control (not crosslinked) film prepared by method II with enzyme-solubilized collagen. The film was removed after 7 days

Table III. Resistance to Collagenase[a]

Method	Glutaraldehyde		Dialdehyde Starch		Composite 80% Insoluble Collagen Fiber—20% Soluble Collagen	
	Control	Cross-linked	Control	Cross-linked	Control	Cross-linked
I	0.230	0.028	0.230	0.000	0.473	0.000
II	0.069	0.025	0.069	0.005	0.159	0.000
III	0.218	0.000	0.218	0.004	—	—

[a] In micromoles of amino acids released per milligram of sample.

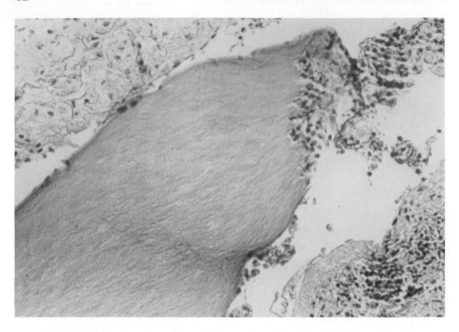

Figure 1.   Control (not crosslinked) collagen film removed after 7 days sub-cutaneous implantation in rabbits. ×128

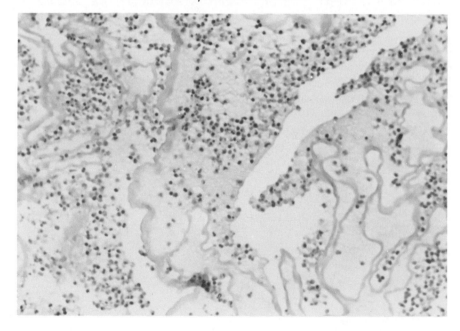

Figure 2.   Control (not crosslinked) collagen film removed after 14 days intra-muscular implantation in rabbits. ×128

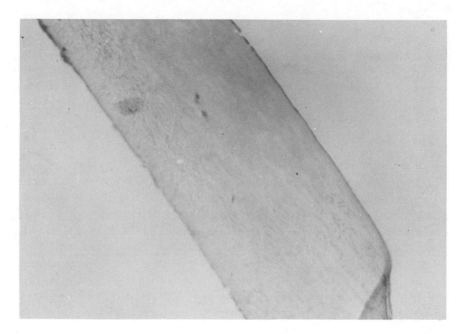

Figure 3. Method II glutaraldehyde crosslinked collagen film removed after 90 days subcutaneous implantation in rabbits. ×128

of subcutaneous implantation. Note the foreign body type of tissue reaction with histiocytes and beginning resorption of the collagen.

Figure 2 is the same type of film removed after 14 days of intra-muscular implantation. Here the collagen is fragmented and infiltrated with inflammatory cells. After 21 days, all control films disappeared except for the ones prepared from a mixture of insoluble collagen fiber and enzyme-solubilized collagen. Some of these remained for 30 days.

The inflammatory response with crosslinked films was minimum and not dissimilar to the sham operations. Grossly, there was little in-flammation at day 7. What inflammatory reaction appeared, peaked around day 14. This was characterized by an enlarged cuff of cells sur-rounding the implant occasionally accompanied by a fluid exudate.

All of the crosslinked implants lasted for at least 60 days. At this time, a few of the enzyme-solubilized collagen films crosslinked with either glutaraldehyde or dialdehyde starch appeared to be thinner. Some of the films crosslinked with glutaraldehyde appeared to be somewhat thinner after 90 days, although most were intact.

Figure 3 is a method II enzyme-solubilized collagen film crosslinked with glutaraldehyde and implanted subcutaneously. It was unchanged after 90 days.

The films prepared from a mixture of collagen fiber and soluble collagen crosslinked with glutaraldehyde were all intact after 90 days implantation.

Figure 4 is a method II fiber film that was implanted intramuscularly for 90 days. The thin, delicate surrounding fibrous tissue can be seen partially adhering to the surface of the film. The 180-day films have not yet been removed.

Discussion

These results indicate that the biologic degradation of collagen can be controlled and delayed for at least 90 days. It is relatively unimportant whether collagen is crosslinked in the wet or the dry state or whether the films are precipitated or not prior to crosslinking. Both glutaraldehyde and dialdehyde starch are effective in stabilizing collagen molecules and retarding their bio-degradation for at least 90 days and perhaps longer. The inflammatory response has been minimum, reaching a peak at 14 days and then subsiding. The films are eventually covered by a thin, fibrous membrane which in some cases adheres to the surface of the film.

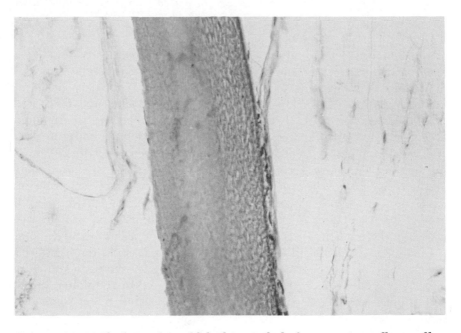

Figure 4. Method II glutaraldehyde crosslinked composite collagen fiber-soluble collagen film removed after 90 days intramuscular implantation in rabbits. ×128

These studies provide a base for further development of collagen materials for specific biomaterial applications. Further studies are being directed to the effect of collagen film on blood in terms of platelet and white cell adhesion, protein absorption, and thrombogenicity.

References

1. Nishihara, T., Miyata, T., *Collagen Symp. Jap.* (1962) **3**, 66-84.
2. Drake, M. P., Davidson, P. F., Bump, S., Schmitt, F. O., *Biochemistry* (1966) **5**, 301-312.
3. Gallop, P. M., Blumenfeld, O. O., Seifter, S., *Ann. Rev. Biochem.* (1972) **41**, 617-672.
4. Bailey, A. J., "Comprehensive Biochemistry," M. Florkin and E. H. Stotz, Eds., **26**, pp. 297-423, American Elsevier, New York, 1968.
5. Traub, W., Piez, K. A., *Advan. Prot. Chem.* (1971) **25**, 243-352.
6. Piez, K. A., *Ann. Rev. Biochem.* (1968) **37**, 547-571.
7. Yannas, I. V., *J. Macromol. Sci. Rev. Macromol. Chem.* (1972) **C7** (1), 49-104.
8. Stenzel, K. H., Miyata, T., Rubin, A. L., *Ann. Rev. Biophys. Bioeng.* (1974) **3**, 231-253.
9. Rubin, A. L., Pfahl, D., Speakman, P. T., Davidson, P. F., Schmitt, F. O., *Science* (1973) **13**, 37-38.
10. Tanzer, M. L., *Science* (1973) **180**, 561-566.
11. Steffen, C., Timpl, R., Wolff, I., *Immunology* (1968) **15**, 135-144.
12. Timpl, R., Wolff, I., Wick, G., Furthmayr, H., Steffen, C., *J. Immunol.* (1968) **101**, 725-729.
13. Baumgartner, H. R., Handenschild, C., *Ann. N.Y. Acad. Sci.* (1972) **201**, 22-36.
14. Wilner, G. D., Nossel, H. L., LeRoy, E. C., *J. Clin. Invest.* (1968) **47**, 2616-2621.
15. *Ibid.* (1968) **47**, 2608-2615.
16. Jaffe, R., Deykin, D., *J. Clin. Invest.* (1974) **53**, 875-883.
17. Kefalides, N. A., *Biochem. Biophys. Res. Commun.* (1971) **45**, 226-231.
18. Kefalides, N. A., *Int. Rev. Conn. Tiss. Res.* (1973) **6**, 63-104.
19. Rosenberg, N., Lord, G. H., Henderson, J., Bothwell, J. W., Gaughran, E. R. L., *Surgery* (1970) **65**, 951-956.
20. Reichle, R. A., Stewart, G. J., Essa, H., *Surgery* (1973) **74**, 945-960.
21. Bothwell, J. W., Lord, G. H., Rosenberg, N., Burrowes, C. B., Wesolowski, S. A., Sawyer, P. N., "Biophysical Mechanisms in Vascular Homeostasis and Intravascular Thrombosis," P. N. Sawyer, Ed., pp. 306-313, Appleton-Century-Crofts, New York, 1965.

RECEIVED July 7, 1974. Work was supported in part by the National Science Foundation and John A. Hartford Foundation.

Proteins as Substrates: Introduction

ROBERT E. BAIER

One area of applied chemistry where protein-dominated surfaces are of enormous importance is that of food science. Contributions representing this globally critical science have not been included in this volume, however, because a dedicated press has already published up-to-date resource material on the subject (1, 2, 3, 4, 5). A noteworthy recent article emphasizes the specific role of food proteins as substrates which can adsorb and preferably orient materials at their own interfaces (6).

Here, we focus first on the most expansive organ of the animal body, the skin, which is dominated in its interactions with all environments by the flat stratum corneum cells which represent more than 99% of the skin surface. This section opens with a major contribution from a military laboratory concerned for decades with the protective function of skin which provides its resistance to chemical threats. This up-to-date review article by M. M. Mershon makes it clear that the skin surface is dominated by protein and proteoglycan components. Yet, the outer cell membranes resist enzyme digestion, protein solvents, and strong alkali. Further research in this subject must be fostered to explain the surface properties of this chemically resistant envelope which, nonetheless, absorbs greater than five times it own weight of water. Mershon's review also reiterates the point that protein-dominated surfaces (especially of stratum corneum) can present lipophilic surfaces to lipophilic solvents and hydrophilic surfaces to water and aqueous solutions. Thus, the concept of proteinaceous interfaces as responsive interfaces, adjusting spontaneously to minimize gradients in interfacial tension, is endorsed.

The second paper in this section reviews and summarizes studies of an important industrial group working with skin and other protein-dominated surfaces. This well-illustrated contribution provides important fundamental information on skin, comparing its properties with both wool and hair. Data included demonstrate the dominance of α-helical protein molecules in the barrier layers, the role of water as the plasticizer of interfacial structural proteins, the resistance of such films to penetration by strong solvents or acids, and the admixture of lipids with natural proteins. Data acquired on simple polyamide models for proteins, such as the nylons, are also reported. The physical/chemical techniques used by

Figure 1.   Direct analysis of naturally moist protein layers in their functional state. Forearm of subject rests comfortably on internal reflection prism mounted in recording spectrophotometer.

the group from Johnson & Johnson to characterize stratum corneum are similar to those used (as described later in this volume by Malcolm) by chemists working with protein films at gas/liquid interfaces.

A paper from the Lever Brothers Research Center describing a technique for studying water vapor interactions with excised skin surfaces is the only method-focused article in this volume. It is presented here because of the general applicability of the method described to all protein surfaces under consideration and even to nonprotein surfaces, such as poly(hydroxyethyl methacrylate) hydrogels discussed later in this volume by Horbett and Hoffman. This article by El-Shimi and co-workers points up the enormous importance of adsorbed water in the function (and structure?) of natural materials.

In this regard, it is also important to dispel the notion that highly hydrated materials, such as proteins, in their natural functional stages might not be amenable to direct analysis. Many investigators believe that the natural state of hydration of protein-dominated boundaries must prevent the direct, in-place analysis of such layers by water-sensitive techniques such as IR. This is not true. Figure 1 illustrates the simplicity of direct analysis of naturally moist protein layers in their functional state, in this case providing the active barrier function of human skin as discussed in detail in the following chapters. Figure 2 presents an internal reflection IR spectrum which illustrates the typical spectral contrast which can be obtained with hydrated protein films including skin. In this case, the sample is a gelatin product acceptable for the preparation of photographic emulsions.

Photographic products probably make greater use of, and depend more crucially on, the surface properties of protein preparations than do any other commercial product. The art of preparing acceptable gelatin batches for photographic end uses is closely held. Although most such research remains proprietary, the subject prompts intensive experimental efforts. We are fortunate in having a contribution on this subject in the next section of this volume from a leading industrial laboratory, but our scientific curiosity about the surface properties of the protein preparations actually used remains piqued. Figure 3 and Table I serve the dual purpose here of introducing the contact angle technique which is the subject of two basic chapters in this section and characterizing the wettability of one commercial gelatin preparation which has been used widely in photographic products. We are indebted to Henry E. Wilkie, recently retired after 55 years as chief chemist of the now-closed Chalmers Gelatin Co. of Williamsville, New York, for providing samples of the product gelatin which has been used for decades in fine quality photographic emulsions. Figure 3 plots the contact angle values, measured for a variety of highly purified wetting liquids on the surfaces of gelatin samples (as cast in thin sheet form at the site of manufacture, as re-cast from hot water in the laboratory, and as cast from hot water using a granulated and blended version of the product also prepared at the manufacturing site). Table I lists the actual wetting liquids used and their liquid/vapor surface tensions at 20°C along with the average contact angles obtained

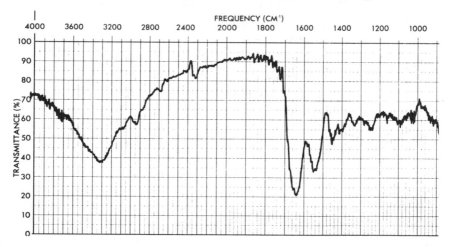

Figure 2.   Internal reflection infrared spectrum of hydrated gelatin (from water buffalo hide) used in photographic emulsions

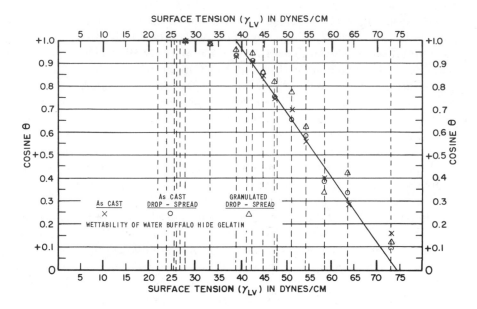

Figure 3.   Contact angle data characterizing the wettability of a commercial gelatin preparation used in photographic emulsions

on these three versions of the same protein stock material. Throughout this volume, the terms "critical surface tension" and "Zisman plot" are used. The extrapolated intercept at the cosine-contact-angle = 1 axis is defined as the critical surface tension. As chapters near the end of this section will discuss more completely, the simple contact angle behavior

Table 1.   Contact Angles on Water Buffalo Hide Gelatin

	Average Contact Angle (θ)		
Wetting Liquid and Surface Tension (γ_{LV}), dynes/cm, 20°C	*As Cast (Smooth Flat Sheet)*	*As Cast Drop-Spread at >80°C*	*Granulated and Drop-Spread at >80°C*
Water	72.8	81	84	83
Glycerol	63.4	73	70	64
Formamide	58.2	66	67	70
Thiodiglycol	54.0	55	54	50
Methylene iodide	50.8	46	48	38
sym-Tetrabromoethane	47.5	42	41	34
1-Bromonaphthalene	44.6	31	30	29
o-Dibromobenzene	42.0	25	24	19
1-Methylnaphthalene	38.7	22	21	14
Dicycolhexyl	33.0	10	8	9
n-Hexadecane	27.7	0	0	0

revealed in Figure 3 and Table I for this photographic quality gelatin is not necessarily typical of highly purified protein materials. In the case of water-buffalo-hide gelatin with acceptable photographic characteristics, solution of the manufactured material at temperatures above 80°C does not sufficiently randomize the polymeric structure to produce evidence for increased surface polarity or surface-hydrogen-bonding interactions in the contact angle data. Does this apparent resistance to surface chemical change differentiate gelatin preparations acceptable for photographic end use from the large majority of gelatin preparations which are not?

The paper by El-Shimi and Goddard in this section provides an important demonstration of the variety of interactions which protein surfaces can exhibit, as if in direct response to the variety of challenging environments. This contribution also follows closely some of the concepts introduced in the earlier reports of the study of skin, as well as the earlier model studies with simple polyamides like nylon. Following El-Shimi and Goddard's summary of theoretical descriptions of wetting and spreading phenomena on the chemically heterogeneous surfaces of proteins, a paper by W. A. Zisman and me reports the changes of such wetting properties with collagen and gelatin preparations which correlate with molecular conformation changes and hydration of the proteins. Closing this section is a description of the surface modifications of collagenous surfaces which presents immediately useful grafting techniques. This article, from the Dental Research Section of the National Bureau of Standards, demonstrates direct coupling of modifying reagents to collagen present in powders, films, or in skin, bone, or the dentin fraction of teeth. One example of the direct commercial utility of the techniques described would be in their application to the mold-proofing of a variety of protein-dominated materials.

Literature Cited

1. H. W. Schultz, A. F. Anglemier, "Proteins and Their Reactions," AVI, West-port, Conn., 1969.
2. A. K. Smith, S. J. Circle, *Soybeans: Chemistry and Technology, Vol. I* "Proteins," AVI, Westport, Conn., 1972.
3. G. E. Inglett, "Seed Proteins," AVI, Westport, Conn., 1972.
4. L. P. Hanson, "Vegetable Protein Processing, 1974," *Food Technol. Rev.* No. 16, Noyes Data Corp., Park Ridge, New Jersey, 1974.
5. Gutcho, M., "Textured Foods and Allied Products, 1973," *Food Technol. Rev.* No. 1, Noyes Data Corp., Park Ridge, New Jersey, 1973.
6. E. Berlin, B. A. Anderson, M. J. Pallansch, "Hydrocarbon Binding by Particles of Bovine Casein," *J. Colloid Interface Sci.* (1974) **48**, 470.

Barrier Surfaces of Skin

M. M. MERSHON

Biomedical Laboratory, Edgewood Arsenal, Aberdeen Proving Ground, Maryland 21010

Stratum corneum, the nonliving layer of skin, is refractory as a substrate for chemical reactions, but it has a strong physical affinity for water. The chemical stability of stratum corneum is evident in its mechanical barriers which include insoluble cell membranes, matrix-embedded fibers, specialized junctions between cells, and intercellular cement. The hygroscopic properties of stratum corneum appear to reside in an 80 A-thick mixture of surface-active proteins and lipids that forms concentric hydrophilic interfaces about each fiber. This combination of structural features and surface-active properties can explain how stratum corneum retains body fluids and prevents disruption of living cells by environmental water or chemicals.

The stratum corneum, the outermost layer of skin, controls water flux and thereby protects our tissues from fatal drying and from osmotic damage by bathing. This tough, flexible layer restricts the passage of most substances to a degree that varies with the substance and the conditions of exposure. Permeation is limited by interrelated contributions of structural elements, time factors, spatial arrangements, chemical compositions, and physical properties.

Differences Among Barriers

Size. Stratum corneum surfaces that impede diffusion fall into four sizes. The largest structures, such as hair and the cornified surface, are visible. Cell surfaces, their interfaces, and their elaborations are microscopic. Keratinized elements are ultramicroscopic. Finally, interactions between solvents, solutes, and barrier surfaces occur on the molecular level. The functions of these structures are integrated into an overall barrier capability.

PERCUTANEOUS ABSORPTION OF SOLUTIONS WITH TIME AND SITE

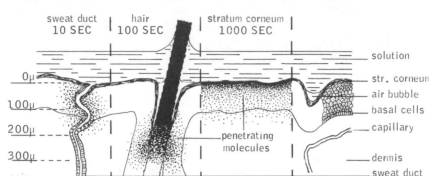

Figure 1.   Schematic drawing showing relative sizes of sweat ducts, hair follicles, and cornified layers of human abdominal skin. Approximate transient concentrations of polar molecules in each site at different times are shown by variations in density of stippling. Adapted from Ref. 1.

Routes of Entry. Microscopic sections show the stratum corneum (SC) of the abdomen as thin layers of dead, flattened cells arrayed over a much thicker layer of epithelial cells. Both layers are pierced at intervals by hair follicles and sweat ducts (Figure 1) (*1*). Sebum flows into, lubricates, and tends to fill the space between each hair shaft and its surrounding conical sheath (*2*). Sweat ducts are cellular tubes that spiral through epidermis with increasing radius and decreasing pitch (*3*). Therefore, they approach the surface at an acute angle and empty through slit-like pores (*2, 3*).

Conflicting evidence and divergent views exist concerning the relative capacities of hair follicles, sweat ducts, and SC as parallel routes for admission of chemicals (*1, 2, 4–21*). Preferential penetration of follicles and their associated sebaceous glands is suggested by tracer studies using dyes, heavy metals, and sulfonamides (*4*). This indication is reinforced by results with applications of histamine or naphthazoline base (*5*), hydrocortisone or organophosphorus pesticides (*6*), or other organophosphates (*7*). These drugs produce enhanced responses in areas where sebaceous glands are most numerous. However, responses to ethyl nicotinate (*5*) and radioactive tracer studies with tributyl phosphate (*8*) show no difference between hairy and hair-free areas. Furthermore, Maibach and associates suspect that differences in SC, rather than in follicular penetration, account for increased absorption in hairy sites (*6*). Fredriksson (*9*) suggests that chemical affinities can explain the high concentrations of labeled pesticides that he found in ducts and follicles (*10*).

The available data are similarly divergent for penetration *via* sweat ducts. Van Kooten and Mali (*11*) estimate that sweat ducts contain 70% of the precipitate formed when potassium ferrocyanide and ammonium ferrisulfate diffuse into skin from opposite sides (*10, 12*). This result suggests preferential sweat-duct conduction of strongly ionized compounds. Wahlberg (*13*) came to the opposite conclusion, *i.e.*, that sodium chloride passes through SC faster than it does through ducts or follicles (*6, 12*). Such strongly ionized or polar compounds generally penetrate skin very poorly (*4, 14, 15, 16*).

Scheuplein and associates report that polar steroids penetrate preferentially through sweat ducts and hair follicles while more lipid-soluble steroids and primary alcohols penetrate rapidly in SC (*16, 17*). Surfactant properties of compounds facilitate their penetration of ducts and follicles (*18*). However, massage displaces air and improves penetration through follicular structures (*19*). Lindsey (*14*) reconciles such diverse evidence by stating that barrier properties do not depend on structures, but they vary with the properties of the penetrant.

Hair shafts also constitute routes of entry; water and other chemicals tend to diffuse through hair shafts more readily than through SC (*20*). Some organic solvents enter skin through unprotected hair that protrudes through an impervious coating over SC, pores, and follicular openings (*22*).

Biphasic Diffusion. Much of the confusion about routes of entry has been resolved by Scheuplein (*1*). His work shows that any one of the routes can be dominant under appropriate conditions. For instance, dominance may vary with time (Figure 2). Scheuplein observed biphasic diffusion with rapid onset and initial dominance of flux *via* ducts, hair,

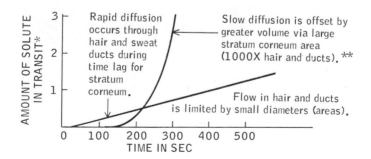

Figure 2.   Relative time course and capacities of stratum corneum and other routes for transfer of polar solvents through skin. Data from Ref. 1.

  * *(Amount of substance)/(thickness of membrane) × concentration × 10³*
** *Human abdominal skin*

and follicles. The short lag time is explained by data on diffusion constants (*1*). These show that some polar molecules travel 50 times as fast in hair and follicles and up to 1000 times as fast in sweat ducts as in hydrated SC. The small diameters (70 μ) and high flow rates (*1*) of these tiny tubes permit them to fill quickly and deliver solute to the dermis with short lag times.

In the second phase, SC becomes saturated and begins to deliver a greater volume of penetrant to the dermis than is carried by all ducts and follicles combined (Figures 1 and 2). This condition occurs because the surface area of SC may be 1000 times greater than the combined surface areas of all ducts and follicles (*1*). This ratio and differences in permeability vary with numbers of pores and hairs in various skin locations (*7*). SC resembles a large pipe with slow rates of filling and flow, permitting it to carry volumes that greatly exceed the total capacity of the small pipes.

Stratum Corneum Structure. Reviewers agree that for most compounds the rate-limiting barrier properties of skin are located within the SC (*2, 10, 12, 15, 16, 21*). The texture and cohesiveness of this tissue are familiar to anyone who has ever peeled bits of it from sunburned skin; large sheets of SC can be separated from skin (*2*). Such sheets look like used polyethylene film, and their resistance to water diffusion approximates that of Mylar film of similar thickness (*16*). However, each sheet is a mosaic made of individual cells (*2*).

A top view of SC cells (Figure 3) shows some features that are essential to barrier functions. Their flat surfaces represent about 99% of the exterior skin surface. Attachment sites (desmosomes) and overlapping interfaces hold cells together. About 15 layers of SC cells, total-

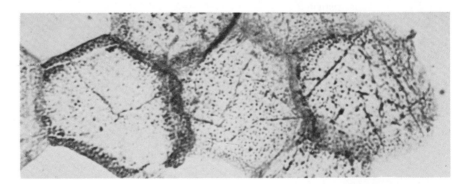

The Epidermis

Figure 3.   Cornified cells on adhesive slide show hexagonal shape and dark bands of overlap. Areas of intercellular attachment (desmosomes) appear as dark stippling (after modified Gram staining). Magnification: x1065 (2).

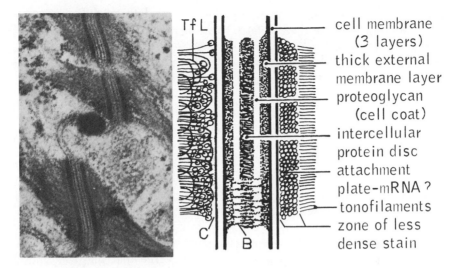

Ultrastructure of Normal and Abnormal Skin

Figure 4.   Desmosomes within the granular layer and in a schematic diagram. The electron microphotograph (x93,000) is from Ref. 44.   The diagram is adapted from Refs. 26 and 35.   The tonofilament loops (TfL), cement layer (C), and bridging material (B) of Kelly (26) are represented on the left side of the diagram.

ling about 12μ in thickness, can be counted in SC from the human back (*2*). Individual cells vary from 25 to 45μ in diameter (*23*) but they are only 0.8 to 1.0μ thick; thus each cell is at least 30 times as wide as it is thick (*2*). There is some tendency for these cells to be stacked in vertical columns (*2, 24*). Loss of this stacking pattern is observed in certain diseases such as psoriasis; diffusion of water and other substances is increased in such diseased skin and in the unstacked normal callus of palms and soles (*16, 24*).

Desmosomes and Filaments. The cohesiveness and internal structure of SC cells are dependent on the structures that are called desmosomes (*25, 26, 27*). These structures unite SC cells of different layers (Figure 3) and of the same layer (Figures 4 and 5). The sequence of embryonic development is a thickening of opposed cell membranes followed by intercellular disc formation and intracellular formation of the attachment plate (*25, 29*). This sequence is blocked if normal protein production in rough endoplasmic reticulum is altered by virus activity (*28*). When complete, the desmosomes of basal cells sprout filaments from their attachment plates (*25, 27, 30*). These filaments become longer, thicker, and more numerous as cells differentiate and migrate toward the skin surface (*31, 32*). Filaments arise either from the attachment plate (*27*) or they form loops that are anchored there (*26*). The smallest filaments form complexes with each other, then add newly synthesized molecules to become tonofilaments (32, 33). These filaments anastomose to unite various desmosomes and provide the cell with an internal meshwork skeleton (30, 31, 34). According to Mercer (25), "they are quite literally tonofibrils—they occur in precisely the situations where support is demanded."

Desmosomes and Intercellular Material. The extracellular structure of desmosomes (Figure 4) includes two proteoglycan layers (acid mucopolysaccharide) that are continuous with the surface coats of each cell (26, 35). Staining responses suggest that the intercellular disc and the thick external layer of the tripartite cell membrane both contain protein and acid proteoglycans (35). Kelly (26) suggests that the disc is formed by overlapping of protein strands originating in the thick membrane layer (Figure 4).

Studies of blistering (acantholysis and vesiculation) show that disc proteins are attacked by papain, trypsin, and by reagents that split sulfur or hydrogen bonds (36, 37). Desmosomes and cells are also separated by extraction of calcium ions (35, 37), presumably from proteoglycans. Desmosomes are completely dissolved *in vitro* by chloroform:methanol mixtures (38) that dissolve proteolipids (39). The separated cells become reaggregated after evaporation of the solvent leaving an intercellular deposit with normal appearance by electron microscopy (38).

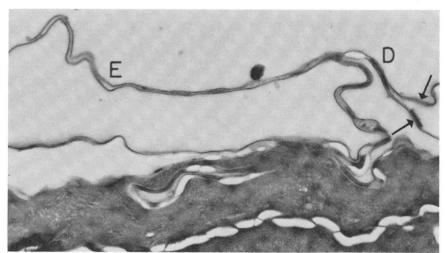

Journal of Investigative Dermatology

Figure 5. Cross section through an external surface of stratum corneum after 0.5-hr digestion with subtilisin. The thick membrane envelopes (E) of the digested cells and of lower cell layers remain attached by desmosomes (D). The arrows indicate the residues of a separated desmosome. Magnification: x17,770
(67).

The above results suggest intimate association of proteins and proteo-glycans in desmosomes and in surface coats of cells. A possible example of this kind of association is provided by studies of films made with various proportions of protein and proteoglycan. Mixtures of 2 parts gelatin and 1 part hyaluronic acid are much more resistant to diffusion of various substances than other combinations of these ingredients (*40*). Perhaps intercellular discs represent mechanically dense and viscid complexes of similar nature. The disappearance of desmosomal components into re-constituted intercellular cement suggests that they have similar chemical components but different stoichiometry.

Variable viscosity and separation of desmosomal surfaces by enzymatic activity are implicit in results showing that labeled epidermal cells occasionally trade neighbors (*41, 42*). Rather than being irreversibly united at desmosomes, cells appear to slide laterally over each other (*30*). Some cells reach the skin surface in 1 week while others straggle along for 6 weeks (*41*). Stoughton concluded that demosomal protein includes a proteolytic enzyme system that requires oxygen and that can be artificially activated with cantharidin to separate cells (*36*). Such a system may explain the infrequently observed separation of facing desmosomal halves in normal tissue (*42*). Half desmosomes and cells slide by each other after enzymatic softening of the protein-proteoglycan adhesive (*36*). Then new pairings occur with secretion of fresh proteoglycan and protein (*30, 41*). Some material for desmosome formation apparently originates within the rough endoplasmic reticulum (*28*).

Intercellular Volume and Penetration. The surface coats of epidermal cells occupy intercellular space and these gelatinous layers probably act as watery diffusion channels for nutrients (*35*). Surface coats are quite different from keratinized intercellular cement (*32, 43*); keratinization modifies desmosomes and intercellular materials to resist diffusion and premature separation in SC (*32, 43, 44*).

Although SC appears as loosely separated layers in micrographs (Figure 5), this appearance is false. It results from loss of intercellular cement during preparation for sectioning (*2*). However, such views do show some of the interdigitations that cooperate with desmosomes to lock cells together (*44*). The intercellular cement probably fills all space between these mechanical junctions in SC (*32*).

The top view of SC cells in Figure 3 (*2*) shows both the horny plates and their overlapping interfaces. Chemicals could enter SC through the intercellular spaces, the cellular plates, or both. But opposing opinions on the subject have been expressed. In 1964 Kligman (*2*) said,

It is to be emphasized that, structurally, the horny cell is very simple, containing essentially nothing more than filaments and matrix. . . . This dense and total fibrous packing would offer tremendous resistance to penetration by chemicals. That agents enter skin by direct diffusion through such "solid" cells is improbable. The failure of aniline dyes to stain the contents of monolayers of horny cells is practical evidence of this. On the other hand, these become deeply stainable after being damaged by protein denaturants such as anionic detergents. We hold that substances permeate the horny layer via intercellular spaces. For quite different reasons, Tregear (8) insists that agents pass between the cells, not through them.

In 1971 Scheuplein and Blank (16) stated,

Current electron-microscopic evidence shows that intercellular regions in SC are filled with lipid-rich amorphous material (43, 45, 46). In the dry membrane the intercellular diffusion volume may be as large as 5% and at least 1% in the fully hydrated tissue . . . this intercellular volume is at least an order of magnitude larger than that estimated for the intra-appendageal pathway and allows the possibility of significant intercellular diffusion. Diffusion between cells cannot be ruled out, but various data show that diffusion cannot be primarily intercellular.

These authors continue by pointing out that polarity and molecular weight influence the route of penetration.

Such differences in opinion continue to exist. These differences could be resolved by separate measurements of flux *via* the respective routes. We remain unable to make such measurements.

Intercellular Area and Penetration. The relationship of SC and resistance to penetration somewhat resembles that of asphalt tile and protection of flooring. The tiles are impervious to water; therefore, any water spillage that reaches the flooring must penetrate the joints between tiles. However, various organic solvents will dissolve in and diffuse through the tile. We can now draw some new conclusions because the areas of cellular tiles and joints have been measured.

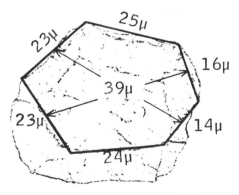

Such cornified cells as this one from a male abdomen are nearly 40 μ long and 1000 μ² in area. Given this 125 μ perimeter and 150–400 Å interspaces, the ratios of interspace to surface areas range from 1:1000 to about 1:250.

Journal of Investigative Dermatology

Figure 6.   Cell removed from the abdominal skin of a man by washing with a detergent solution shows average dimensions of 200 measured cells. Magnification: x1018 (23).

A rough estimate of the relative surface areas projected by intercellular space and by cell membranes can be made from Plewig and Marples data on the length, width, and surface area of cells from a number of sites (23). They also show microphotographs of measured cells including a typically hexagonal cell from an abdominal site. The linear perimeter of this cell has been related to their values for average dimensions and areas of abdominal cell surfaces (Figure 6). Cellular interspaces are about 150 A wide in most tissues but may reach 400 A between keratinized cells (35). Half of the surface area for a 125-μ perimeter of 150-A width is about 1.0 μ^2, one-thousandth of the membrane area (23). The corresponding ratio for the 400-A width is 1:250.

Scheuplein (1) concluded that diffusion rates are high but transport capacities are limited in hair and ducts because these structures occupy only one-thousandth of the total skin surface area. The present data suggest that intercellular cement and desmosomes occupy 0.001–0.004 of the remaining area. Accordingly, penetration through the interspaces will be limited by their relatively small area for exposure. However, the interspace route could be significant for molecules that penetrate SC cells poorly. As with asphalt tile, properties of the solvent and composition of the cellular tile must be considered.

Ultramicroscopic Surfaces

Keratin Structure. Keratin is a nonspecific term applied to various insoluble aggregations in hair, nail, skin, and mucosa. Rothberg (33) says of skin, ". . . what has been called and analyzed as keratin in the past is the total products of epidermal metabolism which are not returned to the metabolic pools but are instead excreted with the cornified epidermal cells." Much has been learned by selective extraction and analysis of these complex products, but the question often arises—what was analyzed? Biologists have tried to determine keratin composition by observing synthesis, assembly, and differentiation of the composite parts.

Products of Differentiation. The distinguishing features of basal cells are their hemidesmosomal attachments to the basement membrane and their formation of filaments at desmosomes (Figure 7); otherwise, these cells resemble other reproducing cells (44). Migration away from the basement membrane is associated with a spinous appearance of tonofilaments at desmosomes and early signs of membrane-coating granules (32, 47). The granular cell layer is named for masses of irregular keratohyalin that surround and infiltrate spaces between tonofibrils (44). Keratohyalin (KH) granules vary in size, shape, and homogeneity in various tissues and species (48, 49). Dense homogeneous deposits are observed in some KH granules (49). The rest of the KH consists of spherical particles 20 A (50) or 50–100 A in diameter (51). Numerous ribosomes surround (49) the 20 A macromolecular aggregates of amorphous structural protein (50). KH and endoplasmic reticulum proteins that infiltrate the filaments probably constitute about half of the horny cell content (47).

Many epithelial cells are connected by tripartite junctional complexes (Figure 7) that include desmosomes, intermediate gap junctions of 20 A, and tight junctions that unite opposing cell membranes into a pentalaminar unit (52, 53, 54). Gap junctions have low electrical resistance and are suspected of providing communication between cells (55). Work with autolyzed cells suggests that gap junctions bind ribosomes and cal-

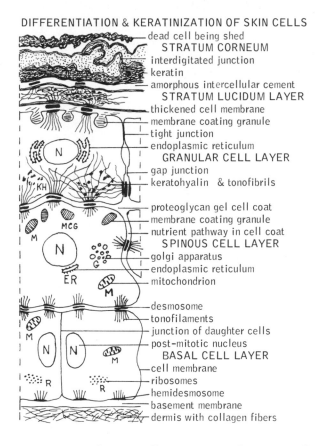

DIFFERENTIATION & KERATINIZATION OF SKIN CELLS

dead cell being shed
STRATUM CORNEUM
interdigitated junction
keratin
amorphous intercellular cement
STRATUM LUCIDUM LAYER
thickened cell membrane
membrane coating granule
tight junction
endoplasmic reticulum
GRANULAR CELL LAYER
gap junction
keratohyalin & tonofibrils

proteoglycan gel cell coat
membrane coating granule
nutrient pathway in cell coat
SPINOUS CELL LAYER
golgi apparatus
endoplasmic reticulum
mitochondrion

desmosome
tonofilaments
junction of daughter cells
post-mitotic nucleus
BASAL CELL LAYER
cell membrane
ribosomes
hemidesmosome
basement membrane
dermis with collagen fibers

Figure 7. Schematic illustration of ultrastructural changes at successive stages of epithelial cell keratinization. Adapted from Refs. 25 and 32.

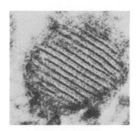 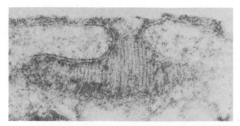

Ultrastructure of Normal and Abnormal Skin

Figure 8.   These electron micrographs by Robert Scoggins show an intact membrane-coating granule (x196,000) and another granule (x105,000) within an invagination of the trilaminar cell membrane (44).

cium ions after injury or lysis to seal the cell membrane (56). Tight junctions form a continuous belt that obliterates intercellular space (52) and prevents diffusion to or from the cell surface, as demonstrated with lanthanum (57). Such sealing occurs after cells lose substance during lysis (47). This change and changes in desmosomes of cells in transition (44) suggest that the proteoglycan surface coat of the cell is lysed, possibly by enzymes of membrane-coating granules.

Extracellular Changes. Membrane-coating granules (Figure 8) were named following observation of the extrusion of their contents into spaces between granular cells and SC (46). These granules are also called keratinosomes because they resemble lysosomes in several ways (58, 59, 60). Both originate in the Golgi apparatus and contain acid phosphatase plus other hydrolases (59). They have similar proteolytic responses to ultraviolet radiation (60). Both show enhanced staining with osmium zinc iodide (OZI) (61). The stained material that lies between keratinosome lamellae is associated with sterols and other nonpolar lipids because extraction with hexane prevents OZI staining without disrupting tissue patterns (61). The visible lamellae probably contain chondroitin sulfate B or other proteoglycans, cholesterol–hexosamine compounds, and ordered phospholipids like those in myelin structures (32, 59, 62).

As granular cells mature, their membranes fuse with those of keratinosomes, which must extrude their lamellar content through both membranes (32, 44, 61). This process (Figure 8) may involve either selective lysis or opening of a pore in the fused membranes (54). Authorities suggest that the coating material becomes part of intercellular cement which unites SC cells, seals the interspace, and becomes a diffusion barrier (32, 43, 46). Meanwhile, the extracellular portions of desmosomes become opaque, and tight junctions are formed around the lateral cell margins (44, 57). This sequence probably represents lysis of the outermost proteoglycan surface coats followed by mixing of the residues with the membrane-coating materials (32, 47).

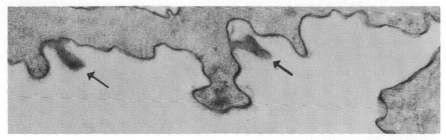

Ultrastructure of Normal and Abnormal Skin

Figure 9. Flakes of intercellular cement (arrows) can be seen attached to the membrane of a cell separating from oral mucosa of a mouse. Magnification: x26,400 (32).

The modified desmosomes and intercellular cement of SC are much more resistant to mechanical and chemical disruption than their counterparts between living cells; yet they come apart on schedule for desquamation (*2, 26, 58, 62*). The intercellular material includes resistant membranous residues (*46*) and an amorphous substance that does not resist keratolytics (*32*). Normal shedding is dependent on proper cholesterol metabolism and may involve slow lysis of a cholesterol–hexosamine compound and chondroitin sulfate B (*58, 62, 63*). The persistence of membranous bodies between SC cells (*46*) suggests that specialized lysosomes may be present to initiate desquamation. Desmosomal separation at the distal membrane surface has been reported (*64*) and is observable (Figure 5) in membranes treated with subtilisin (*67*). However, other pictures show disintegration of the entire intercellular disc (*44*). Flakes of intercellular cement, absence of desmosomes, and convoluted shapes can be observed among desquamating SC cells (Figure 9).

Membrane Thickening. Cell envelopes are usually neglected in discussions of skin penetration, although Matoltsy (*65*) wrote, "the cell membrane can be looked on as the specific protective element of each cornified cell, since it is much more resistant than the epidermal keratin." He also observed that loss of barrier function is maximum when SC is treated with reagents that affect membranes (*66*).

The tough, reinforced cell membrane is the first organized barrier surface presented to most impacting molecules. The nature of these cornified cellular envelopes (Figure 5) has been clarified by work with subtilisin, one of the proteolytic enzymes used in laundry products (*67*). Rupture of cell membranes opens them to lysis of their keratin content, but the membranes resist subtilisin, trypsin, pronase, protein solvents, and keratolytic agents (*32, 67*). Thickened membranes also resist strong alkali, remaining as 5% of the original dry weight after the rest of the cell dissolves at pH 13 (*2, 66*).

Cell membranes abruptly change from 70 to 150 or 200 A in thickness shortly after discharge of the membrane-coating material into the extracellular space immediately below SC (*32, 43, 46*). Amorphous keratinosome material may be deposited (*32, 43*), but most of the thickening occurs in the intracellular leaflet of the tripartite membrane (*58, 61*). This increase is attributed to deposits of cystine-rich KH particles and other cytoplasmic substances (*32, 48, 68*). A similar situation is observed in autolyzed cells when calcium ions (presumably released from lysed proteoglycan) modify gap junctions (*58*). The result is binding of ribosomes at these sites. Lysis of intercellular proteoglycan by keratinosome enzymes may have a similar effect. Such binding of KH particles or the ribosomes that synthesize these cystine-rich and histidine-rich granules could lead to deposition or synthesis of a chemically-resistant envelope layer (*47, 48, 49*).

Cell Lysis. Transitional cells have been described (*31*), but Lavker and Matoltsy (*47*) give the most explicit description of the sudden change from live cells to keratinized plates as follows:

. . . transformation is initiated by the release of hydrolytic enzymes . . . As lysozymes increase in number, the nucleus, ribosomes, mitochondria, Golgi apparatus, and mucous granules are gradually degraded. Marked changes occur in permeability of the plasma membrane as . . . lysed cell components pass through and accumulate in the intercellular space. . . . Filaments, KH granules, and the content of the endoplasmic reticulum (ER protein) are not lysed . . . filaments become displaced toward the cell periphery . . . and KH granules disperse and mix with the ER protein in the cell center. Subsequently, the KH–ER protein complex infiltrates the filament network. . . . Loss of fluids through the

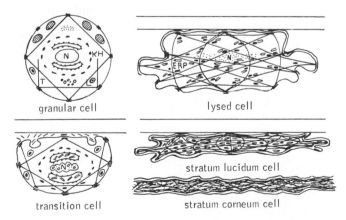

granular cell   lysed cell

transition cell   stratum lucidum cell

                         stratum corneum cell

Figure 10.   Diagram shows sequence of changes in distribution of tonofilaments (T), keratohyalin (KH), lysosomes (L), and endoplasmic reticulum protein (ERP) that occur as the cell is cornified

SEQUENCE AND TIME COURSE OF KERATINIZATION

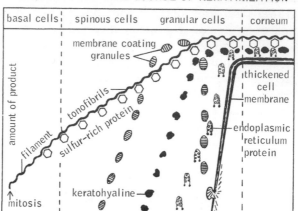

Figure 11. The sequential formation of products during differentiation of epithelial cells. Adapted from Ref. 32.

cell membrane leads to reduction of cell volume and consolidation of the remaining cell content.

Most of the filaments in basal cells lie perpendicular to the skin surface, but in the stratum spinosum they gradually shift into a perinuclear network (*31, 64*). The above concept is diagrammed in Figure 10. As cells become flattened, the radial pattern of filaments about the nucleus (*31, 34*) is preserved; but the keratinized fibers come to lie parallel with the skin surface, like spokes in a bicycle wheel (*31, 64*). Collapse of the fibrous cage produces tension on desmosomes and deforms the cell envelope so that, "the deep interdigitations formed between

Table I.   Amino Acid Residues

		Membrane	
Acid	*Significance*	*Man* (68)[c]	*Man* (32)
Proline	non-helix[d]	13.7	12.9
Cystine (1/2)	-S-S-[e]	4.9	7.5
Methionine	first on pept.[f]	0	—
Leucine ⎰	⎰ basal cell	5.8	6.2
Phenylalanine ⎱	⎱ label[g]	2.3	3.1
Histidine ⎰	⎰ granular	2.1	3.7
Glycine ⎱	⎱ cell label[g]	14.1	8.7

[a] Fibrous protein with 50% α helix content (*71*).
[b] Protein isolate of unknown function but distinctively high histidine content (*49*).
[c] Numbers in parentheses refer to literature citatations.
[d] Amorphous or non-helical protein. Proline is incompatible with α helix structure (*32*).

the cells ultimately interlock the outer part of the epithelium into a cohesive and protective SC" (*47*). Each cornified cell contains roughly twice as much fibrous protein as a basal cell, but the volume is only about one-tenth as large (*69*).

The sequence of appearance, relative amounts, and final relationships of cell components that survive lysis are presented diagrammatically in Figure 11. The smallest filaments that have been isolated (35 A thick) have low sulfur content (*32*). They thicken to 60–90 A with addition of sulfur-rich protein (demonstrated by heavy metal staining) (*32, 64*) and histidine-rich protein (demonstrated with radioactive labeling) (*49*). Five to 10 of these thickened filaments aggregate to form fibrils that average 250 A in diameter (*70*). Meanwhile, KH and ER protein accumulate until the cell is lysed when they are mixed and dispersed (*47*) to coat the 250-A fibrils (*70*). The coated fibrils are submerged in a matrix that includes nucleoproteins and nonfibrous proteins; these incorporate about 10 times more sulfur than the fibrils (*32*). The insoluble fibrils and matrix constitute about 65% of the cornified cell (*66*); other components include 10% soluble keratin, 10% dialyzable substances (amino acids, etc.), 7–9% lipids, and about 5% membrane protein (*65, 66*).

Chemical and Physical Composition of Keratin

Amino Acid Composition. Values vary substantially with differences of species, body location, and technique of isolation, purification, or analysis. However, distinctive patterns are associated with some membrane and fibrous protein isolates (Table 1). High proportions of proline and cystine are consistent with observed chemical and mechanical resistance of membranes (*32, 68, 71*). Similarly, increased methionine and

per 100 Residues in Epidermis

α Protein[a]		Keratohyalin		Histidine-Rich[b]	
Man (71)	Rat (72)	Rat (72)	Rat (51)	Rat (76)	Cow (77)
2.7	0.7	3.8	12.9	0	1.6
1.0	0.5	1.6	9.2	trace	0.8
1.2	0.8	1.0	—	trace	0.5
9.8	7.8	7.5	10.0	10.0	—
4.3	2.7	2.9	3.9	0.4	1.0
1.0	1.4	3.3	1.0	6.4	7.0
18.5	20.2	10.6	9.9	13.6	13.9

[e] A relatively large proportion of cystine suggests stabilization with disulfide bonds (*32*).

[f] Methionine is the first amino acid of most peptide chains (*73*).

[g] Tendency of radiolabeled amino acid to be incorporated in basal or granular cells (*74, 84*).

leucine suggest active synthesis of peptide chains in basal cells (73, 74). A preponderance of glycine is also associated with filamentous protein (72).

Hydrolysis with peracetic acid reveals a membrane composition of about 2/3 protein and 1/3 lipid, plus small amounts of carbohydrates (32). This protein incorporates nearly 13% proline, an amino acid that prevents normal helix formation (68, 75). Membranes also contain about 7½% of half-cystine units; this composition may permit substantial disulfide bonding (68). The formation of cross linkages like those in insect cuticle is postulated (32). A common origin of membrane and KH proteins is suggested by their similarly high contents of proline and half-cystine residues (75).

Studies of KH clearly indicate complexity that is only partially resolved (49). Differential staining reveals small, dense, homogeneous particles within amorphous KH masses, usually associated with tonofibrils (32, 48, 49). Amino acid analyses of supposed KH materials show at least three distinctive patterns (see Table I). The amorphous material of Tezuka and Freedberg (72) has much less proline and cystine than KH studied by Matoltsy (51). Other workers have associated histidine with KH in granular cells (48, 49, 76). Tezuka's histidine values (72) fall between those of Matoltsy (51) and Hoober (76) and conceivably represent an analysis of mixed components. Ugel's bovine material is a nucleoprotein that may be either a ribosomal product or still another KH component (71, 77).

Assembly of Polypeptides. Attempts to visualize the construction of keratin are presented (Figures 12, 13, and 14). The first fibrous protein is seen as fine filaments in ectodermal cells, but only thicker tonofilaments are observed opposite fully differentiated desmosomes (28): The attachment of tonofilaments at desmosomes may be essential for normal keratinization (27). Authorities believe that large molecules of prekeratin are synthesized on polyribosomes and assembled into polypeptide chains (32, 69, 73). Adjacent chains probably associate by hydrogen bonding at peptide linkages to form filaments (71, 78). Polyribosomes are collections of ribosomes that are thought to read simultaneously several different coded sequences of messenger ribonucleic acid (mRNA) (73, 79). Such ribosomes may be bound to fibers or membranes while the mRNA is moving and the code is being read (73, 80).

Polyribosomes may or may not be localized in the desmosomal attachment plate, as represented in Figure 12. Such localization is inferred from the apparent embryological and anatomical origin of filaments at desmosomes (25, 27, 28, 32). Some authors suggest that desmosomes are necessary to anchor filaments during their synthesis and to orient fibrils within each cell (27, 29). Also, there is evident economy in form-

MECHANISTIC DIAGRAM OF POLYPEPTIDE ASSEMBLY

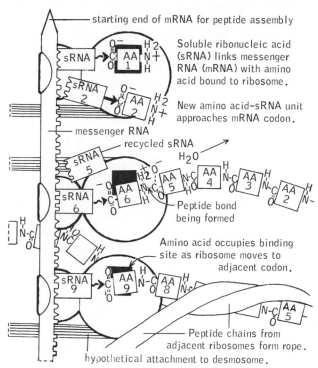

*Figure 12.   Schematic concept of a polyribosome show-
ing stepwise growth of a polypeptide chain and assembly
of protein strands. Soluble ribonucleic acid is also called
transfer RNA (tRNA). Adapted from Refs. 73, 79, and
80.*

ing filaments near amino acid supplies at the membrane surface. However, the filaments probably increase in diameter (Figure 13) by aggregation with pre-keratin molecules released from other ribosomes that
are observed alongside fibrils (*51, 78*).

Ribosomes synthesize pre-keratin units about 1000 A long and 35 A
thick (*32*). These aggregated peptide chains (Figure 12) mutually
align laterally and longitudinally in staggered ranks to make very long
strands (*32, 71*). These filaments are anchored at their sites of origin
then become looped through other attachment plates (*25, 26, 27*).

This arrangement explains Brody's (*31*) observation that, "Tonofibrils form concentric rings around the nucleus. The tonofibrils anastomose within these rings and from ring to ring, and also show connections
with the fibrous part of the desomosomes."

Tonofilament Assembly. The available information suggests that keratin is sequentially assembled around primary fibers that originate within the attachment plates of desmosomes (27). The dense cores of 50-A filaments (stained with uranyl acetate) represent such fibers (64, 84). These cores and their surrounding fibrous protein probably contain about 50% α helix (71). They contain even less sulfur (32, 64) than the high methionine (1.4 residues/100 amino acid residues), low cystine (1.1 residues/100 amino acid residues) fraction obtained by Baden after partial enzymatic hydrolysis (82). Studies with tritiated amino acids suggest that basal cells preferentially incorporate methionine, leucine, and phenylalanine within their elements (73, 74). In Figure 13A the primary rope is identified with the 35-A diameter of the smallest filaments that have been isolated (32).

The second layer within 50-A filaments can be visualized as two coils of peptide chains wound concentrically about the primary rope (Figure 13D), in the concept of Swanbeck (70). Alternately, the second layer may be pictured as a single layer of six ropes formed from triads of polypeptide chains (Figure 13A).

Electron microscopists see the third layer as a moderately osmiophilic sheath that probably contains some crosslinked sulfur (64). The sheathed filaments look like tubes 70 A in diameter; their radial orientation within the plane of SC cells probably accounts for the keratin pattern observed by Brody (31) with the electron microscope (64, 81). Baden's (82)

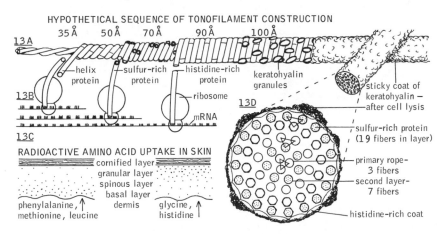

Figure 13. A composite of three drawings: Model of tonofilament assembly (13A) based on the ribosome model (13B) (Ref. 79), the distribution of radioactive amino acid uptake shown by silver grains (13C) (Ref. 74), and the cross section model (13D) Ref. (70). Second and fifth layers (dotted in 13D) are omitted from 13A. The triangles in 13D locate three possible layers of tripeptide chains.

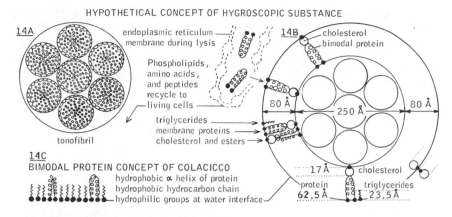

Figure 14.   Proposed structure and composition of hygroscopic elements within stratum corneum. A composite of three drawings: filaments aggregate to form tonofibrils of 250-A diameter (14A), adapted from Ref. 70; lysed lipids and proteins form an 80-A coating (14B) around the tonofibrils, adapted from Refs. 47, 96, and 97. The depicted relationship of bimodal protein and lipids to form hydrophilic surfaces (14C) is adapted from Ref. 98.

low methionine (0.3 residue/100 amino acid residues), moderate cystine (1.6 residues/100 amino acid residues) fractions may represent this third layer.

As cells move through the spinous layer, these maturing filaments acquire a highly ordered fourth layer that appeared speckled to Brody (*31*) and increases their diameter up to 90 A (*44, 64*). These filaments stain poorly in the granular layer (*31*) after they take up radiolabeled histidine, glycine, and arginine (*49, 64, 84*). The patterned differences in staining suggest that regularly spaced elements of this layer form a sheath around the helical proteins (*31*). This histidine-rich layer is distinguished from underlying protein by results of alkaline hydrolysis of SC material fixed with glutaraldehyde (*83*).

Keratohyalin and Tonofibrils. The investment of tonofibrils with a histidine-rich protein is followed by their aggregation into bundles of filaments (*64, 84*). X-ray diffraction studies and electron microscopy indicate that tonofilaments tend to associate in bundles of 5 to 10 filaments (Figure 14), each with a collective diameter of about 250 A (*70, 81*). These bundles correspond in size to the tonofibrils of the light microscopist (*81*).

The histidine-rich sheath seems to act as a priming coat for distribution of intensely osmiophilic globules (*49, 64*). This osmiophilic matrix or cementing substance is probably sulfur-rich (*64, 82*) and has been observed as patchy deposits (*49*) that embed the 100-A filaments within 250-A fibrils (*64, 70*). Various evidence suggests that this osmiophilic material is derived from sulfur- and proline-rich KH material such as the 20-A spheres described by Matoltsy and associates (48, 49, 50, 51, 75, 83, 84, 85, 86).

These sticky spheres probably become the globules of chemically resistant glue that spread over the histidine-rich priming coat and bind the filaments into bundles (64, 70, 75). They are also associated with nuclear inclusions, reinforcement of cell membranes, and dense homogeneous deposits (DHD) in the periphery of KH granules (49, 75, 76). The distribution of cystine-^{3}H and DHD suggests that Matoltsy's (75) 20-A spheres are a common source of protein, rich in cystine and proline, for KH and membrane reinforcement (48, 49, 83). Matoltsy suggests that KH granules are wholly composed of these spheres (51, 75). However, it is more likely that they only represent DHD which appear in electron micrographs as much the smaller of two KH components (49).

Completed fibrils form a network that progressively accumulates amorphous KH, especially where fibrils intersect (44, 75, 81). Extensive histochemical and radiolabeling studies indicate that KH granules are rich in histidine, glycine, and arginine (74, 75, 76, 77, 84, 87). Hoober and Bernstein (76) separated one SC fraction that contained 42% of the available histidine-^{3}H and 20% of the glycine-^{3}H; their other fraction contained 14% of the histidine-^{3}H and 17% of the glycine-^{3}H. If the bulk of the labeled histidine and glycine are derived from the large KH granules, the more resistant glycine-rich (76) fraction may represent the smaller amount of histidine-rich protein needed to sheathe the tonofilaments. The basophilic KH of light microscopy is probably histidine-rich material manufactured on free ribosomes (49, 76, 86); it may be the fraction with 42% of the histidine-^{3}H.

After deposits of chemically refractory layers encase major structural elements, the fully differentiated cell is ready for lysis and restructuring, as described by Lavker and Matoltsy (47). The histidine-rich KH is probably associated with ribonuclear (49, 77) and endoplasmic reticulum proteins during lysis (47). Another consequence of lysis is formation of disulfide crosslinks in cystine-rich protein; these are formed after conformational changes or enzymatic cleavages that expose sulfhydryl groups (68, 71). Cleaved and sticky KH (Figure 13) is probably crosslinked in the stratum lucidum (Figure 7) in the final stage of keratinization (75).

The onset of differentiation and its subsequent control is attributed to inhibitory feedback by varying concentrations of cellular products or epidermal chalone (30, 88). However, the specific means for initiating cell lysis of desmosomes and intercellular cement to accomplish desquamation is unexplained, although it is known that normal cholesterol metabolism is required for proper separation (2, 62).

Hygroscopic Substance. Perhaps the most distinctive and vitally important surface property of SC is its capacity to absorb up to six times its own weight in water (*16, 89, 90*). Attempts have been made to associate this property with protein surfaces (*20, 91*), protein–lipid interfaces (*38, 63, 66*), or with lipids alone (*66, 92*). Recently, investigators have emphasized the joint importance of lipids and proteins, or proteoglycan complexes, in hygroscopic properties (*16, 66, 92*).

Analysis of lipids from various skin layers shows that most phospholipids are degraded during cell lysis while other lipids are retained for reassembly (*93*). Normal SC contains lipids that constitute 2–9% of the dry weight (*66, 94*). The composition of SC lipids was investigated by Wheatley and Flesch (*62, 63*). They washed pooled SC specimens with petroleum ether and water to remove sebum and related lipids and then extracted lipids composing 6.4% of this dry residue. After extraction with chloroform:methanol (2:1 by volume), they separated 0.3% fatty acids, 0.4% neutral lipids and 2.5% proteolipids. Such proteolipids contain fatty acids, sterols, sterol esters, phosphatides, hydrocarbons, waxes, and triglycerides, together with amino acids and hexosamines (*94*). Flesch and Esoda (*95*) extracted up to 2.5% of an amorphous substance that melted at 55°C; they reported it as a glycoproteolipid complex that includes proteins, choline, phosphorus, and hexosamine. Wheatley also found cholesterol–hexosamine compounds (*62*).

Swanbeck and Thyresson (*96*) account for the location of the described lipids, protein residues, and other products of lysis as constituents of an 80 A thick coating around the 250-A fibrils (*93*). Their x-ray diffraction studies suggest that this noncrystalline coating is radially oriented with molecules aligned perpendicular to the fibrils (Figure 14). Diffraction of ether extracts of their tissue produced repeats at 34 A (paired cholesterol molecules), 47 A (attributed to pairs of triglycerides), and 123 A (unidentified). X-ray diffraction studies by Wilkes *et al.* (*97*) confirm the radial orientation and the presence of three diffracting lipids in human SC. Goldsmith and Baden (*71*) also reported a chloroform:methanol soluble material that is oriented perpendicular to the fibrils.

Lavker and Matoltsy (*47*) report that ER protein is formed in large amounts in granular cells and is set free during lysis to mix with dispersed KH between fibrils. The distribution of ER protein and its resistance to lysis is such that it may be an intrinsic bimodal protein, as described by Colacicco (*98*). He claims that membrane proteins have short sequences of hydrophilic amino acids and long hydrophobic chains of α helix. He shows how such proteins associate with lipids to create aqueous interfaces (Figure 14C). Vandeheuvel (*99*) has created models that demonstrate the large water-binding capacity of lipid–protein interfaces (Figure 15) of proteolipids. He also observes that amphiphilic solvents, such as chloroform:methanol, are needed to solvate concurrently the coexisting polar and nonpolar regions of proteolipid molecules. Therefore, the complexes reported by Wheatley et al. (94) and the bimodal proteins of Colacicco (98) fit the model proposed by Vandeheuvel (99) and can be expected to have hygroscopic properties.

If Swanbeck's 80-A layer is assumed to represent most of the hygroscopic substance, its water-binding capacity may be attributed to concentric hydrophilic interfaces (Figures 14B and 15). A myelin-like packing with the substitution of bimodal ER protein for the phospholipids of

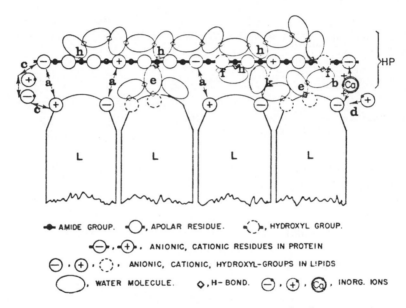

Figure 15.   Potential sites for binding of water molecules at interfaces of hydrated protein (HP) and lipids (L). Adapted from Ref. 99.

Vandeheuvel's model is likely with polar lipids aligned as surfactants at the water interface. The reported proteolipids and glycoproteolipids may be ER protein with lipids incorporated into the hydrophobic portion that contacts the hydrocarbon chains of other polar lipids. Lipids may be combined with the protein during lysis.

The Hygroscopic Barrier

Properties. Blank (90) studied callus (thickened SC) from human soles by varying relative humidity and measuring the parallel changes in tissue weight. These studies associated the water-binding capacity of callus with material that can be extracted in water if the tissue is first

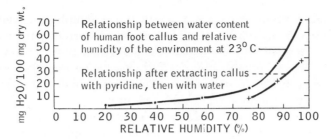

Figure 16.   Demonstration of hygroscopic properties of cornified tissue and loss of such properties after extraction with pyridine and water. Adapted from Refs. 90 and 91.

damaged with an organic solvent such as pyridine (Figure 16) (*91*). Buettner (*100*) subjected SC to 24-hr extraction in water and obtained 20 to 30% of the dry weight as a light-refractive, sticky, very hygroscopic residue.

Blank's work has been extended and confirmed by Scheuplein (*20*) (Figure 17) and others (*38, 89, 101*). The later work shows that chloroform:methanol mixtures very effectively remove hygroscopic substance from SC (*38, 94, 101*). Extracted SC dries rapidly as if it were paper (*16, 89*). Blank (*91*) observed that water is the only substance that softens SC. The hygroscopic substance is the apparent receptor for the plasticizing water.

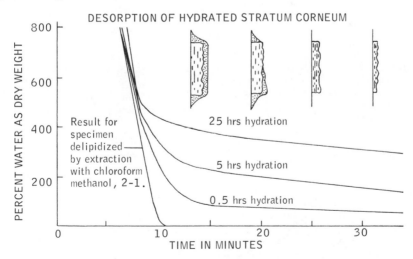

Figure 17.   Desorption curves show rapid evaporation of surface water followed by progressively slower losses with longer duration of hydration. The drawings in the upper right hand corner illustrate stages in desorption. Adapted from Ref. 89.

Surfaces. Scheuplein's three different levels of residual water (associated with 0.5, 5, and 25 hrs of hydration in Figure 17) can be related, in terms of Figure 14, to the postulated concentric water interfaces. The outermost interface probably absorbs and desorbs water rapidly, perhaps to hold most of the water that is absorbed within 0.5 hr. Similarly, the lipid–fibril water interface may hold the increment absorbed between 0.5 and 5 hrs. The least accessible sites may be located within the fibrils. Another possible explanation is that swelling of the 80 A coating may expose new hydrophilic surfaces of ER protein and lipid micelles. These sites may take up water between 5 and 25 hrs.

An alternate or complementary explanation for different levels of water retention is offered by Scheuplein (Figure 18). He cites evidence of different structures and properties in water bound at protein interfaces as compared with free water (89). His concept is that of SC fibrils and lipid matrix arranged to form capillary channels that have water bonded completely, partially, or indirectly to protein surfaces depending on their spacing (20, 69). This concept does not readily explain the reversibility of water retention at ambient temperatures (Figure 15), but it may explain the observations of other workers (Figure 18). Baden and Goldsmith have a similar structural concept (71).

Scheuplein's HOH:HOH bonds probably correspond to free water. Bulgin (102) found that differential thermal analysis (DTA) yielded the same 103°C peak for SC of man or rat and also for wet sand (38). Walkley (103) correlates this peak with frozen free water (which yields latent heat while melting during differential scanning calorimetry). His data show that 30% of SC is bound water that remains after evidence

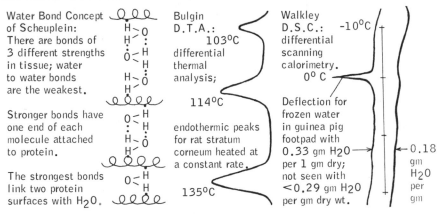

Figure 18. Co-presentation of postulated bonding concept, data on endothermic transitions observed during DTA of skin, and evidence that water:water bonds are absent in skin containing less than 0.29 g water per g dry tissue. Adapted from Refs. 20, 38, and 103.

of melting ice is gone. He cites similar findings by Hansen and Yellin (*104*), who used NMR and IR to differentiate two bound water fractions that have about sixfold differences of molecular mobilities.

Bulgin's data show loss of DTA peaks at 114° and 135°C if SC is heated for 2 hrs at 100°C or extracted with chloroform:methanol (2:1) (*103*). Rehydration restores the 114°C peak of the heated SC, but the 135°C peak is irreversibly lost. The reverse is true for extracted SC; rehydration restores the 135°C peak but not the 114°C peak. These results suggest association of the 114°C peak with water in lipid-rich hygroscopic substance; heating would drive off the water without destroying the lipids. However, the lipids would be removed by the chloroform:methanol. The 135°C peak may be associated with water bound by protein surfaces which resist chloroform:methanol yet are denatured by heat. Perhaps dehydrated protein surfaces become bonded together thereby eliminating hydrophilic sites.

All this evidence supports identification of the free water in SC with HOH:HOH bonds, the 103°C DTA peak, and Walkley's freezable water (Figure 18). These data also suggest correlation of the hygroscopic substance, the 114°C peak, and the more mobile of Hansen and Yellin's bound water fractions. Their less mobile fraction probably relates to the 135°C peak and to protein-water bonding. Perhaps only one of Scheuplein's hypothetical protein:water bonds exists when SC is dehydrated during DTA.

Permeability. As a diffusion barrier, SC is most effective when dry, less effective when hydrated, and still less effective when treated with solvents such as dimethylsulfoxide (DMSO) (*16, 92*). The hydrating effect of increased relative humidity, occlusion, or immersion can be visualized as separation of hygroscopic and protein elements to create diffusion channels containing free water (*20, 69, 71*). Obviously, water and its solutes should be more mobile in free-water channels than in bound water. The degree of hydration also can be influenced indirectly by organic solvents that hold water (glycol, DMSO) (*18*) or that modify surfaces (surfactants) (*16, 18*).

The ultrastructural evidence of alternating hydrophilic and hydrophobic regions is supported by measurements of organic solvent partitioning (Figure 19) in SC (*16, 69, 105*). Results with dilute alcoholic solutions show increasing permeation by straight-chain alcohols in parallel with increasing carbon number and lipophilia (*105*). This says that these alcohols tend to move from water into the lipophilic elements of SC. The low water solubility of the higher alcohols (C^{7-10}) reduces their aqueous concentrations and their fluxes (Figure 19). Decreasing permeation is observed with undiluted alcohols, of which methyl alcohol is most likely and decyl alcohol is least likely to leave homologous molecues to enter SC (*105*). These results suggest that SC presents lipophilic surfaces to lipophilic solvents and hydrophilic surfaces to water and its solutes (*16*). However, SC lipophilia is shown to be less than that of the higher alcohols (*105*).

The partitioning data (*16, 105*) and empirical results with various solvents (*18, 106*) suggest that lipophilic solvents form diffusion channels among lipoidal elements of keratin in the same way that water forms hydrophilic channels (*15, 16, 105*). In terms of Figure 14, it appears that hydrophobic substances can diffuse through Swanbeck's 80-A layer. This view is in accord with observed properties of solvents that facilitate diffusion of drugs. Such solvents are low in surface tension, dielectric constant, and melting point; they are extracted by ether from water and mix with other organic solvents (*106*). Some solvents (cresols, amines) react destructively with SC (*101, 106*), and others dissolve or split off moieties of the hygroscopic substance, thereby lowering barrier resistance (*63, 95, 101, 105*).

Examples. Interactions between solvents and SC are shown in Figure 20 (*106*). The rapid and effective transfer of nonionized scopolamine from chloroform into SC suggests that cell membranes and the 80-A coatings are saturated with the nonpolar solvent to become diffusion channels for the drug. Both the base and the salt of scopolamine are absorbed into SC much more slowly from water or *N,N*-dimethylformamide than from chloroform. These polar solvents are unlikely to dissolve in the 80-A layer or to form diffusion channels for the lipid-soluble scopolamine base.

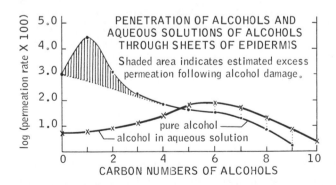

Figure 19. Comparison of permeation rates for pure alcohols and aqueous alcoholic solutions. Dotted lines indicate more rapid permeation of higher alcohols from dilute aqueous solution than from pure alcohol. Adapted from Ref. 105.

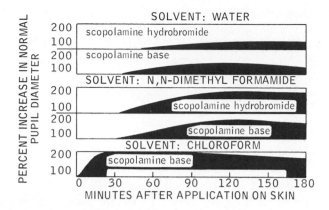

Figure 20. Relative contributions of diverse solvents to penetration of skin by ionized and nonionized forms of a mydriatic drug measured as changes in pupil diameters
(106)

The drugs in water or N,N-dimethylformamide appear to enter SC by partitioning according to their relative solubilities. The ionized salt is more soluble than the base in water; therefore, the base moves out of water and enters SC more readily than the salt does. N,N-Dimethylformamide is less polar than water. It therefore dissolves the base more readily than water, and the base moves less readily out of the solvent or into SC.

Blank measured 20- to 50-fold increases in water permeability of cornified tissue after he extracted hygroscopic substance into a mixture of 8% ethyl alcohol and 92% ether (*90*). Vinson *et al.* measured slight water losses from the skin of guinea pigs washed with pyridine, larger losses with tetrahydrofuran washing, and marked losses with chloroform:methanol (2:1) application (*101*). These barrier losses appear to parallel the effectiveness of the solvents in dissolving proteolipids, which are resistant to most solvents (*63, 95*). Tissues containing epithelial cells are the most common sources of proteolipids (*39*).

The nature of solvent damage to SC is clarified by Buettner (*100*). He removed 5% of SC as amino acids, including histidine, during 5 min of extraction in water. After 24 hrs of soaking and water extraction, he obtained hygroscopic substance constituting up to 30% of SC. His extraction of the hygroscopic substance in water did not affect x-ray diffraction patterns of SC although its resistance to water vapor penetration was drastically reduced. Soaking SC in ether or ethyl alcohol did destroy x-ray diffraction patterns without affecting barrier properties. It now appears that the amino acids dialyze readily from free water distributed between the coated fibrils. Replacement of the hygroscopic substance by free water probably explains the observed loss of vapor-barrier resistance and the retention of spatial relationships after extraction with water. Leaching of ether, or ethyl alcohol-soluble lipids, probably results in disorder of the 80-A coating. Spatial dislocations of its elements and of the interposed fibrils should destroy x-ray diffraction patterns. However, the residues would remain lipophilic and resistant to water vapor.

Summary

Stratum corneum is composed of physically and chemically transformed cells that are locked into interdigitating sheets by desmosomes and intercellular cement. Reinforced cell membranes constitute most of the exterior surface of skin. Less than 1% of the surface is occupied by microscopic spaces between membranes and by the gross openings of sweat ducts and hair follicles. These spaces and openings may admit various substances more readily than do the SC cells. The less permeable membranes of SC cells enclose a mixture of complex surface-active proteins and lipids that form concentric hydrophilic interfaces about the closely picked fibrous elements. These hygroscopic interfaces reversibly exchange water with cells and ambient water sources; this water acts as a plasticizer of the fibrous mass. These surfaces physically attract water with the result that they retard diffusion of water and hydrophilic substances. Lipophilic substances tend to dissolve in and diffuse through the proteolipid–lipid combinations in cell envelopes and hygroscopic coatings. The various diffusion barriers are either products of cell differentiation or modifications of normal structures that are transformed during lysis of granular cells.

Literature Cited

1. Scheuplein, R. J., "Mechanism of Percutaneous Absorption. II. Transient Diffusion and the Relative Importance of Various Routes of Skin Penetration," *J. Invest. Dermatol.* (1967) **48**, 79–88.
2. Kligman, A. M., "The Biology of the Stratum Corneum," in "The Epidermis," W. Montagna, W. C. Lobitz, Eds., pp. 387–433, Academic, New York, 1964.
3. Montagna, W., Hu, F., Giacometti, L., "General Morphology," in "Ultrastructure of Normal and Abnormal Skin," A. Zelickson, Ed., pp. 1–20, Lea and Febiger, Philadelphia, 1967.
4. McKee, G. H., Sulzberger, M. B., Hermann, F., Baer, R. L., "Histologic Studies on Percutaneous Penetration With Special Reference to the Effect of Vehicles," *J. Invest. Dermatol.* (1945) **6**, 43–61.
5. Cronin, E., Stoughton, R. B., "Percutaneous Absorption. Regional Variations and the Effect of Hydration and Epidermal Stripping," *Brit. J. of Dermatol.* (1962) **74**, 265–272.
6. Maibach, H. I., Feldmann, R. J., Milby, T. H., Serat, W. F., "Regional Variation in Percutaneous Penetration in Man," *Arch. Environ. Health* (1971) **23**, 208–211.

7. Sim, V. M., Biomedical Laboratory, Edgewood Arsenal, personal communication.
8. Tregear, R. T., "Relative Permeability of Hair Follicles and Epidermis," *J. Physiol. (London)* (1961) **156**, 303–313.
9. Fredriksson, T., "Studies on the Percutaneous Absorption of Parathion and Paraoxon. II. Distribution of ^{32}P-labeled Parathion Within the Skin," *Acta Dermatol.-Venereol.* (1961) **41**, 353.
10. Grasso, P., Lansdowne, A. B. G., "Methods of Measuring, and Factors Affecting, Percutaneous Absorption," *J. Soc. Cosmet. Chem.* (1972) **23**, 481–521.
11. Van Kooten, W. J., Mali, J. W. H., "The Significance of Sweat Ducts in Permeation Experiments on Isolated Cadaverous Human Skin," *Dermatologica* (1966) **132**, 141.
12. Nicoll, P. A., Cortese, T. A., "The Physiology of Skin," *Ann. Rev. Physiol.* (1972) **34**, 177–203.
13. Wahlberg, J. E., "Transepidermal or Follicular Absorption," *Acta Dermatol. Venereol.* (1968) **48**, 334–336.
14. Lindsey, D., "Percutaneous Absorption: An Informal Review," C.R.D.L. Technical Memorandum **20-28**, Edgewood Arsenal, Md., Feb. 1962.
15. Tees, T. F. S., "The Permeability Characteristics of the Skin: A Literature Survey," Porton Note No. **388**, Chemical Defense Experimental Establishment, Porton, Wiltshire, England, June, 1968.
16. Scheuplein, R. J., Blank, I. H., "Permeability of the Skin," *Physiol. Rev.* (1971) **51**, 702–747.
17. Scheuplein, R. J., Blank, I. H., Brauner, G. J., MacFarlane, D. J., "Percutaneous Absorption of Sterols," *J. Invest. Dermatol.* (1969) **52**, 63–70.
18. Idson, B., "Biophysical Factors in Skin Penetration," *J. Soc. Cosmet. Chem.* (1971) **22**, 615–634.
19. Hadgraft, J. W., Somers, G. F., "Percutaneous Absorption," *J. Pharm. Pharmacol.* (1956) **8**, 625–634.
20. Scheuplein, R. J., "Molecular Structure and Diffusional Processes Across Intact Epidermis," Final Comprehensive Report, No. 7, Contract No. DA 18-108-AMC-148A. AD 822 655, National Technical Information Service, Springfield, Va., 22151.
21. Barrett, C. W., "Skin Penetration," *J. Soc. Cosmet. Chem.* (1969) **20**, 487–499.
22. Mayer, W. H., Biomedical Laboratory, Edgewood Arsenal, personal communication.
23. Plewig, G., Marples, R. R., "Regional Differences of Cell Sizes in the Human Stratum Corneum. Part I," *J. Invest. Dermatol.* (1970) **54**, 13–18.
24. Menton, D. N., Eisen, A. Z., "Structural Reorganization of the Stratum Corneum in Certain Scaling Disorders of the Skin," *J. Invest. Dermatol.* (1971) **57**, 295–306.
25. Mercer, E. H., "Keratin and Keratinization," Pergamon, New York, 1961.
26. Kelly, D. F., "Fine Structure of Desmosomes, Hemidesmosomes and a Adepidermal Globular Layer in Developing Newt Epidermis," *J. Cell. Biol.* (1966) **28**, 51–72.
27. Wilgram, G., Caulfield, J. B., Madgic, E. B., "A Possible Role of the Desmosome in the Process of Keratinization," in "The Epidermis," W. Montagna, W. Lobitz, Eds., pp. 275–301, Academic, New York, 1964.
28. Barrett, C. P., Donati, E. J., Petrali, J. P., "A Possible Relationship of Intercellular Cohesiveness in Various Murine Neoplasms and the Presence of Intracisternal Particles," *Anat. Rec.* (1971) **169**, 272–273.
29. Breathnach, A. S., "Embryology of Human Skin. A Review of Ultrastructural Studies," *J. Invest. Dermatol.* (1971) **57**, 133–143.

30. Mercer, E. H., "Protein Synthesis and Epidermal Differentiation," in "The Epidermis," W. Montagna, W. Lobitz, Eds., pp. 161–178, Academic, New York, 1964.
31. Brody, I., "The Ultrastructure of the Tonofibrils in the Keratinization Process of Normal Human Epidermis," *J. Ultrastr. Res.* (1960) **4**, 264–297.
32. Matoltsy, A. G., Parakkal, P. G., "Keratinization," in "Ultrastructure of Normal and Abnormal Skin," A. Zelickson, Ed., pp. 76–104, Lea and Febiger, Philadelphia, 1967.
33. Rothberg, S., "Biosynthesis of Epidermal Protein," in "The Epidermis," W. Montagna, W. Lobitz, Eds., pp. 351–364, Academic, New York, 1964.
34. Charles, A., Smiddy, F. G., "The Tonofibrils of the Human Epidermis," *J. Invest. Dermatol.* (1957) **29**, 327–338.
35. Mercer, E. H., Jahn, R. A., Maibach, H. I., "Surface Coats Containing Polysaccharides On Human Epidermal Cells," *J. Invest. Dermatol.* (1968) **51**, 204–214.
36. Bagatell, F., Stoughton, R. B., "Vesication and Acantholysis," in "The Epidermis," W. Montagna, W. Lobitz, Eds., pp. 601–612, Academic, New York, 1964.
37. Kahl, F. R., Pearson, R. W., "Ultrastructural Studies of Experimental Vesiculation. I. Papain," *J. Invest. Dermatol.* (1967) **49**, 43–60.
38. Vinson, L. J., Masurat, T., Singer, E. J., "Basic Studies in Percutaneous Absorption," Final Report, No. 10, Contract No. DA 18-108-CML 6573, Lever Brothers Co. AD 627 810, National Technical Information Service, Springfield, Va., 22151.
39. Folch-Pi, J., Stoffyn, P. J., "Proteolipids From Nervous Membranes," *Ann. N.Y. Acad. Sci.* (1972) **195**, 86–107.
40. Feinsilver, L., Grainger, M. M., "Do Mucopolysaccharides Have a Role in Skin Penetration," in C.R.D.L. Special Publication **2-56**, F. N. Marzulli, M. M. Mershon, Eds., pp. 124–143, Jan. 1964. AD 434 294, National Technical Information Service Center, Springfield, Va. 22151.
41. Epstein, W. L., Maibach, H. I., "Cell Renewal in Human Epidermis," *Arch. Dermatol.* (1965) **92**, 462–468.
42. Overton, J., "Experimental Manipulation of Desmosome Formation," *J. Cell. Biol.* (1973) **56**, 636–646.
43. Brody, I., "Intercellular Space in Normal Human Stratum Corneum," *Nature* (1966) **209**, 472–476.
44. Odland, G. F., Reed, T. H., "Epidermis," in "Ultrastructure of Normal and Abnormal Skin," A. Zelickson, Ed., pp. 54–75, Lea and Febiger, Philadelphia, 1967.
45. Brody, I., "Observations on the Fine Structure of the Horny Layer in the Normal Human Epidermis," *J. Invest. Dermatol.* (1964) **42**, 27–31.
46. Matoltsy, A. G., Parakkal, P. F., "Membrane-Coating Granules of Keratinizing Epithelia," *J. Cell. Biol.* (1965) **24**, 297–307.
47. Lavker, R. M., Matoltsy, A. G., "Formation of Horny Cells. The Fate of Cell Organelles and Differentiation Products in Ruminal Epithelium," *J. Cell. Biol.* (1970) **44**, 501–512.
48. Jessen, H., "Two Types of Keratohyalin Granules," *J. Ultrastr. Res.* (1970) **33**, 95–115.
49. Fukuyama, K., Epstein, W. L., "Heterogenous Ultrastructures of Keratohyalin Granules. A Comparative Study of Adjacent Skin and Mucous Membranes," *J. Invest. Dermatol.* (1973) **61**, 94–100.
50. Lavker, R. M., Matoltsy, A. G., "Substructure of Keratohyalin. Granules of the Epidermis as Revealed by High Resolution Electron Microscopy," *J. Ultrastr. Res.* (1971) **35**, 575–581.

51. Matoltsy, A. G., Matoltsy, M. N., "The Amorphous Component of Keratohyalin Granules," *J. Ultrastr. Res.* (1972) **41**, 550–560.
52. Farquhar, M. G., Palade, G. E., "Junctional Complexes in Various Epithelia," *J. Cell. Biol.* (1963) **17**, 375–412.
53. Cavoto, F. V., Flaxman, B. A., "Communication Between Normal Human Epidermal Cells *in Vitro*," *J. Invest. Dermatol.* (1972) **59**, 370–374.
54. Lagunoff, D., "Membrane Fusion During Mast Cell Secretion," *J. Cell. Biol.* (1973) **57**, 252–259.
55. Flaxman, B. A., Cavoto, F. V., "Low-Resistance Junctions in Epithelial Outgrowths From Normal and Cancerous Epiderm *in Vitro*," *J. Cell. Biol.* (1973) **58**, 219–223.
56. David-Ferriera, J. F., David-Ferriera, K. L., "Gap Junction-Ribosome Association After Autolysis," *J. Cell. Biol.* (1973) **58**, 226–230.
57. Hashimoto, K., "Intercellular Spaces of the Human Epidermis as Demonstrated With Lanthanum," *J. Invest. Dermatol.* (1971) **57**, 17–31.
58. Bonneville, M., Weinstock, M., Wilgram, G. F., "An Electron Microscope Study of Cell Adhesion in Psoriatic Epidermis," *J. Ultrastr. Res.* (1968) **23**, 15–23.
59. Weinstock, M., Wilgram, G. F., "Fine Structural Observations on the Formation and Enzymatic Activity of Keratinosomes in Mouse Tongue Filiform Papillae," *J. Ultrastr. Res.* (1970) **30**, 262–274.
60. Wilgram, G. F., Kidd, R. L., Krawczyk, W. S., Cole, P. L., "Sunburn Effect on Keratinosomes," *Arch. Dermatol.* (1970) **101**, 505–511.
61. Wilgram, G. F., Krawczyk, W. S., Connolly, J. E., "Extraction of Osmium Zinc Oxide Staining Material in Keratinosomes," *J. Invest. Dermatol.* (1973) **61**, 12–21.
62. Wheatley, V. R., "Identification and Role of Non-Sebaceous Keratin Lipids in Skin," Final Report, No. 8, Contract No. DA 18-035-AMC-304A, Biomedical Laboratory, Edgewood Arsenal, APG, Md., 21010.
63. Wheatley, V. R., Flesch, P., "Horny Layer Lipids. II. Further Studies on the Overall Chemical Composition of Lipids from Normal and Pathological Human Stratum Corneum," *J. Invest. Dermatol.* (1967) **49**, 198–205.
64. Zelickson, A. S., "Normal Human Keratinization Processes as Demonstrated by Electron Microscopy," *J. Invest. Dermatol.* (1961) **37**, 369–379.
65. Matoltsy, A. G., Balsamo, C. A., "A Study of the Components of the Cornified Epithelium of the Human Skin," *J. Biophys. Biochem. Cytol.* (1955) **1**, 339–360.
66. Matoltsy, A. G., Downes, A. M., Sweeney, T. M., "Studies of the Epidermal Water Barrier. Part II. Investigation of the Chemical Nature of the Water Barrier," *J. Invest. Dermatol.* (1968) **50**, 19–26.
67. Loomans, M. E., Hannon, D. P., "An Electron Microscopic Study of the Effects of Subtilisin and Detergents on Human Stratum Corneum," *J. Invest. Dermatol.* (1970) **55**, 101–114.
68. Matoltsy, A. G., Matoltsy, M. N., "The Membrane Protein of Horny Cells," *J. Invest. Dermatol.* (1966) **46**, 127–129.
69. Scheuplein, R. J., "Mechanism of Percutaneous Absorption. I. Routes of Penetration and the Influence of Solubility," *J. Invest. Dermatol.* (1965) **45**, 334–346.
70. Swanbeck, G., "A Theory for the Structure of α-Keratin," in "The Epidermis," W. Montagna, W. Lobitz, Eds., pp. 339–350, Academic, New York, 1964.
71. Baden, H. P., Goldsmith, L. A., "The Structural Protein of Epidermis," *J. Invest. Dermatol.* (1972) **59**, 66–76.

72. Tezuka, T., Freedberg, I. M., "Epidermal Structural Proteins. I. Isolation and Purification of Keratohyalin Granules of the Newborn Rat," *Biochim. Biophys. Acta* (1972) **261**, 402–417.
73. Freedberg, I. M., "Pathways and Controls of Epithelial Protein Synthesis," *J. Invest. Dermatol.* (1972) **59**, 56–65.
74. Bernstein, I. A., "Relation of the Nucleic Acids to Protein Synthesis in the Mammalian Epidermis," in "The Epidermis," W. Montagna, W. Lobitz, Eds., pp. 471–431, Academic, New York, 1964.
75. Matoltsy, A. G., Matoltsy, M. N., "The Chemical Nature of the Keratohyalin Granules of the Epidermis," *J. Cell. Biol.* (1970) **47**, 593–603.
76. Hoober, J. K., Bernstein, I. A., "Protein Synthesis Related to Epidermal Differentiation," *Proc. Nat. Acad. Sci. U.S.A.* (1966) **56**, 594–601.
77. Ugel, A. R., Idler, W., "Further Characterization of Bovine Keratohyalin," *J. Cell. Biol.* (1972) **52**, 453–464.
78. Matoltsy, A. G., "Soluble Prekeratin," in "Biology of the Skin and Hair Growth," A. G. Lyne, G. F. Short, Eds., pp. 291–305, Angus and Robertson, Sydney, Australia, 1965.
79. Watson, J. D., "Molecular Biology of the Gene," W. A. Benjamin, New York, 1965.
80. Krebich, G., Sabatini, D. D., "Microsomal Membranes and the Translational Apparatus of Eukaroytic Cells," *Fed. Proc.* (1973) **32**, 2133–2138.
81. Odland, G. F., "Tonofilaments and Keratohyalin," in "The Epidermis," W. Montagna, W. Lobitz, Eds., pp. 237–249, Academic, New York, 1964.
82. Baden, H. P., "Enzymatic Hydrolysis of the α-Protein of the Epidermis," *J. Invest. Dermatol.* (1970) **55**, 184–187.
83. Fukuyama, K., Buxman, M. M., Epstein, W. L., "The Preferential Extraction of Keratohyalin Granules and Interfilamentous Substances of the Horny Cells," *J. Invest. Dermatol.* (1968) **51**, 355–364.
84. Fukuyama, K., Epstein, W. L., "Ultrastructural Autoradiographic Studies of Keratohyalin Granule Formation," *J. Invest. Dermatol.* (1967) **49**, 595–604.
85. Fukuyama, K., Wier, K., Epstein, W. L., "Dense Homogeneous Deposits of Keratohyalin Granules in Newborn Rat Epidermis," *J. Ultrastruct. Res.* (1972) **38**, 16–26.
86. Suzuki, H., Kurosumi, K., Myata, C., "Electron Microscopy of Spherical Keratohyalin Granules," *J. Invest. Dermatol.* (1973) **60**, 219–223.
87. Fukuyama, K., Epstein, W. L., "Synthesis of RNA and Protein During Epidermal Cell Differentiation in Man," *Arch. Dermatol.* (1968) **98**, 75–79.
88. Bullough, W. S., "The Control of Epidermal Thickness," *Brit. J. Dermatol.* (1972) **87**, 187–199, 347–354.
89. Scheuplein, R. J., Morgan, L. J., "Bound Water in Keratin Membranes Measured by a Microbalance Technique," *Nature* (1967) **214**, 456–458.
90. Blank, I. H., "Factors Which Influence the Water Content of the Stratum Corneum," *J. Invest. Dermatol.* (1952) **18**, 433–440.
91. Blank, I. H., "Further Observations on Factors Which Influence the Water Content of the Stratum Corneum," *J. Invest. Dermatol.* (1953) **21**, 259–269.
92. Sweeney, T. M., Downing, D. T., "The Role of Lipids in the Epidermal Barrier to Water Diffusion," *J. Invest. Dermatol.* (1970) **55**, 135–140.
93. Nicolaides, N., "Lipid, Membranes and the Human Epidermis," in "The Epidermis," W. Montagna, W. Lobitz, Eds., pp. 511–538, Academic, New York, 1964.

94. Wheatley, V. R., Flesch, P., Esoda, E. C. J., Coon, W. M., Mandol, L., "Studies of the Chemical Composition of the Horny Layer Lipids," *J. Invest. Dermatol.* (1964) **43**, 395–405.
95. Flesch, P., Esoda, E. C. J., "Isolation of a Glycoproteolipid From Human Horny Layers," *J. Invest. Dermatol.* (1962) **39**, 409–515.
96. Swanbeck, G., Thyresson, N., "A Study of the Stage of Aggregation of the Lipids in Normal and Psoriatic Horny Layer," *Acta Dermatol.-Venereol.* (1962) **42**, 445–457.
97. Wilkes, G. L., Nguyen, A., Wildnauer, R., "Structure-Property Relations of Human and Neonatal Rat Stratum Corneum. I. Thermal Stability of the Crystalline Lipid Structure as Studied by X-ray Diffraction and Differential Thermal Analysis," *Biochim. Biophys. Acta* (1973) **304**, 267–275.
98. Colacicco, G., "Surface Behavior of Membrane Proteins," *Ann. N.Y. Acad. Sci.* (1972) **195**, 224–261.
99. Vandeheuvel, F. A., "Structural Studies of Biological Membranes. The Structure of Myelin," *Ann. N.Y. Acad. Sci.* (1965) **122**, 57–76.
100. Buettner, K. J. K., Charlson, R., "Biochemical Studies with Stratum Corneum," in C.R.D.L. Special Publication **2-56**, F. N. Marzulli, M. M. Mershon, Eds., pp. 104–107, January 1964. AD 434 294, National Technical Information Center, Springfield, Va. 22151.
101. Vinson, L. J., Singer, E. J., Kroehler, W. R., Lehman, M. D., Masurat, T., "The Nature of the Epidermal Barrier and Some Factors Influencing Skin Permeability," *Toxicol. Appl. Pharmacol., Toxicol. Suppl.* II (1965) **7**, 7–19.
102. Bulgin, J. J., Vinson, L. J., "The Use of Differential Thermal Analysis to Study the Bound Water in Stratum Corneum Membrane," *Biochim. Biophys. Acta* (1967) **136**, 551–560.
103. Walkley, K., "Bound Water in Stratum Corneum Measured by Differential Scanning Calorimetry," *J. Invest. Dermatol.* (1972) **59**, 225–227.
104. Hansen, J. R., Yellin, W., "Abstracts of Papers," 161st Nat. Mtg., Amer. Chem. Soc., March-April 1971, COLL 21.
105. Scheuplein, R. J., Blank, I. H., "Mechanisms of Percutaneous Absorption. IV. Penetration of Non-Electrolytes (Alcohols) From Aqueous Solutions and From Pure Liquids," *J. Invest. Dermatol.* (1973) **60**, 286–296.
106. Mershon, M. M., "Chemical Basis for Selecting Skin Penetrants," in C.R.D.L. Special Publication **2-56**, F. N. Marzulli, M. M. Mershon, Eds., pp. 90–107, January 1964. AD 434 294, National Technical Information Center, Springfield, Va. 22151.

RECEIVED January 29, 1974. Edgewood Arsenal Clearance 74-155.

4

A Physicochemical Approach to the Characterization of Stratum Corneum

RICHARD H. WILDNAUER, DAVID L. MILLER, and
WILLIAM T. HUMPHRIES

Department of Skin Biology, Johnson & Johnson Research,
New Brunswick, N. J. 08903

*Stratum corneum is the principal diffusion barrier of the
skin to molecules and is also a protective surface against
mechanical insults; their function depends on morphological and macromolecular organization of the membrane.
Its physical behavior is a two-phase system of oriented,
amorphous, and crystalline regions, principally fibrous proteins associated with lipids. Thermally induced viscoelastic,
dimensional, enthalpic, spectral, and diffusional changes
occurring at 30°–50°C and 190°–220°C suggest transformations in both amorphous and crystalline regions, respectively.
The magnitude and temperature of these transitions depend
on moisture content, solvent exposure, chemical crosslinking,
and degree of orientation. The fundamental and empirical
parameters derived from the physical characterization methods discussed here aid in understanding the influence of
physical and chemical factors on stratum corneum properties.*

Skin, the most expansive human organ, envelops the entire surface of
the body such that its epithelium is continuous with the epithelia
of the external orifices of the digestive, sweat, sebaceous, respiratory, and
urinogenic systems. As a result of its anatomical location, the skin functions as the physical interface between the body tissues and the environment. The physiological functions of the skin are protection, containment,
and thermoregulation.

The skin is two discrete tissue layers, both polymeric but differing
in protein composition, morphology, and thickness (*1, 2*) (Figure 1).
Epidermis, the outer layer, is cellular and is composed primarily of the
intracellular fibrous protein keratin associated with lipids. In contrast,

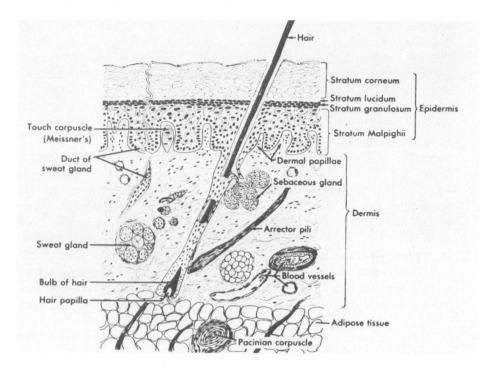

Figure 1. Diagramatic transverse section of full thickness human skin (2)

the dermis, or inner layer is an order of magnitude thicker, principally acellular, and composed of extracellular proteins such as collagen and elastin complexed with mucopolysaccharides. The dermis also contains the vascular system and nerve endings as well as protein-synthesizing cells. It is the principal load-bearing tissue of the skin. Physicochemical characterization of the polymeric nature of dermis has been studied extensively and will not be considered in this work (3, 4, 5). Few systematic studies have been made to characterize thoroughly the epidermis in terms of its functionally related physical and chemical properties (6–15). This form of characterization is the topic of this discussion.

Stratum corneum, the outermost layer of mammalian epidermis, functions physiologically as the principal diffusion barrier to molecules penetrating the skin and as a protective physical barrier to mechanical insults at the skin surface. Data suggest that these functions are critically dependent on the specific morphological and macromolecular organization of the membrane mosaic (16, 17, 18, 19, 20). Thus, alterations of biophysical properties arise from environmental factors acting directly on the membrane or upon the keratinization process, and they affect significantly biological performance (21). Similarly, the smoothness and flexibility of stratum corneum are important to cosmetic aspects. It is thus reasonable that certain requirements of stratum corneum strength and elasticity are essential properties to maintain a contiguous membrane and to permit adequate physiological function.

Practical and fundamental information come from studies of the physical and chemical properties of isolated samples of stratum corneum. First, the gross manifestations of many skin disorders are altered physical properties of the corneum such as cracking, scaling, roughness, inflexibility, and increased permeability. Studies can separate factors which influence keratinization from those which affect the membrane directly. Second, the quality of stratum corneum is an indicator of epidermal function, since the stratum corneum is the final differentiated product of epidermopoesis. The composition and structural organization of the keratinized stratum corneum cells contain the inscribed history of their biological formation and are a potential source of practical clinical information. Recognition of the clinical implications of physical and chemical studies of stratum corneum to dermatology was pioneered by Kligman (17).

The physical properties affecting physiological function are principally determined at the macromolecular level by the three-dimensional network structure of component long-chain polymeric molecules complexed with small molecules such as lipids and polysaccharides. These network structures are best characterized by physicochemical measurements of isolated stratum corneum to establish a profile of parameters relating structure and properties to functions.

Characterization of the keratinized cells by classical histological and biochemical approaches has been difficult because of the intractable nature of the tissue. Yet it is precisely these properties of mechanical strength, insolubility, macromolecular character, and lack of metabolic activity along with its ease of isolation which makes stratum corneum amenable to analysis by physical methods. The extreme complexity of composition, molecular structure, and organization of stratum corneum make interpretation of these macroscopic properties in terms of molecular structure and events dependent heavily on analogous studies of model synthetic polymer systems and the more thoroughly characterized, keratin-containing wool.

Sample Preparation

Stratum corneum used in these studies was isolated from human, newborn rat, callus, and guinea pig foot pad. The methods of isolating the various tissues have been discussed elsewhere (10, 11). Even though the various corneum tissues differ somewhat in morphology and chem-

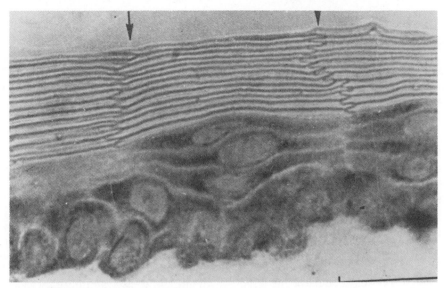

"Epidermal Wound Healing"

Figure 2.   Dilute NaOH-swollen transverse section of mouse ear epidermis illustrating the morphological organization of the corneum (22)

istry, physical characterization data obtained on one tissue can often be extrapolated to the others. Significant differences in physical or chemical behavior among the various corneum tissues studied will be noted where they occur. Solvent-extracted samples were prepared by immersing dry corneum samples in the specified solvent for 90 min unless otherwise stated. Samples were not tested until at least 48 hrs following extraction to allow for evaporation of residual solvent. Samples were prestretched by mounting dry strips of corneum in a stretching device and submerging the entire apparatus in water for 1 hr. The strips were then stretched by manual adjustment of the device to the required extension and allowed to dry prior to testing. Hydration increased sample length by 4–8%, and this is taken into consideration when calculating strain (*18*).

The dry weight used to calculate the extent of sorbed water (wt %) is based on the sample weight at 110°C in an atmosphere of dry air (less than 10 ppm H_2O). Desiccated samples refer to corneum stored over calcium sulphate at room temperature and containing approximately 5 wt % H_2O.

Morphology and Biology

Stratum corneum is a multicellular membrane of acutely-flattened, metabolically-inactive cells stacked in vertical columns (*22*). This stratified morphological organization is demonstrated in a freshly frozen transverse section of epidermis (Figure 2) where the swelling aids the visualization. The morphological architecture is such that the cells in one column interdigitate with those in adjacent columns to form a con-

tiguous and coherent membrane. The compactness of this cellular arrangement is shown in a differential interference micrograph (Figure 3).
On general body areas, the membrane is composed of 10–15 stacked cells
and is about 10μ thick when dry. Callus areas of the hands and feet are
atypical, being considerably thicker ($\sim 200\,\mu$) with much less regular
stacking. In many of the scaling diseases, a thickened corneum with a
poor cell-stacking pattern is observed clinically (16).

The stratum corneum is dynamic. The cells at the surface are con-
tinually lost through desquamation (intercellular fracture of small cell
aggregates) and replaced by differentiated epidermal cells, maintaining
a reasonably constant number of cell layers. Stratum corneum cells are
the final differentiated product of the keratinization process which is
initiated by mitotic divisions of cells located in the germinative basal
layer. Following division, one of the daughter cells begins to differentiate
and joins the stream of viable cells proceeding to the surface of the skin.
During this transit to the skin surface, a number of important biochemical
and biophysical events occur which result in the dense, acutely flattened
corneum cells (24, 25). In the early stages of this transit from basal layer
to the stratum corneum (lasting 14 days), the major metabolic activity
is the synthesis of fibrous protein. In the upper layers of the epidermis,
the cells gradually dehydrate, lose metabolic activity, and flatten along
the plane parallel to the skin surface resulting in a biaxial orientation of

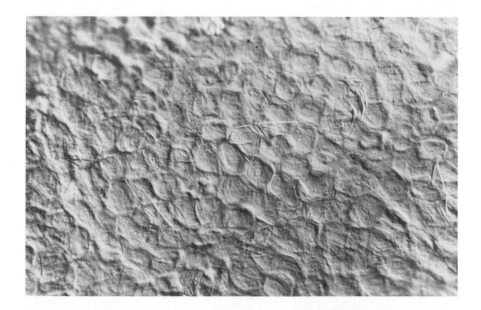

Figure 3.   Interference micrograph of human stratum corneum surface (23)

the fibrous proteins (*26*). The polygonal (principally hexagonal) shaped cells vary in size with body location ranging from a diameter of 34μ on the forehead and hands to 46μ on the thigh axilla (*27*). The cells are generally about 0.8μ thick when dry.

In spite of the fact that stratum corneum cells are metabolically inert, changes in keratin structure and organization occur as each cell transits through the stratum corneum prior to desquamation (*28*). This suggests some asymmetry in physical and chemical properties through the thickness of the corneum. One demonstration of this is the swelling of fresh frozen transverse sections of corneum in dilute acid or base. The most mature surface cells swell considerably more slowly and to a lesser extent than the lower layers of the corneum (*18*). Such asymmetry is of particular importance in studying the diffusion and mechanical properties of this membrane.

Chemistry and Supramolecular Structure

The detailed chemistry of the stratum corneum is complicated by the membrane's composition, formation, and structure. Some gross chemical characterizations have determined the primary chemical components of the tissue which are shown in Table I (*29*). The tissue is primarily cellular with approximately 10% extracellular components which are lipid and mucopolysaccharides. The bulk of the tissue is densely packed intracellular fibrous protein associated with lipids, resulting in a dry general body corneum density of 1.35–1.40 gm/cm^3 as determined by a gas displacement technique (*30*).

Table I.   Composition of Stratum Corneum (27)

Tissue Component	Chemical Composition	% of Stratum Corneum
Cell membrane	lipid, protein	5
Intercellular	lipid, protein, and mucopolysaccharides	10
Intracellular	lipid, 20% fibrous protein, 70% non-fibrous protein, 10%	85

Electron microscopic studies further substantiate the presence of highly-ordered macromolecular structures with the corneum cells (*24, 31*). The fibrous keratin structure has been described as low-density filaments, low in sulfur but embedded in a dense sulfur-rich amorphous interfilamentous matrix. The keratin fibrils organize into lipid-covered bundles preferentially oriented in the longitudinal plane of the acutely flattened cell (*16*). These filaments terminate at the periphery of the cell

Biochimica et Biophysica Acta

Figure 4. Wide angle x-ray diffraction patterns of newborn rat stratum corneum with beam normal and parallel (edge) to the corneum plane (13)

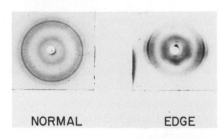

NORMAL   **EDGE**

adjacent to desmosomes, the specialized intercellular attachment plaques (*32*). In addition to the desmosomes, mucopolysaccharides form the ground substance that fills the intercellular spaces and aids in intercellular cohesion (*33*).

A hypothetical model for the molecular organization of the fibril unit of keratin with lipids consists of protein cylinders, surrounded by a lipid layer, with the lipid chains arranged radially on the protein cylinder. Support for this model comes from wide- and small-angle x-ray diffraction studies (*13, 19, 34*) as well as small-angle light-scattering studies (*35*). The relative contribution of this complex to stratum corneum properties and functions is not well understood. These lipids are present in the lower epidermal cells associated with 1000g precipitate of homogenized epidermis (*36*). These findings suggest an early association of lipids with the protein filaments which may have a role in determining their final organization. Swanbeck has suggested on the basis of x-ray diffraction studies that one of the defects in the dermatological conditions of psoriasis and ichthyosis is related to the lack of proper lipid–protein filament complex formation (*37*).

Figure 4 shows the wide angle x-ray diffraction pattern for newborn rat stratum corneum with the beam normal and parallel to the plane of the flattened corneum cell. The pattern shows two sharp reflections of lipid origin at 4.2 A and 3.7 A and two diffuse halos at 4.6 A and 9.8 A attributable to protein. The intensity of the azimuthal reflections suggests that the lipids are associated with the proteins which are oriented parallel to the long axis of the flattened cell. The reflections are not present in samples which have been extracted with chloroform–methanol (3/1 by volume) or after long exposure to ether. X-ray diffraction patterns of corneum specimens at various temperatures demonstrate the thermal stability of these lipid components (*13*). Two melt temperatures are observed by the disappearance of the 3.7-A spacing at 40°C and the 4.2-A one at about 70°C. Upon cooling the heated newborn rat corneum samples to room temperature, both reflections reappear. The recrystallization process was not moisture-dependent.

There also appears to be some variation in the reversibility of the melts, number of reflections, and their intensity among the various types of corneum tissues. For example, human corneum exhibited reflections at 3.7, 4.2, and 4.6 A, of which that at 3.7 A did not return when the sample was heated and cooled back to room temperature. Diffraction patterns from psoriatic scales and human callus exhibit weak lipid reflections while human hair exhibits no lipid reflections (*38*).

The 4.2-A spacings of other lipid-containing membrane structures are interpreted as the average interchain separation in directions perpendicular to the long axis of the hydrocarbon portion of the molecules (*39*). This explains the independence of this spacing from the length of the chain and the nature of the polar group.

Lipid Characterization. The lipid extracts of rat stratum corneum recovered from solvents of varying polarities have been characterized further by differential scanning calorimetry (DSC) and shown to display two major melting endotherms: 37°–40°C and 58°–62°C (Figure 5). The fact that the melts of the intact corneum determined by x-ray diffraction are both somewhat higher in temperature than the DSC melts of the extract suggests additional specific interaction of the lipid molecules with membrane proteins. The extracts are quite waxy looking and are composed of a complex mixture of esters of long-chain, water-insoluble alcohols and higher fatty acids (*40, 41*).

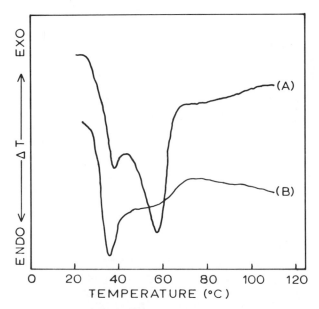

Figure 5. DSC analysis of ether extract from A: newborn rat stratum corneum and B: human hair (N₂ atm). Data from Ref. 18.

The lipid extract from wool or hair exhibits only one melting endotherm at 35°–40°C (Figure 5) and is composed of about 90% esters of
long-chain acids and alcohols with 10% free acids and alcohols. The
acid fraction of this hydrolysate contains principally branched chain and
hydroxy acids melting at 40°–45°C. The long-chain alcoholic fraction
melts at 55°–65°C (40). IR and DSC data of the extracts from hair
and corneum indicate that corneum contains considerably more free
alcohols than wool or hair (42).

Keratin Structure and Orientation. Acute flattening of the fibrous
protein-filled cells in the final stages of keratinization establishes a
biaxial orientation. As would be expected, no birefrigence is observed
normal to the plane of the corneum surface, but significant birefrigence
is observed parallel to the plane of the corneum surface (1, 42). The
x-ray diffraction pattern of this isolated epidermal protein, when highly
drawn, exhibits the classical alpha pattern (7, 43).

The wide angle x-ray diffraction pattern of undeformed corneum
exhibits diffuse halos at 4.6 A and 9.8 A common to proteins (Figure 4).
The lack of the 5.1-A reflection characteristic of alpha-keratin structures
in undeformed corneum suggests that the protein is considerably less
oriented and perhaps of a lower alpha content than wool. This is supported by the fact that the 5.1-A reflection begins to appear in samples
of corneum which were hydrated and stretched to 100% or more (Figure
6) and allowed to dry in the extended state. The increased orientation
of the lipid reflections in the stretched sample demonstrates further their
association with the orienting protein fibrils.

Additional evidence for the presence of alpha-helical fibrous protein
in stratum corneum is provided by IR dichroism studies (42). The
transmission IR of newborn rat corneum (Figure 7) is diffuse because
of a large variety of side-chain bands, but it is characteristically protein.

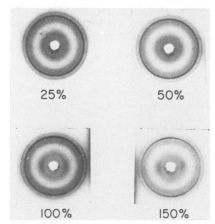

Critical Reviews in Bioengineering

*Figure 6. Wide angle x-ray dif-
fraction pattern of newborn rat stratum corneum stretched to various
elongations while hydrated and allowed to dry in extended state (normal to plane of membrane) (6)*

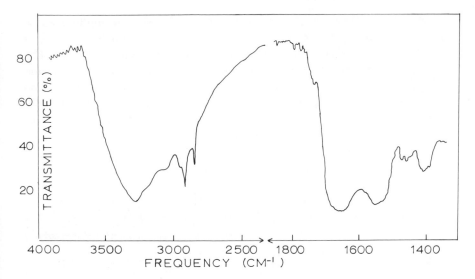

Figure 7. IR transmission spectrum of newborn rat stratum corneum (10μ thick). Data from Ref. 18.

The oriented fibrous protein in samples extended to greater than 100% strain is predominately in the alpha-helical form evidenced by the dichroism of the amide I band at 1658 cm^{-1} and of the amide II band at 1550^{-1} cm rather than the beta configuration which would show dichroism at 1640 and 1525 cm^{-1}. In prestretched samples varying from 0 to 200%, the amide I band was most pronounced when the polymer stretch orientation was parallel to the electrical vector of the radiation, but the amide II band was greatest when the polymer stretch orientation was perpendicular to the electrical vector. This behavior has been shown for model polypeptides in the alpha form. Even in highly stretched samples ($\sim$ 200%) there was no shift in absorption bands characteristic of the beta configuration (44). The lack of any significant alpha-to-beta conversion as a result of stretching hydrated samples to 200% extension was also confirmed by wide angle x-ray diffraction (6).

Thermal Decomposition Studies. Recently thermogravimetric analysis (TGA) has been used to characterize block and random copolymers of alpha-amino acids and to determine some qualitative information about the chemical composition and crosslink density of protein systems (45, 46). The thermal degradation curves of block copolymers display distinct degradation temperature ranges while the random copolymers exhibited a broad degradation temperature region which is between those of the two homopolymers. The thermal decomposition curve for stratum corneum (nitrogen atmosphere) (Figure 8) suggests a random copolymer system. The maximum decomposition temperature of 375°C is midway between that of poly(L-valine) at 405°C and poly(L-phenyl-alanine) at 355°C (45).

Thermal decomposition of proteins in air has been shown to occur by an oxidative mechanism in two successive stages. The first reaches its maximal rate at about 320°C and results in the partial cleavage of peptide bonds with the formation of peptide fragments of low molecular weight (46, 47). The second major decomposition temperature depends on the degree of chemical crosslinking in the remaining fragments and varies from 500° to over 650°C for highly crosslinked fragments.

The decomposition of stratum corneum in air is considerably more complex than in nitrogen (Figure 9). As expected for a crosslinked polymer, about 50% of the corneum protein decomposed at 320°C, and the highly crosslinked fragments decomposed at 600°C. Several other peaks are related to the decomposition of chemical components such as

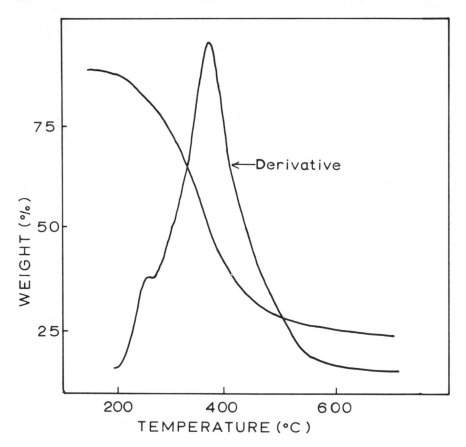

Figure 8.   TGA of desiccated newborn rat stratum corneum in nitrogen (~5% water content, 50°C/min heating rate). Data from Ref. 18.

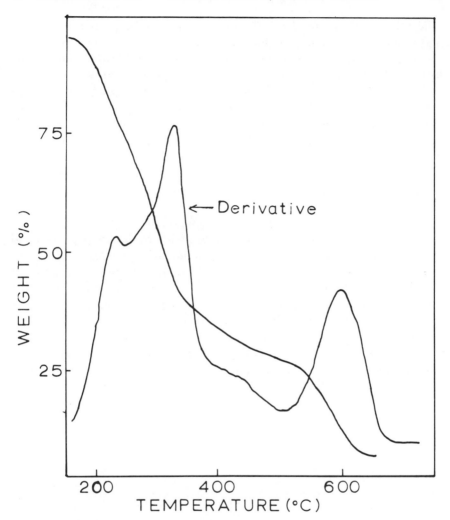

Figure 9.   TGA of newborn rat stratum corneum in air (∼5% water content, 50°C/min heating rate). Data from Ref. 18.

glycosaminoglycans which decompose in the 240° to 270°C range (48). In the thermal degradation of wool, ammonia is lost at about 180°C, and the sulfur-containing amino acids decompose at 240°C (49, 50). No attempt was made in these studies to analyze the various decomposition products.

Sorption of Water in Newborn Rat Stratum Corneum. The physical properties of polymeric materials are markedly dependent on the interaction of the chain network with low molecular weight compounds (plasticizers). Similarly, stratum corneum interactions with water significantly modify its mechanical and diffusion barrier properties. Meaningful interpretation of the influence of water on physical properties requires quantitative characterization of both equilibrium and kinetic aspects of the water–corneum system. The apparent equilibrium properties of the system may be constructed from a sorption isotherm of regain data as a function of relative humidity (RH). The isotherm for newborn rat stratum corneum at 25°C is shown in Figure 10. For comparison the isotherms of some other protein and protein-related polymer–water systems are included. A characteristic sigmoidal shape, indicative of sorption at specific sites at low relative humidities followed by multilayer build up of water molecules at higher humidities, is common to all of the isotherms (51).

Proteins sorb water molecules at low relative humidities through interactions at hydrophilic sites associated with the polar peptide bond and various polar side chains. A correlation of the polar amino acid content and ability to sorb water has been demonstrated for a series of soluble proteins by Bull (52). However, the influence of protein–protein interactions on the availability of particular polar groups complicates similar relationships for insoluble proteins such as keratin.

Table II lists the polar amino acid content of four polyamide materials along with their corresponding water regains at 50% RH. These data suggest that insoluble proteins with polar side chains bind more water than those without (53, 54). However, the wide range of water

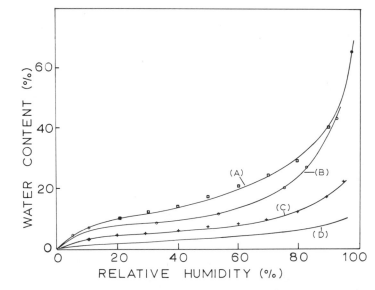

Figure 10. Water vapor adsorption isotherms as a function of % RH for A, collagen; B, RSC; C, silk; D, nylon. Data from Refs. 18, 51, and 62.

Table II. Comparison of Water Content of Polyamides and Proteins with Polar Side Chain Content

Substrate	Polar Group, $\times 10^3$ moles	Ref.	g H_2O/g Dry Tissue at 50% RH	Ref.
Stratum corneum	4.4	50	0.15	26
Collagen	4.0	51	0.179	48
Silk	3.6	51	0.074	48
Nylon	0		0.031	48

regains for quite similar polar amino acid contents indicates that the other factors mentioned above may be involved in determining the equilibrium water uptake. These factors relate to the physical aspects of the system such as backbone chain flexibility and crosslink density (55).

In spite of the composite nature of the stratum corneum, its water sorption isotherm is qualitatively identical to those of the more simple protein systems shown, suggesting that water interacts predominately with the protein components of the corneum. This conclusion is supported further by the results of chloroform–methanol (3/1 by volume) extraction which removed as much as 25% of the original dry weight (lipids and low molecular weight water-soluble components) but did not quantitatively alter the isotherm in the low relative humidities (18). The application of the Zimm–Lundberg cluster theory (56, 57) to the isotherm yields additional information as to the state of the sorbed water in the corneum. The tendency of water to cluster is expressed in this theory by the cluster function C_1G_{11}:

$$C_1G_{11} = (1-\phi_1)\ (d\ \ln\ \phi_1/d\ \ln\ a_1) - 1$$

where ϕ_1 is the volume fraction of water in the polymer and a_1 is the activity of the water. Values of C_1G_{11} greater than -1 indicate the tendency of the water to prefer self association. Figure 11 shows the dependency of the cluster function on relative humidity for newborn rat stratum corneum at 25°C. A dramatic increase in clustering tendency of water occurs over the range 40–60% RH. As a consequence of this transition, water sorbed at low relative humidities is probably associated at isolated sites in the corneum whereas at high relative humidities a cooperative effect in the sorption is observed. The presence of more highly clustered water correlates highly with a rapid decrease in the tensile modulus of the corneum with increasing relative humidity (*see* Figure 18). From this correlation, water clusters are shown to be more efficient plasticizing agents than the individual molecular species. On the other

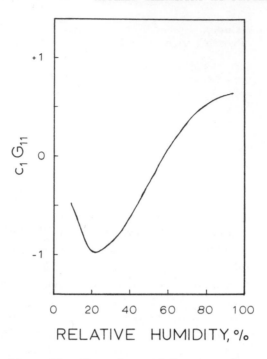

Figure 11.   Dependence of the water cluster function, C_1G_{11}, on relative humidity (newborn rat stratum corneum, 25°C). Data from Ref. 57.

hand, both the clustering and the modulus effects may result from water-induced alteration in the conformation of the protein itself.

Diffusion of Water Vapor in Newborn Rat Stratum Corneum. Measurement and interpretation of diffusion in heterogenous biological systems such as the stratum corneum are difficult compared with similar measurements for well-defined synthetic polymer systems, but studies of water diffusion in stratum corneum are essential to a better understanding of those factors which contribute to the barrier function of the corneum. Water diffusion measurements under both equilibrium and non-equilibrium conditions are useful to probe the influence of temperature and other factors on stratum corneum macromolecular structure.

Thermal desorption is a dynamic (non-equilibrium) technique in which a sample of hydrated corneum is heated at a constant rate in a dry atmosphere. The water desorption rate is plotted as a function of temperature. The general shape and temperature maxima of the desorption rate vs. temperature curves (Figure 12) are characteristic of the material's diffusion and equilibrium sorption behavior as well as experimental conditions such as heating rate. In a simple desorption process where there are no significant thermally induced changes in the membrane matrix, there is a gradual increase in the desorption rate with temperature. The desorption rate reaches a maximum and begins to decrease when the gradient of sorbed water in the membrane is reduced to the extent that it overshadows the influence of the increase in diffusion coefficient with temperature. In more complex systems such as stratum corneum, heating produces structural changes in the matrix which are reflected in the shape of the resultant thermal desorption curves.

Examples of the resulting plots of the desorption rate *vs.* temperature are shown in Figure 12. Untreated corneum samples exhibit one maximum at about 80°C whereas ether extraction (90 min) produces a second lower temperature maximum in addition to the higher temperature peak. The low temperature peak perhaps indicates the presence of loosely bound water which can diffuse out of the corneum more easily than the primary sorbed water. The thermogram for the chloroform–methanol-extracted corneum reveals a single, broad peak indicative of a more general solvent damage to the corneum matrix. The thermal de-

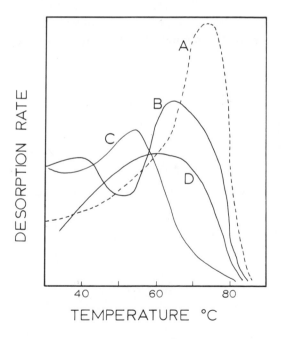

Figure 12. Thermal desorption of water from newborn rat stratum corneum. A: control newborn rat, B: extracted 90 min with ether, C: formaldehyde crosslinked, D: extracted 90 min with chloroform-methanol (3:1); heating rate 5°C/min.

sorption curve for formaldehyde crosslinked corneum also peaks at a lower temperature than the control. This effect is ascribed to the sample's reduced water-holding capacity coupled with a higher initial desorption rate. At this early stage of technique development, characterization of stratum corneum by thermal desorption is useful only to survey qualitatively the effect agents have on the thermal behavior of corneum.

The more conventional method for studying the energetics of diffusion in membranes is to perform permeation experiments as a function of equilibrium temperature. Figure 13 illustrates the effect of temperature on the apparent diffusion coefficient calculated from the water vapor permeation time lag established by steady-state permeation with a 75 to 0% RH gradient across the membrane. The principles of the time lag permeation method are adequately discussed elsewhere (58). The lower curve corresponds to a sample which was not mechanically supported and was observed to deform into a hemispherical shape. This deformation is the combined result of a small pressure difference across the membrane and a decrease in modulus of stratum corneum as the temperature is increased. The upper curve corresponds to a supported sample. Previous to the experiment, both samples had identical thermal histories. Stresses accompanying deformation of the unsupported cor-

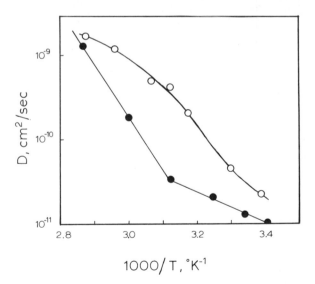

Figure 13. Temperature dependence of apparent diffusion coefficient of water vapor in newborn rat stratum corneum, 0 to 75% RH vapor gradient. Open circles, supported membrane; closed circles, unsupported membrane. Thickness assumed constant for the purposes of calculation.

**Table III. Diffusion Coefficients for Water in
Newborn Rat Stratum Corneum**

Method	Conditions	Nominal Water Content, $g\ H_2O/g$ Dry Tissue	D, cm^2/sec
Thermal desorption	bulk water	4.80	1.9×10^{-11}
	vapor	0.35	9.5×10^{-12}
Permeation lag time	bulk water[a]	1.5	1.4×10^{-10}
	vapor	0.35	1×10^{-11}
Steady state flux and sorption values	bulk water[a]	1.5	1.2×10^{-10}
	vapor	0.35	2.1×10^{-11}

[a] ^{3}H-H_2O radio tracer method.

neum greatly alter the diffusion behavior. The deformed sample has a diffusion coefficient an order of magnitude lower than that of the undeformed sample at 50°C. Upon cooling the samples in the diffusion cell to room temperature, the diffusion coefficients of neither the supported nor the unsupported samples return to the initially observed magnitude. Thermalsorption hysteresis from accumulated stresses created as the sample cools in the mechanical constraints of the diffusion cell could account for the altered diffusion coefficients. The irreversible change in the apparent diffusion coefficient for the unsupported membrane is caused by alterations in the corneum matrix resulting from the deformation. The mean energy of activation for water diffusion in the supported membrane over the range studied is 18 kcal/mole. This agrees well with that reported for bulk diffusion of water in highly swollen human corneum (59). Table III lists diffusion coefficients calculated *via* three different methods: initial desorption rate, permeation time lag, and steady-state flux combined with equilibrium distribution values calculated from well known formulas (58). For each method, a comparison is made between the diffusion coefficient for water from vapor and that from bulk liquid. As observed by others (15), the diffusion coefficient of wet corneum is somewhat higher than that of dry corneum—*i.e.*, the presence of large amounts of imbibed water appears to plasticize the matrix. Except for the desorption value, the results represent good agreement among the various methods explored and good agreement with a previously reported value of 5×10^{-11} cm²/sec for newborn rat corneum (60). The low value obtained from the desorption of the wet corneum most likely results from its highly swollen condition. As has already been shown, water tends to cluster in this system at vapor pressures well below saturation. Because the clustered water is not dissolved in the matrix, it cannot contribute to the driving force of the diffusion. Correcting for the cluster effect will effectively increase the magnitude of the diffusion coefficient. The apparent depression of the diffusion coefficient of vapors in synthetic polymer systems has been attributed to a similar clustering effect (*61*).

Physical and Chemical Properties

Thermal Behavior. The most characteristic parameter of an amorphous polymer is the glass transition temperature (T_g). In the glass transition region, a viscoelastic transition occurs as a result of the onset of motions of chain segments in the amorphous region of the polymer which transforms the material from a rigid state to a rubbery one. The mechanism of deformation response by the material is dependent on T_g, and it determines the ductility and brittleness of the polymer. The usual method for determining T_g is to measure the temperature at which the specific volume–temperature plot shows an inflection indicating an increased thermal expansion coefficient. In general, all physical properties of amorphous polymers which are dependent on segmental relaxation rate such as viscous flow, mechanical and dielectric relaxation, creep, and diffusion show a major change on heating through the glass transition region. Similarly, T_g can be determined by differential scanning calorimetry (DSC) from the temperature at which there is a sudden change in the specific heat of the sample. T_g depends also on the molecular weight of the polymer, on internal strain, and to a lesser extent on heating rate. Quenched, highly amhorphous polymers generally display a more pronounced inflection when heated through the T_g region (*63*).

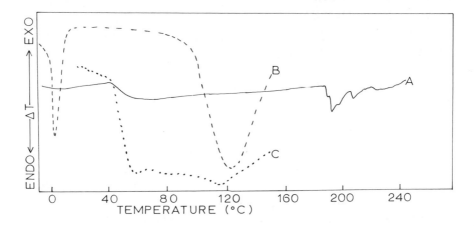

Figure 14.   DSC scan of guinea pig footpad: A, desiccated control; B, hydrated in water 45 min; and C, high sensitivity of dry control (hermetically sealed N_2 atm, 20°C/min heating rate). Data from Ref. 42.

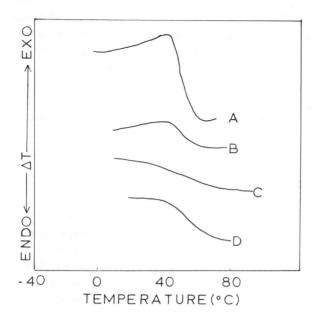

Figure 15. DSC scan of guinea pig footpad: A, control; B, rerun of rapidly cooled; C, rerun of slowly cooled; and D, rerun of C after 48 hrs N_2 atm (samples of comparable weight, 20°C/min heating rate). Data from Ref. 42.

A DSC scan of stratum corneum (Figure 14) indicates several thermally induced transformations from 0° to 250°C. As standard conditions, all samples were cooled to −40°C and heated at 20°C/min in a stream of dry nitrogen. To avoid the endothermic contribution of water evaporation, all samples were hermetically sealed and weighed before and after the scan. The thermogram for desiccated corneum (~ 5 wt % H_2O) exhibits an abrupt change in specific heat at 48°C suggesting a glass transition. This temperature region is generally accepted for the T_g of various nylons and other polyamides (63, 64).

This 48°C transition in dry corneum is anomolous in that it does not fit all the criteria usually associated with classical glass transitions. For example, once the sample has been heated through the T_g region, there is no corresponding specific heat change on cooling. Immediate reheating after a slow cooling cycle displays little or no glass transition (Figure 15). The transition appears at slightly lower temperatures on the second cycle when the sample is rapidly quenched below T_g before reheating. In this case, the specific heat change is considerably reduced from that observed in the first cycle, and a second glass appears at 92°C. The 48°C transition does begin to reappear for the slow-cooled samples after a few hours but at slightly lower temperatures.

Figure 16 demonstrates the effect on the glass transition of annealing stratum corneum at 75°C for 18 hrs. When the sample is scanned immediately following slow cooling from the annealing temperature, only the 92°C glass transition appears. Annealed samples allowed to rest in a desiccator for 72 hrs exhibited glass transitions at both 42° and 92°C. The lowered T_g generally observed on the second cycle may arise from induced internal stress on cooling through T_g.

One possible interpretation of this anomolous glass transition in terms of molecular structure and reactivity is based on the suggestions from simpler polyamide systems such as nylon. It is generally held that this transition originates with the rupture of interchain hydrogen bonds by the motion of long-chain segments in the amorphous regions. The unusual aspects of the glass transition is believed in part to be related to the unique structure of polyamides with alternating nonpolar chain segments and strongly hydrogen-bonding sites along the polymer chain (65). Since the hydrogen-bonding sites occur only at intervals along the chain, steric factors hinder the formation of the network and may explain the time dependence of its return. This explanation is further substantiated by the fact that this transition has been observed in the 40° to 50°C region for polyamides in which the number of methylene groups between

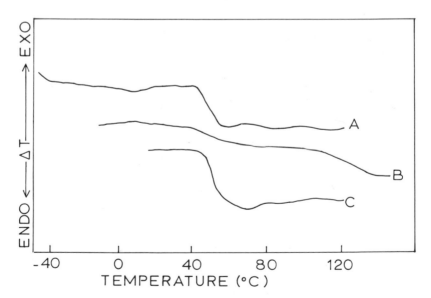

Figure 16. DSC scan of guinea pig footpad: A, control; B, annealed at 75°C 18 hrs; and C, rerun of B after 72 hrs N_2 atm (samples of comparable weight). Data from Ref. 42.

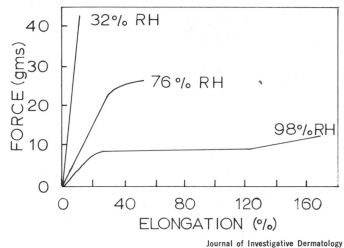

Journal of Investigative Dermatology

Figure 17. Force-extension curves at various RH levels for human stratum corneum (9)

potential hydrogen-bonding sites on the polymer backbone was varied from 4 to 11 (63). Hence, the transition observed at 40°–50°C in the polyamide structures appears to depend not on the flexibility of the network but on the existence and disruption of hydrogen bonds at points throughout the amorphous regions of the polymer.

Since the primary stabilizing forces in the glass appear to be hydrogen bonds, this transition is highly sensitive to the presence of water and other hydrogen-bonding molecules in the amorphous regions. Water then acts as a plasticizer of the amorphous protein regions of corneum as noted by the relationship between the extent of water found and the degree to which T_g is reduced. Also, there appear to be two amorphous regions which differ in their accessibility by water at ambient temperature. The T_g for the most accessible region shifts to $-18°C$ at 40% H_2O while the less accessible is only lowered to approximately 35°C (82). The T_g of corneum at water contents less than 5 wt % is difficult to determine since the remaining water is tightly held by primary sorption sites and technically difficult to remove without introducing structural alterations. Heating the corneum above 100°C to remove this fraction of water results in a T_g at 92°C.

It is not clear whether this shift in T_g to higher temperature is attributable to the lower water content or to the annealing process. There is an abrupt decrease in the magnitude of the associated specific heat change at a sample water content of about 15 wt %. In this water content range (15–20 wt %) force-extension curves for stratum corneum first begin to display a yield phase (Figure 17) as well as the initiation of a dramatic drop in tensile modulus (Figure 18). These data as well as cluster calculations (57) strongly suggest that water sorbed at these water contents plasticize the amorphous regions associated with the observed T_g.

An additional major transition which is observed in DSC scans of stratum corneum is a doublet endotherm which peaks at 194°C and 210°C in dry samples and at 120°–130°C in wet samples (Figure 14). These transitions are also characteristic of the more extensively investigated keratin-containing wool (49, 65). Polyamides such as the various nylons also show melting endotherms above 200°C (63).

Below 200°C, the heat-induced changes in dry wool structure are confined to the amorphous parts of the protein. Above 200°C two melting endotherms are present, a small one at 215°C from the melting of a low crosslinked fraction of the helix and a major one at 235°C as a result of the melting of the higher crosslinked fraction of the helix (49). In wool, the most direct evidence for this interpretation of helix melting is provided by x-ray diffraction patterns at the various temperatures.

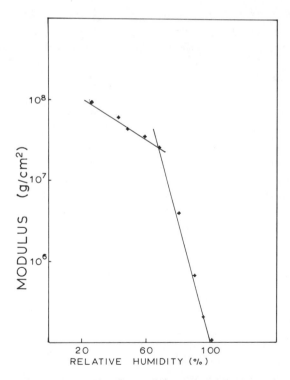

Figure 18.   Tensile modulus as a function of RH for newborn rat stratum corneum. Data from Ref. 82.

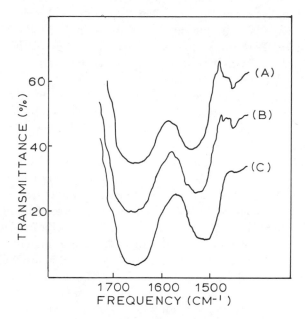

Figure 19. IR transmission spectra for newborn rat stratum corneum at various temperatures: A, 25°C; B, 125°C; C, 250°C. Data from Ref. 18.

Both the 5.1-A spacing characteristic of the spacing of helix turns and the 9.8 A of the lateral helical spacing decrease in intensity at 210° and 230°C (*49*). The fiber also loses its birefringence in this temperature range. These reflections are not present in fibers which have been super-contracted or otherwise rendered amorphous. Similarly, the x-ray diffraction pattern of wool heated to 130°C while immersed in water shows the total disappearance of the alpha-keratin reflections and the appearance of a disordered beta pattern (*66*). When either the dry or the wet fiber is heated above these melting temperatures, the transformation is irreversible.

Influence of Water. Even in desiccated corneum there is a weak, broad endotherm centering at about 120°–130°C as well as the higher temperature doublet melts at 194°C and 210°C. Both melts appear to be somewhat lower in temperature for corneum than for wool, which is consistent with the accepted higher degree of orientation and helical content of wool. The 130°C transition has been reported in other polypeptide systems by DSC, IR, and x-ray diffraction where it was found to be highly moisture sensitive (*67*). It has been suggested that this endothermic process which occurs between 110° and 150°C in a number of polypeptide systems is a partial conversion from the alpha-helical form to a beta configuration. This is supported by the appearance of a 4.7-A spacing and a shift in the amide I and II bands in the IR spectra (68, 69). Both these changes are characteristic of the conversion from the intrachain hydrogen-bonded alpha to the interchain hydrogen-bonded beta configuration.

Transmission IR of dry corneum at various temperatures from 25° to 250°C are shown in Figure 19. From about 120°C to 250°C there is a shift in amide I and II bands from 1660 to 1640 cm^{-1} and from 1550 to 1520 cm^{-1}, respectively, which is consistent with an alpha-to-beta transformation (44). This transition coincides with the broad but weak endotherm at 120°C in the DSC scans of dry corneum (Figure 14).

The intensity of the 120°–130°C endotherm increases as stratum corneum moisture content increases; there is a corresponding decrease in the high temperature meltings at 194° and 210°C (Figure 20). Highly hydrated corneum displays the 120°–130°C endotherm while below 15 to 20% water content, the transition is quite small. These DSC measurements were performed in hermetically sealed pans to avoid the endothermic loss of water. The transition in the presence of water is a cooperative one in that water facilitates the magnitude of the observed transition. The heat change associated with the 120°–130°C transition

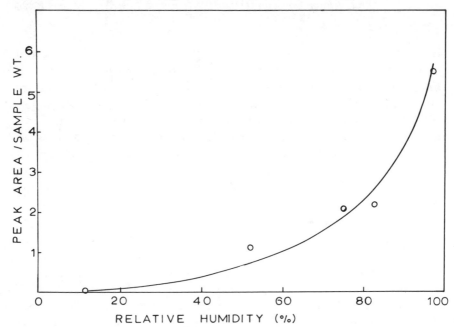

Figure 20.   Relative energy associated with DSC 120°C endotherm in newborn rat stratum corneum at various RH levels (hermetically sealed system). Data from Ref. 42.

is a maximum at 98% RH and is not increased further by immersion in water.

Influence of Nonpolar Solvent. The influence of various solvents on the physical and chemical properties of corneum has been studied widely (*70, 71*) with particular interest in their influence on water binding. In addition to the extraction of lipids by the nonpolar solvents, the samples are subsequently more susceptible to further damage by water. Extracted samples have lowered water-binding affinity and altered mechanical behavior (*9, 70*). Lipid-soluble materials removed by nonpolar solvents perform a protective role in preventing the loss of water-soluble components responsible for the water binding at high relative humidities.

DSC scans of ether and chloroform–methanol (3/1 by volume) extracted samples exhibit broader and less detailed melting endotherms at 194° and 210°C than untreated samples. An additional difference is shown by chloroform–methanol-extracted samples when quench cooled and rerun. In reruns of control and ether-extracted samples, T_g is still rather sharp although reduced in amount and temperature while the chloroform–methanol-extracted sample is quite broad with a second apparent glass at about 90°–100°C (*42*). Either the water-soluble materials act as plasticizers or their loss through solvent extraction causes structural changes in the proteins which inhibit reforming of the original glass. It will be shown later that dynamic spectroscopy demonstrates a higher tensile modulus for chloroform–methanol-extracted samples than the control suggesting a structural reorganization has occurred (*14*).

Influence of Orientation. Induced orientation in stratum corneum was achieved by stretching hydrated samples to varying degrees and allowing them to dry in that elongated state. DSC scans of these oriented samples display a number of alterations in the melt endotherms and glass transition which vary with the extent of prestretching and may provide some insight into the molecular mechanisms responsible for elasticity.

There is a gradual reduction in T_g with some broadening and loss of definition of the high temperature endotherms near 200°C as the amount of prestretch increases (Figure 21). The orientation produced by low deformations appears to produce an internal stress in the amorphous regions which acts as an external load to lower the softening temperature (*72*).

Secondary Transitions. In addition to the anomalous glass transition in the 40°–50°C region because of the motion of large segments of the polymer chain, stratum corneum like many other polymers displays smaller secondary transitions at lower temperatures. These arise generally from the motion of side chains or the small segments of the backbone. The tan δ behavior from dynamic mechanical studies of dry

stratum corneum indicate small peaks *(14)* at $-10°$ and $-60°C$ suggesting secondary transitions. The tan δ for the dielectric properties of
newborn rat stratum corneum also demonstrates a transition at $-10°C$
with an apparent activation energy of 7 kcal/mole *(6)*.

Similar relaxation phenomena are observed in other polyamide systems such as the protein collagen and the synthetic polypeptide 50/50

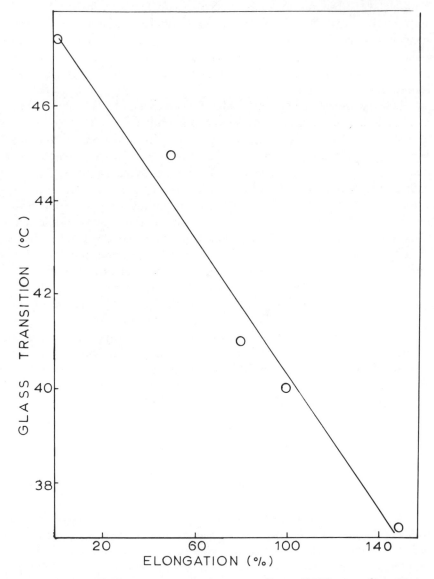

*Figure 21.   Glass transition temperature (from DSC) vs. elongation
of prestretched guinea pig footpads (N₂ atm). Data from Ref. 42.*

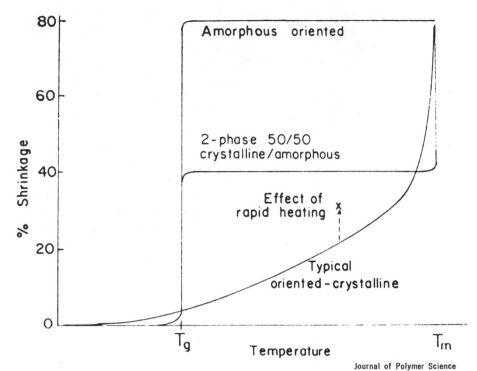

Journal of Polymer Science

Figure 22.   Diagramatic representation of the longitudinal shrinkage behavior of oriented fibers (77)

L-glutamic acid–L-leucine. The dynamic mechanical spectra of these materials display highly moisture-sensitive relaxations at $-13°$ and between $-60°$ and $-90°C$ (73, 74).

Spontaneous Dimensional Changes. One technique used to approximate the degree of orientation in opaque amorphous materials is to measure the amount of shrinkage or magnitude of retractive force developed when they are heated. When an oriented amorphous plastic sheet is heated above T_g, it shrinks back to the approximate shape it had before orientation. Thermodynamically this represents an attempt by the oriented polymers to attain a state of maximum disorder (entropy). If the film is held at constant length, it generates a retractive force when heated above T_g such that an increased retractive force means a greater degree of orientation in the sample. In the latter case, the retractive force generated is also dependent on the modulus of the material (4, 72, 75, 76).

The shrinkage behavior of a hypothetical, oriented fiber composed of varying proportions of amorphous and crystalline phases is shown in Figure 22. Shrinkage first occurs in the region of T_g. A model fiber consisting of oriented amorphous regions and oriented crystalline regions would undergo partial shrinkage around T_g but would show no further shrinkage until the crystalline melt. A completely amorphous oriented system would only display shrinkage at T_g (77).

It is apparent from Figure 23 that the thermally induced spontaneous contractions of stratum corneum are quite similar to the hypothetical model fiber composed of a two-phase system of oriented amorphous and crystalline phases. The initial contraction begins at 50°C as predicted from the DSC determination of T_g. As would be expected for a crystalline melt, there is a rapid loss of modulus (Figure 24), a large increase in transverse thickness (Figure 25), and an endothermic heat process accompanying the 196°C contraction. Similar behavior is observed in the α-keratin-containing hair (78). An energy of activation of 110 kcal/mole was calculated from the frequency dependence of dynamic mechanical spectra of corneum in the 207°C region. This high activation energy is indicative of the motion of rather large segments of the polymer chain and is consistent with an alpha-to-beta transformation (14).

Additional evidence for this two-phase model for stratum corneum can be demonstrated by measuring shrinkage as a function of tension. The degree of shrinkage corresponds to disorientation of the amorphous phase and the successive decrease in shrinkage with increased loads corresponds to an extension from that disoriented state (77). As would be predicted for a simple two-phase amorphous–crystalline system above

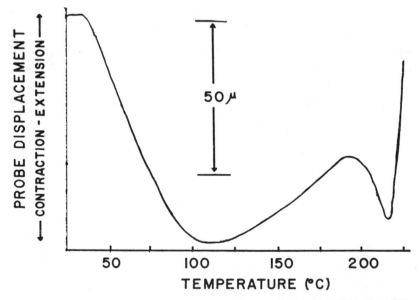

Figure 23.   Thermally induced longitudinal shrinkage in dry newborn rat stratum corneum (He atm, 20°C/min heating rate) (10)

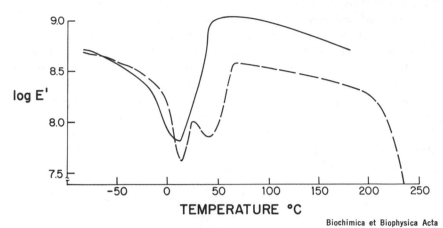

VIBRON-DRY N$_2$ STRATUM CORNEUM (HUMAN)

Biochimica et Biophysica Acta

Figure 24. Dynamic mechanical spectrum of dry human stratum corneum. Dashed line, control; solid line, chloroform–methanol extracted (N$_2$ atm, 1°C/ min heating rate) (14).

T_g, the force–extension curve constructed in this manner for the first contraction displays a rubber-like behavior at low strains as the amorphous regions become oriented (*42*).

The magnitude of the two thermally induced contractions is quite uniform among samples of newborn rat corneum, but considerable variation is encountered with human corneum, particularly for the 196°C contraction. The contraction values for human corneum samples varied from 1 to 5% but were quite consistent for any given subject. The increased magnitude for this contraction in some human specimens could indicate a higher degree of orientation of the crystalline fibrous material (*11*).

The degree of longitudinal contraction at 196°C and the temperature at which stratum corneum begins to melt are influenced by exposure to formaldehyde vapor (*11*). Formaldehyde is highly reactive and capable of forming methylene bridges between and within polypeptide chains. The increased melting temperature and increased degree of contraction (Figure 26) caused by the additional crosslinks imposed on an already oriented system agree with the results predicted by Flory for crosslinked polymer chains (*76*).

Figure 27 indicates the influence of choloform–methanol (3/1 by volume) or ether extraction on the magnitude of the 196°C contraction as a function of exposure time. Chloroform–methanol also appears to interact with the amorphous regions as indicated by the decrease in the 50°C contraction. These findings are consistent with their influence in the DSC studies where solvent exposure broadened both T_g and the melting endotherms at 190°C.

Induced orientation produced by prestretching hydrated stratum corneum samples demonstrates a decrease in the degree of the 50°C contraction with increasing elongation (0–200%). The amount of the 196°C contraction increases with increasing elongation, showing the greatest effect at high extensions (> 100%). These data suggest that elongation is accomplished by decreasing the degree of orientation in the amorphous regions and producing increased orientation in the crystalline regions (72). The proposed mechanism agrees with the increased dichroism of the IR amide I and II bands of prestretched samples (~ 100% extension) discussed earlier.

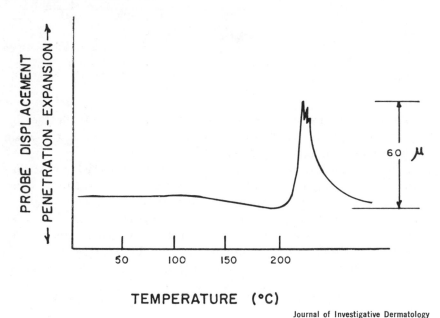

Journal of Investigative Dermatology

Figure 25.   Thermally induced transverse swelling of dry newborn rat stratum corneum (He atm) (10)

The temperature dependence of the longitudinal dimensions of hydrated strips of corneum are shown in Figure 28. Untreated samples begin to contract between 65° and 70°C and continue until about 85° to 90°C. Chloroform–methanol (3/1 by volume) extracted samples differed in that the samples begin to contract at about 60°C and continue beyond 90°C. When a slight positive load is applied to the sample strip, an apparent decrease in modulus is observed at 35°C regardless of solvent pretreatment. Solvents such as dimethyl sulfoxide or strong acids which alter the stability of the organized portions of the protein lower this

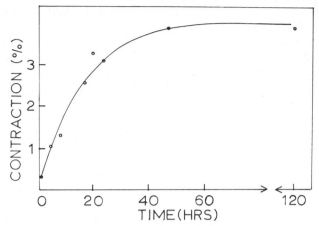

Journal of Investigative Dermatology

Figure 26. Increased shrinkage at 200°C with exposure to formaldehyde vapor in newborn rat stratum corneum (load 2.8 gm, He atm) (11)

initial contraction temperature and are useful in probing the factors determining stratum corneum stability.

Isometric Contraction. One of the earliest observations of the thermal properties of epidermal protein was made by Rudall when he reported that cow snout epidermis contracted in water when heated (79). This thermally induced contraction was accompanied by a change in structure from alpha to cross-beta as exhibited in x-ray diffraction pat-

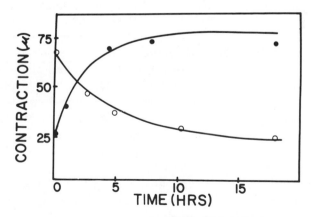

Journal of Investigative Dermatology

Figure 27. Upper curve, effect of chloroform–methanol and ether on 196°C longitudinal shrinkage in newborn rat stratum corneum. Lower curve, chloroform–methanol reduction of 50°C longitudinal shrinkage (He atm) (11).

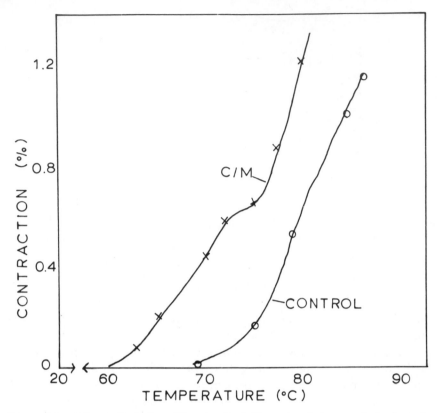

Figure 28.   Thermally induced longitudinal dimensional changes in hydrated guinea pig footpad (samples run in water, 1°C/min heating rate) chloroform–methanol and control. Data from Ref. 42.

terns. More recently, Baden has used this isometric contraction technique to characterize the influence of solvents, pH, salts, and diseases states on hydrated stratum corneum thermal stability (7, 8). Upon heating samples of general body skin epidermis, an initial decrease in tension with little change occurs until about 82°–85°C after which a major continuous rise in tension appears (Figure 29). Prior extraction of the tissue with chloroform–methanol (3/1 by volume) had no effect on the inflection temperature but resulted in a flattening of the curve at about 90°C. X-ray diffraction showed a loss of the 5.15 A alpha reflection and the development of a sharp reflection at 4.65 A (accentuated on the meridian) when specimens were heated above 85°C regardless of whether or not a change in tension was observed or applied. The absence of this transition in the abnormal corneum of Harlequin Fetus was explained by the fact that the x-ray diffraction pattern demonstrated that the cross-beta configuration was present in its native state. Similarly, stratum corneum from palms and soles differed from general body epidermis by displaying a continuous decrease to zero tension with heating. Prior treatment with hexane had no effect on the result, but treating with chloroform–methanol (3/1 by volume) induced behavior similar to that of general body epidermis.

Thermomechanical. Water principally acts as a plasticizer to decrease chain interactions, increase chain mobility, and soften stratum corneum. Similarly, softening can be achieved by introducing enough kinetic energy into the corneum to cause increased segmental motion of the molecular chains and therefore decrease their interactions with adjacent chains.

A representative softening thermogram determined by thermomechanical analysis (TMA) for dry newborn rat corneum is shown in Figure 30. Upon heating, there are transverse softenings of the tissue at 45°C and 155°C followed by a substantial transverse swelling above 180°C (*10*). These transition temperatures are highly load dependent but are reproducible when done at constant load and heating rate. The 45°C modulus decrease appears to be a combination of both lipid melting and T_g softening since it is not entirely removed by chloroform–methanol extraction. The transverse swelling above 100°C is the volume change associated with melting of the α-helical structure. The fundamental

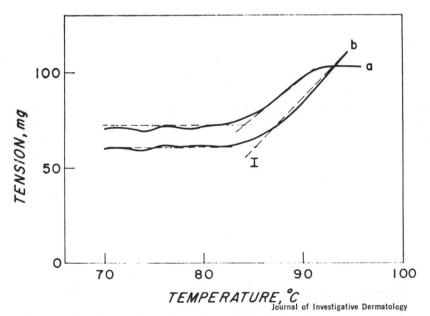

Figure 29.   Isometric contraction of human stratum corneum: A, chloroform–methanol extracted; B, control (7)

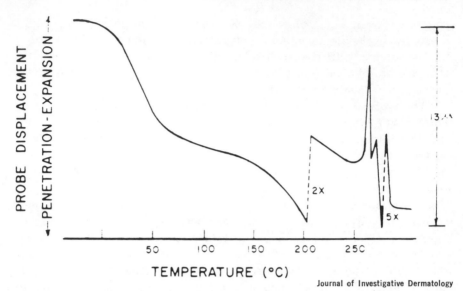

Journal of Investigative Dermatology

Figure 30.   Thermally induced transverse softening of newborn rat stratum corneum (He atm, 3 gm load, 20°C/min heating rate) (10)

molecular events responsible for the softening which occurs at 155°C in newborn rat corneum and 177°C in human corneum have not been elucidated.

Table IV shows a comparison of various viscoelastic parameters for newborn rat and human stratum corneum determined by TMA. The most significant differences relate to the 155°C softening and the degree of the expansion at 213°C. These two parameters appear to be the most sensitive and useful indicators for characterizing normal, extracted, and diseased corneum. For example, exposure of corneum to formaldehyde resulted in little change in the low temperature softening but caused a significant increase in the 155°C softening temperature (Figure 31). Psoriatic corneum exhibited a softening at 192°C which is consistent with the theory that it is more highly crosslinked than normal corneum. It also exhibits a transverse expansion at 40°C which is reversible, suggesting that it is associated with T_g (42). Similarly, guinea pig footpad exhibits expansions at both 40° and 155°C.

Mechanical Properties. The important influence of sorbed water on the general mechanical behavior of stratum corneum was first recognized by Blank (80), and it is best demonstrated by force–elongation curves at various eqiuilibration relative humidities (Figure 17). When dry, stratum corneum is rather inextensible with modulus values of polymeric glass, and it is unable to function properly physiologically. However, as the membrane water content increases, plasticization occurs (T_g lowered) and corneum behaves mechanically as a highly extensible,

Table IV. Viscoelastic and Dimensional Transition Temperatures for Human and Newborn Rat Stratum Corneum Samples as Well as the Degree of the Dimensional Change

| | *Temperature, °C* | | | *Displacement, %* | |
Transition	*Rat*	*Human*	*p*[a]	*Rat*	*Human*
Lenetration	45	53	.25	18	20
Penetration	155	177	.05	52	53
Expansion	213	229	.40	400	200
Contraction	50	60	.05	1	1
Contraction	196	200	.40	1	1–5
Penetration	285	290	.80	—	—

[a] Probabilities obtained from the *t*–test are shown for human *vs.* rat transition temperatures. Further statistics on measurements in Refs. *10, 11,* and *78.*

elastic material (*9, 81, 82, 83, 84*). This plasticizing influence of sorbed water is demonstrated by the significant drop in elastic modulus as a function of RH (Figure 18) where the observed thousandfold drop in modulus is characteristic of a glass transition (*82*). There is an apparent change in the softening effectiveness of water sorbed above 60% RH (15–20 wt %), and it corresponds to the clustering of the sorbed water as discussed previously. Similarly at this RH and water content, there is a change in deformation behavior (appearance of a yield phase) and an abrupt decrease in magnitude of specific heat change at 40°–50°C as discussed earlier.

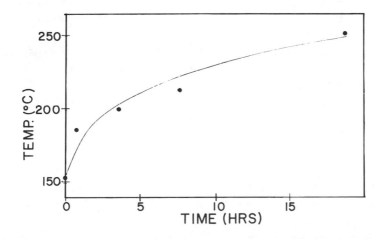

Journal of Investigative Dermatology

Figure 31. Effect of time exposed to HCHO vapors on the 155°C transverse softening in newborn rat stratum corneum (He atm) (11)

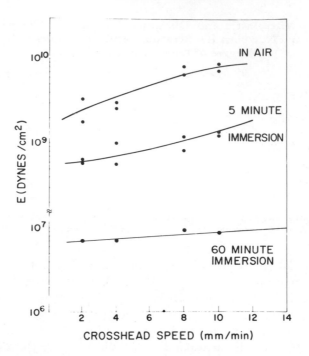

Figure 32. Effect of strain rate on the elastic modulus of newborn rat stratum corneum as a function of time in water. Data from Refs. 6, 85.

The viscoelastic character of stratum corneum is further demonstrated by the dependence of elastic modulus on strain rate (Figure 32). Even the dry corneum ($\sim$ 10 wt % H_2O) showed significant viscoelastic behavior over the narrow range of strain rates studied. In the case of hydrated membranes, there is also a change in shape of the force–elongation curve (*6, 85*).

Stratum corneum breaking strength decreases fourfold over the 0–100% RH range reaching a minimum at approximately 90% RH which is not lowered further by immersion in water. Of fundamental importance is the morphological location within the stratum corneum where failure occurs under a uniaxial load. Scanning electron microscopy (SEM) and conventional analysis of fractured samples indicate that the samples predominately fracture within the intercellular junctions rather than intracellularly (*9*).

Further demonstration of the influence of water on mechanical properties is shown in Figure 33 where the elastic modulus of stratum corneum is plotted as a function of immersion time in water. The elastic modulus begins to decrease after about a 5-min immersion time corresponding to a membrane water content of 15–20 wt % as determined from water uptake studies (*62*). The full decrease in modulus is achieved at 50–60-min immersion time and a water content of about 40–50 wt %. As shown in Figure 10, this is also the water content achieved when equilibrated at 95–100% RH. In contrast, the breaking strength in uniaxial testing is rapidly lowered (75–80%) during the first 5–10 min of immersion time to a saturation value equivalent to that at 98–100% RH (*18*).

The influence of temperature at constant immersion time (2 min) shows a similar plasticizing effect at lower water content, suggesting a cooperative plasticizing effect of water and temperature. There is an abrupt decrease in modulus between 30° and 40°C to that achieved for a fully hydrated membrane (*6, 85*). A similar softening was observed between 30° and 40°C for samples of corneum in studies on the effect of temperature on longitudinal dimensions of hydrated samples discussed earlier. Since this softening occurs in both control and extracted samples, the softening appears to be principally protein related (Figure 34).

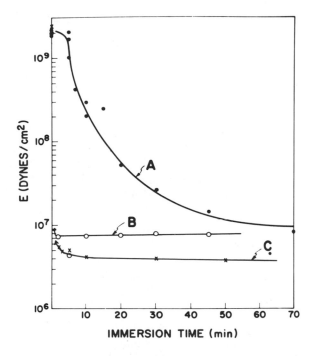

Figure 33. Influence of water on elastic modulus of newborn rat stratum corneum. A, control; B, chloroform–methanol followed by water; C, water followed by chloroform–methanol. Data from Refs.
6, 85.

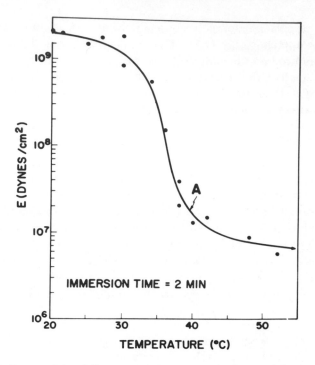

Figure 34. Effect of temperature on the modulus of hydrated newborn rat stratum corneum. Data from Refs. 6, 85.

Influence of Solvents. The stress–strain curves of untreated and ether-extracted corneum in water show marked differences (*81*). Untreated corneum, extended 5% and relaxed, shows hysteresis similar to that observed for other keratinaceous structures (Figure 35). The deformation mechanism is completely reversible, and hydrogen-bond breakdown and slow reformation may be the major factors determining the stress–strain relationships. With ether-extracted samples, complete recovery is observed from 5% extension but with little or no hysteresis. The more rapid swelling and lack of hysteresis of ether-extracted corneum in water may be related to the breakdown of hydrogen bonds normally shielded from the effects of water by the lipid-like materials removed by ether.

Ether and chloroform–methanol extraction have significant effects on the breaking strength of stratum corneum. Both solvents increase the absolute value of the breaking strength in the presence of water vapor or when immersed in water (*9, 81*). In addition, the breaking strength for extracted samples has a lower dependence on RH than do untreated ones. These data suggest a solvent interaction with the intercellular materials responsible for cohesion.   Extraction of stratum corneum by chloroform–methanol (3/1 by volume) produces significant alterations in the influence of sorbed water on the tensile modulus (*84*). Figure 36 demonstrates that the tensile modulus at any given RH is always higher for extracted than for untreated corneum. Also note that the rapid softening effect of water does not occur until about 90–95% RH as compared with 60–65% RH for untreated corneum. This increased modulus for chloroform–methanol-extracted samples is consistent with the results of dynamic mechanical studies (*14*).

The morphological location of the fibrous protein principally responsible for the deformation and viscoelastic behavior is uncertain. Both the cell membrane and intracellular regions are composed of fibrous proteins which differ considerably in amino acid composition. Since the alpha-keratin within the cells shows few orientation properties until high elongations, it has been suggested that the membrane proteins determine the viscoelastic behavior at low deformations (*84*).

X-ray diffraction and IR dichroism studies suggest that the long-range elasticity of wool is related to a reversible molecular transformation of the alpha-keratin to an extended beta form (*66*). No convincing evidence supports this mechanism in stratum corneum viscoelasticity. In fact, the available evidence suggests that the elastic behavior of corneum is primarily entropic in origin. At low deformations, the mechanical properties of hydrated stratum corneum is best described as the behavior of a lightly-crosslinked rubber.

Several additional observations suggest different mechanisms are responsible for the deformation and viscoelastic properties of stratum

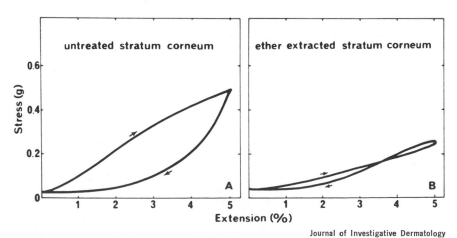

Journal of Investigative Dermatology

Figure 35.   Hysteresis behavior of ether-extracted and untreated stratum corneum in water (81)

corneum and wool. The tensile modulus of wool is only slightly lowered by sorbed water as compared with the dramatic decrease observed in corneum. Similarly, in wool the amount of extension when yielding occurs is almost independent of water content whereas it is dramatically different in corneum (Figure 17) (9).

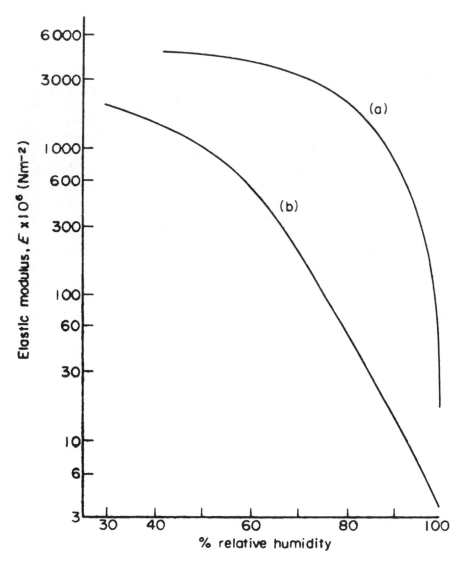

Journal of the Society of Cosmetic Chemists

Figure 36.   Elastic modulus of stratum corneum at various RH levels. A, solvent-exposed; B, control (84)

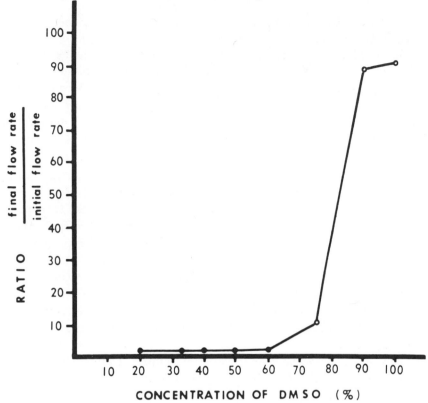

Journal of Investigative Dermatology

Figure 37.   Permeability of water through stratum corneum in the presence of various concentrations of DMSO (86)

Stress Relaxation. In a stress–relaxation measurement, the specimen is deformed by a fixed strain and the stress required to maintain this deformation is measured for a specific time. The degree of stress relaxation is highly temperature-dependent in the region of T_g and very sensitive to interacting solvents as well as other factors which alter chain mobility. Therefore, stress relaxation is useful in determining the influence of solvents on the viscoelastic properties of corneum.

Polar organic liquids such as dimethylformamide (DMF) enhance the permeability of molecules through the stratum corneum. Figure 37 is a plot of water transmission rate through stratum corneum in the presence of varying dimethyl sulfoxide (DMSO)–water concentrations (*86*). The increased water permeability at high DMSO concentrations (> 80%) arises presumably from the structural alterations caused by its ability to interact with polar and ionic groups as well as hydrophobic regions which stabilize the structure.

The increased stress relaxation as a function of DMSO concentration above 80% is indicative of structural alterations resulting in more mobility of side chain segments (Figure 38) (18). This increased chain mobility would be expected to lead to the observed increased permeability by increasing the apparent diffusion constant. The considerable increase of the modulus of the corneum (Figure 39) with increasing DMSO concentration suggests that the strong hydrogen bond-forming ability of DMSO stabilizes the region involved in these low deformations.

Dynamic Mechanical (Low Strain Deformation). When a cyclic strain of small amplitude is applied to a strip of material, a cyclic stress will be generated in response by the sample. If the material is ideal (Hookian) and stores all the input energy, the cyclic stress is in phase with the applied cyclic strain. Viscous components cause a finite phase lag or phase angle, δ, between the stress and strain. E' represents the elastic, real, or storage modulus while E'' is the viscous, imaginary, or loss modulus. Tan δ is equal to the ratio E'/E'' and is related to the molecular relaxations that occur in the sample. Transition temperatures and associated activation energy can be determined (72) by varying the frequency of oscillation at a fixed temperature or the temperature at a fixed frequency.

The dynamic mechanical spectra for human stratum corneum are shown in Figure 24. There is a drop in modulus (E') starting at 0°C

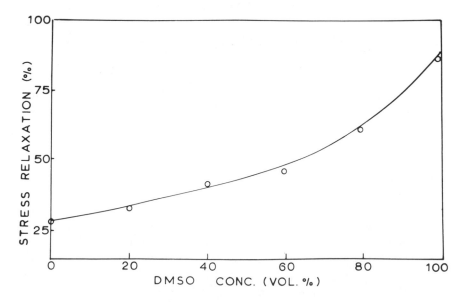

Figure 38. Effect of DMSO on the stress relaxation of newborn rat stratum corneum at 10% extension. Data from Ref. 18.

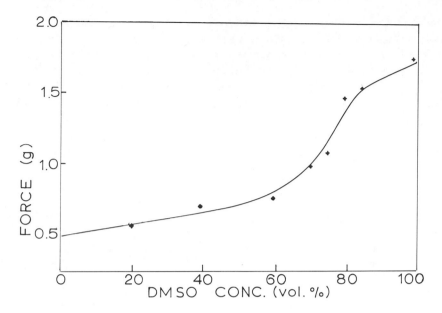

Figure 39. Effect of DMSO on the modulus of newborn rat stratum corneum (force at 10% extension is plotted vs. DMSO concentration). Data from Ref. 18.

accompanied by longitudinal contraction of the sample. Relaxations also occur at 40°C associated with lipid melting and at 207°C associated with the helix melting. The temperature of the major modulus decrease at a low temperature is highly dependent on the sample history (solvent extraction, cooling procedure, and heating rate) (*14*).

Swelling Characteristics. Stratum corneum has the ability to imbibe many times its original volume of liquid without complete dispersion when immersed in certain solvents. Because of the rather dense entanglement of fibrous protein within the cell and cell membrane which cannot dissolve in the surrounding solvent, stratum corneum is subject to osmotic action of the surrounding solvent. When stratum corneum is placed in a swelling agent, liquid is taken up by the tissue with subsequent displacement of the matrix. This displacement of the component polymeric chains bring about elastic retractive forces opposing further expansion of the network. Hence, at equilibrium the elastic retractive forces of the network structure are in balance with the osmotic forces which tend to swell the tissue. The degree of swelling is related to the affinity of the solvent for the tissue as well as the magnitude of the elastic retractive forces stabilizing the network structure.

The dimensional swelling of stratum corneum produced by the sorption of water or formic acid is anisotropic and characteristic of highly oriented films. Highly oriented materials show more swelling perpendicular to the direction of orientation than parallel to it because of the lower cohesive forces between the oriented chains (87).

Figure 40 shows both longitudinal and transverse swelling data for human and newborn rat stratum corneum in formic acid (90 wt %). As would be expected from the biaxial orientation of the fibrous proteins in corneum, transverse swelling is considerably greater than longitudinal swelling for both tissues. The degree of longitudinal swelling is the same for both samples. The magnitude of transverse swelling of newborn rat

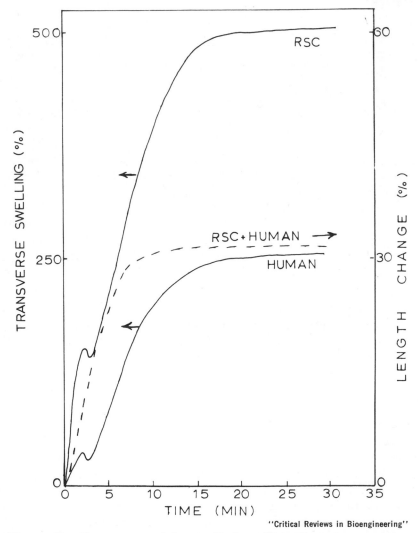

"Critical Reviews in Bioengineering"

Figure 40. Transverse and longitudinal swelling of stratum corneum induced by formic acid 90 wt % (solid lines are transverse swelling) (6)

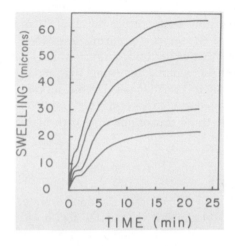

Figure 41. Influence of formaldehyde vapor exposure (crosslinking) on formic acid swelling of newborn rat corneum (sample thickness 12 μ). In decending order the curves represent formaldehyde exposure times of A, control; B, 0.5 hrs; C, 1 hr; and D, 2 hrs. Data from Ref. 18.

corneum is much greater than human, perhaps because of a higher effective crosslink density present in human corneum. The swelling behavior was not affected by previous chloroform–methanol extraction (88).

One of the most widely used measurements of crosslink density in synthetic polymers is based on equilibrium swelling by a good solvent (89, 90). A dramatic example of the influence of chain mobility on the swelling properties of stratum corneum is shown in Figure 41. The degree of transverse swelling of stratum corneum by formic acid is a function of the sample's previous exposure to formaldehyde vapors. The increased crosslink density decreases the degree of chain mobility and hence the equilibrium swelling. However, the degree of longitudinal swelling was unchanged by the increased crosslink density.

The degree of formic acid swelling of various corneum tissues suggests a distinct difference in their apparent crosslink densities (12). The temperatures in Table V are in accord with the structural data suggested by the swelling characteristics of the tissue in formic acid. For example, newborn rat stratum corneum swells considerably more (1800%) than the others, suggesting increased chain mobility and fewer inter- and intra-chain interactions and crosslinks opposing the swelling. Similarly,

Table V.   Physicochemical Parameters for Various Keratin Samples

Sample	Formic Acid Uptake, wt %	Softening T, °C	Expansion T, °C
Newborn rat	1800	155	213
Human (adult)	1100	177	229
Psoriatic	550	192	242
Newborn rat (crosslinked)	450	250	264
Hair	220	242	256

its softening temperature and expansion temperature at 155°C and 213°C, respectively, are the lowest of the group requiring less kinetic energy to alter the structure. The temperatures for psoriatic transitions are significantly higher than for control, again suggesting that its decreased extensibility is partially a result of the decreased chain mobility (increased rigidity) of its macromolecular components. As predicted, samples exposed to formaldehyde have these transitions shifted to higher temperatures than untreated samples.

Summary

Progress has been made in using polymer characterization to understand better the structure–property relationships essential for proper function of the stratum corneum as a biological interface. This has been accomplished primarily through physical techniques and interpretations used in polymer and material sciences for determining fundamental and empirical parameters characteristic of physical behavior. The physical behavior of stratum corneum as a diffusion and mechanical barrier can best be explained by a two-phase system of oriented, amorphous, and oriented crystalline components, principally proteins associated with lipids. The components are highly sensitive to water and to temperature. Both water and temperature behave similarly in their ability to influence diffusion and mechanical properties of stratum corneum by increasing segmental chain motion in the amorphous regions. As expected, many of the physical properties of amorphous polymers which are dependent on segmental relaxation rate such as viscous flow, mechanical and dielectric relaxation, creep, and diffusion, show a major change when sufficient heat or water is added to allow the required segmental motion.

The establishment of similar physical and chemical parameter profiles for normal, altered, and diseased corneum should provide fundamental information and produce clinically useful information as well.

Acknowledgment

The authors express appreciation to Robert Kennedy and Emrick Artz for technical assistance in some of these studies.

Literature Cited

1. Montagna, W., "The Structure and Function of Skin," Academic, New York, 1956.
2. Kent, G., "Comparative Anatomy of the Vertebrates," C. V. Mosby, St. Louis, 1965.
3. "Biophysical Properties of the Skin," H. Elden, Ed., Wiley-Interscience, New York, 1971.

4. Flory, P., "Role of Crystallization in Polymers and Proteins," *Science* (1956) **124**, 53.
5. Humphries, W., Wildnauer, R., "Characterization of Aging Skin *via* Thermal Analysis," in "Polymer Characterization by Thermal Methods of Analysis," J. Chiu, Ed., Marcel Dekker, New York, 1974.
6. Wilkes, G., Brown, I., Wildnauer, R., "The Biomechanical Properties of Skin," in "Critical Reviews in Bioengineering," Vol. I, p. 453, Chemical Rubber Co., 1974.
7. Baden, H., Gifford, A., "Isometric Contraction of Epidermis and Stratum Corneum with Heating," *J. Invest. Dermatol.* (1970) **54**, 298.
8. Baden, H., Goldsmith, L., "The Structural Proteins of Harlequin Fetus Stratum Corneum," *J. Invest. Dermatol.* (1973) **61**, 25.
9. Wildnauer, R., Bothwell, J., Douglas, A., "Stratum Corneum Biomechanical Properties. I. Influence of Relative Humidity on Normal and Extracted Human Stratum Corneum," *J. Invest. Dermatol.* (1970) **56**, 72.
10. Humphries, W., Wildnauer, R., "Thermomechanical Analysis of Stratum Corneum: I. Technique," *J. Invest. Dermatol.* (1971) **57**, 32.
11. Humphries, W., Wildnauer, R., "Thermomechanical Analysis of Stratum Corneum: II. Application," *J. Invest. Dermatol.* (1972) **58**, 9.
12. Wildnauer, R., Miller, D., Humphries, W., "A Physicochemical Approach to the Characterization of Stratum Corneum," *Dermatol. Digest* (1973) **12**, 13.
13. Wilkes, G., Nguyen, An-Loc, Wildnauer, R., "Structure–Property Relations of Human and Neonatal Rat Stratum Corneum: I. Thermal Stability of the Crystalline Lipid Structure as Studied by X-ray Diffraction and Differential Thermal Analysis," *Biochim. Biophys. Acta* (1973) **304**, 267.
14. Wilkes, G., Wildnauer, R., "Structure–Property Relationships of the Stratum Corneum of Human and Neonatal Rat: II. Dynamic Mechanical Studies," *Biochim. Biophys. Acta* (1973) **304**, 267.
15. Scheuplein, R., Ross, L., "Effects of Surfactants and Solvents on the Permeability of Epidermis," *J. Soc. Cosmet. Chem.* (1970) **21**, 853-873.
16. Menton, D., Eisen, A., "Structural Organization of the Stratum Corneum in Certain Scaling Disorders of the Skin," *J. Invest. Dermatol.* (1971) **57**, 295.
17. Kligman, A., "The Biology of the Stratum Corneum," in "The Epidermis," W. Montagna, W. Lobetz, Eds., Academic, New York, 1964.
18. Wildnauer, R., unpublished data.
19. Swanbeck, G., "Macromolecular Organization of Epidermal Keratin," *Acta Dermatol. Venereol.* (1959) **39**, 1.
20. Frost, P., Weinstein, G., Bothwell, J., Wildnauer, R., "Ichthysioform Dermatoses. III. Studies of Transepidermal Water Loss," *Arch. Dermatol.* (1968) **98**, 230.
21. Shahidullah, M., Raffle E., Frain-Bell, W., "Insensible Water Loss in Dermatitis," *Brit. J. Dermatol.* (1967) **79**, 589.
22. MacKenzie, I., "The Ordered Structure of Mammalian Epidermis," in "Epidermal Wound Healing," H. Maibach, D. Rovee, Eds., Yearbook Med. Publish., Chicago, 1972.
23. Rovee, D., unpublished data.
24. Matoltsy, A., Parakkal, P., "Keratinization," in "Ultrastructure of Normal and Abnormal Skin," Chap. 5, A. Zelickson, Ed., Lea and Febiger, Philadelphia, 1967.
25. Wilgram, G., Caulfield, J., Madgic, E., "A Possible Role of the Desmosome in the Process of Keratinization," in "The Epidermis," W. Montagna, W. Lobetz, Eds., Academic, New York, 1964.
26. Odland, G., Reed, T., "Epidermis," in "Ultrastructure of Normal and Abnormal Skin," Chap. 4, A. Zelickson, Ed., Lea and Febiger, Philadelphia, 1967.

27. Plewig, G., Marples, R., "Regional Differences of Cell Sizes in the Human Stratum Corneum," *J. Invest. Dermatol.* (1970) **54**, 13.
28. Brody, I., "An Electron Microscopic Study of the Fibrillar Density in the Normal Stratum Corneum," *J. Ultrastruct. Res.* (1970) **30**, 209.
29. Scheuplein, R., Blank, I. H., "Permeability of the Skin," *Phys. Rev.* (1971) **51**, 702.
30. Miller, D., *Current Peaks* (1973) **6**, Carle Instruments Quarterly Publ., Fullerton, Calif.
31. Brody, I., "The Keratinization of Epidermal Cells of Normal Guinea Pig Skin as Revealed by Electron Microscopy," *J. Ultrastruct. Res.* (1959) **2**, 482.
32. Charles, A., Smiddy, F., "Tonofibrils of the Human Epidermis," *J. Invest. Dermatol.* (1957) **29**, 327.
33. Mercer, E., Jahn, R., Maibach, H., "Surface Coats Containing Polysaccharides on Human Epidermal Cells," *J. Invest. Dermatol.* (1968) **51**, 204.
34. Swanbeck, G., Thyresson, N., "An X-ray Diffraction Study of Scales from Different Dermatoses," *Acta Dermatol. Venereol.* (1961) **41**, 289.
35. Wilkes, G., "Superstructure in Stratum Corneum of Neonatal Rat," *Biochim. Biophys. Acta* (1973) **304**, 290.
36. Goldsmith, L., Baden, H., "A Uniquely Oriented Component of Epidermis," *Nature* (1970) **225**, 1052; Abstract: *Clinical Res.* (1969) **17**, 274.
37. Swanback, G., Thyresson, N., "A Study of the State of Aggregation of the Lipids in Normal and Psoriatic Horny Layer," *Acta Dermatol. Venereol.* (1962) **42**, 445.
38. Wilkes, G., Wildnauer, R., unpublished data.
39. Esfahani, M., Lembrick, A., Knutton, S., Oka, T., Wakel, S., "The Molecular Organization of Lipids in the Membrane of *Escherichia coli:* Phase Transitions," *Proc. Nat. Acad. Sci. USA* (1971) **68**, 3180.
40. Lipson, M., Wool, "Encyclopedia of Chemical Technology," vol. 22, p. 387, John Wiley, New York, 1970.
41. Ansari, M., Nicolaides, N., Fee, H., "Fatty Acid Composition of the Living Layers and Stratum Corneum Lipids of Human Sole Skin Epidermis," *Lipids* (1971) **5**, 838.
42. Humphries, W., Wildnauer, R., unpublished data.
43. Baden, H., Bonar, L., "The Alpha-Fibrous Proteins of Epidermis," *J. Invest. Dermatol.* (1968) **51**, 478.
44. Parker, F., "Applications of Infrared Spectroscopy in Biochemistry, Biology and Medicine," Chap. 10, Plenum, New York, 1971.
45. Boni, R., Filippi, B., Ciceri, L., Peggion, E., "Characterization of Block and Random Copolymers of Alpha-Amino Acids by Thermogravimetric Analysis," *Biopolymers* (1970) **9**, 1539.
46. Bihari-Varga, M., "Thermoanalytical Characterization of Stability of Cross-linked Proteins," *Acta Biochim. Biophys. Acad. Sci. Hung.* (1971) **6**, 265.
47. Bihari-Varga, M., Biro, T., "Thermoanalytical Investigations on the Age-Related Changes in Articular Cartilage, Meniscus and Tendon," *Gerontologia* (1971) **17**, 2.
48. Simon, J., Bihari-Varga, M., Erdey, L., Gero, S., "Thermal Investigations on Structural Glycosaminoglycans and Proteins," *Acta Biochim. Biophys. Acad. Sci. Hung.* (1969) **4**, 273.
49. Menefee, E., Yee, G., "Thermally-Induced Structural Changes in Wool," *Textile Res. J.* (1965) **35**, 801.
50. Crighton, J., Findon, W., Happey, F., "Application of Thermoanalytical Methods in the Study of Keratin and Related Proteins," in "Proceedings of the Fourth International Wool Textile Research Conference," Part II, p. 847, L. Rebenfeld, Ed., Interscience, New York, 1971.

51. Bull, H., "Adsorption of Water Vapor by Proteins," *J. Amer. Chem. Soc.* (1944) **66**, 1499.
52. Bull, H., Breese, K., "Protein Hydration," *Arch. Biochem. Biophys.* (1968) **128**, 488.
53. Mercer, E., "Keratin and Keratinization," p. 9, Pergamon, New York, 1961.
54. Rudall, K., "Comparative Biology and Biochemistry of Collagen," in "Treatise on Collagen," B. Gould, Ed., Vol. 2, Part A, pp. 92, 132, Academic, New York, 1968.
55. Watt, I., Leeder, J., "Effect of Chemical Modifications on Keratin and Water Isotherms," *Trans. Faraday Soc.* (1964) **60**, 1335.
56. Zimm, B., Lundberg, J., "Sorption of Vapors by High Polymers," *J. Phys. Chem.* (1956) **60**, 425.
57. Miller, D., Wildnauer, R., "Clustering of Sorbed Water in Stratum Corneum," *Bull. Amer. Physical. Soc.* (1974) **19**, 265.
58. Crank, J., Park, G., "Diffusion in Polymers," J. Crank, G. Park, Eds., p. 16, Academic, New York, 1968.
59. Scheuplein, R. J., "Analysis of Permeability Data for the Case of Parallel Diffusion Pathways," *Biophysical J.* (1966) **6**, 1.
60. Vinson, L., Singer, E., Koehler, W., Lehman, M., Masurat, T., "The Nature of the Epidermal Barrier and Some Factors Influencing Skin Permeability," *Toxicology Suppl. II* (1965) **7**, 7.
61. Yasuda, H., Stannett, V., "Permeation, Solution and Diffusion of Water in Some High Polymers," *J. Polym. Sci.* (1962) **57**, 907.
62. Wildnauer, R., Humphries, W., Miller, D., unpublished data.
63. Ke, B., "Differential Thermal Analysis" in "Newer Methods of Polymer Characterization," *Polym. Rev.* Vol. 6, Chap. 9, Interscience, New York, 1964.
64. Walters, J., "An Apparatus for the Investigation of Wet Polymer Samples with the Vibron Viscoelastometer," *Textile Res. J.* (1971) **41**, 857.
65. Gordon, G., "Glass Transitions in Nylons," *J. Polym. Sci. Part A-2* (1971) **9**, 1693.
66. Chapman, B., Feughelman, M., "Aspects of the Structure of Alpha-Keratins Derived from Mechanical Properties," *J. Polym. Sci. Part C* (1967) **20**, 189.
67. Puett, D., "DTA and Heats of Hydration of Some Polypeptides," *Biopolymers* (1967) **5**, 327.
68. Green, D., Happey, F., Watson, B., "Conformational Changes in Polyamino Acids and Proteins," in "Proceedings of the Fourth International Wool Textile Research Conference," L. Rebenfeld, Ed., Part I, p. 237, Interscience, 1971.
69. Green, D., Happey, F., Watson, B., "Differential Thermal Analyses Applied to Polypeptides," in "Conformation of Biopolymers," G. Ramachandran, Ed., p. 617, Academic, New York, 1971.
70. Middleton, J. D., "The Mechanism of Water Binding in Stratum Corneum," *Brit. J. Dermatol.* (1968) **80**, 437.
71. Singer, J., Vinson, L., "The Water Binding Properties of Skin," *Toilet Goods Ass.* (1966) **46**, 29.
72. Neilsen, L., "Mechanical Properties of Polymers," Chap. 10, Van Nostrand Reinhold, New York, 1962.
73. Baer, E., Kohn, R., Papir, Y., "Mechanical Relaxation Behavior of Collagen, Polyglycine I. and Nylon 6 in the Range from 77 to 420°K.," *J. Macromol. Sci. Phys.* (1972) **B6(4)**, 761.
74. Hiltner, A., Anderson, J., Baer, E., "Dynamic Mechanical Analyses of Poly-α-Amino Acids—Models for Collagen," *J. Macromol. Sci. Phys.*, in press.
75. Nakajima, A., Scheraga, H., "Thermodynamic Study of Shrinkage and of Phase Equilibrium under Stress in Films Made from Ribonuclease," *J. Amer. Chem. Soc.* (1961) **83**, 1575.

76. Flory, P. J., "Theory of Elastic Mechanisms of Fibrous Proteins," *J. Amer. Chem. Soc.* (1956) **78**, 5222.
77. Bosley, D., "Fiber Length Changes and Their Relation to Fiber Structure," *J. Polym. Sci., Part C* (1967) **20**, 77.
78. Humphries, W., Miller, D., Wildnauer, R., "The Thermomechanical Analysis of Natural and Chemically Modified Human Hair," *J. Soc. Cosmet. Chem.* (1972) **23**, 359.
79. Rudall, K., "The Proteins of Mammalian Epidermis," in "Advances in Protein Chemistry," M. Anson, J. Edsall, K. Bailey, Eds., Academic, New York, 1952.
80. Blank, I. H., "Factors which Influence the Water Content of the Stratum Corneum," *J. Invest. Dermatol.* (1952) **18**, 433.
81. Wolfram, M., Wolejsza, N., Laden, K., "Biomechanical Properties of Delipidized Stratum Corneum," *J. Invest. Dermatol.* (1972) **59**, 421.
82. Papir, Y., Wildnauer, R., "The Mechanical Properties of Stratum Corneum," *Bull. Amer. Phys. Soc.* (1974) **19**, 264.
83. Park, A., Baddiel, C., "Rheology of Stratum Corneum, I. A Molecular Interpretation of the Stress-Strain Curve," *J. Soc. Cosmet. Chem.* (1972) **23**, 26.
84. Park, A., Baddiel, C., "Rheology of Stratum Corneum, II: A Physico-Chemical Investigation of Factors Influencing the Water Content of the Corneum," *J. Soc. Cosmet. Chem.* (1972) **23**, 13.
85. Wilkes, G., Hession, W., unpublished data.
86. Sweeney, T., Downes, A., Matoltsy, A., "The Effect of Dimethyl Sulfoxide on the Epidermal Water Barrier," *J. Invest. Dermatol.* (1966) **46**, 300.
87. Alfrey, T., "Mechanical Behavior of High Polymers," p. 508, Interscience, New York, 1948.
88. Wildnauer, R., Humphries, W., unpublished data.
89. Schreiber, H., Holden, H., Barna, G., "Rapid Determination of Crosslink Densities and Interaction Parameters from Swelling Rate Data," *J. Polym. Sci. Part C* (1970) **30**, 471.
90. Flory, P., "Principles of Polymer Chemistry," Chap. XI, Cornell University, Ithaca, 1953.

RECEIVED December 4, 1973.

5

Water Vapor Sorption, Desorption, and Diffusion in Excised Skin: Part I. Technique

A. F. EL-SHIMI, H. M. PRINCEN, and D. R. RISI

Lever Brothers Research Center, 45 River Rd., Edgewater, N. J. 07020

A gravimetric technique to study the sorption, desorption, and diffusion characteristics of water vapor in excised skin under dynamic conditions is described. The technique features a continuously recording microbalance and a humidity-generating apparatus which provides a stream of air with any given relative humidity. The diffusion coefficient is determined from the kinetics of sorption and desorption. The technique can be used to study other polymeric films, fibers, and powders.

It is generally recognized that water plays an important role in maintaining skin in a healthy state with desirable mechanical properties (1). This work describes a technique for generating information on the state of water in the stratum corneum *in vitro,* with particular emphasis on the mobility of water in the corneum matrix and the effect of stratum corneum components on the characteristics of water diffusion. The characteristics of water diffusion in the stratum corneum are derived from sorption and desorption kinetics by using a gravimetric technique which allows determination of the amount of water vapor sorbed or desorbed continuously from an air stream of any given relative humidity.

NMR has provided the most useful information on the state and mobility of water in keratins (2, 3, 4). It was shown (4) that the water associated with the stratum corneum seems to exist in two distinct states: a bound fraction and a free fraction. This type of information could be obtained conveniently by the new technique using desorption kinetics under a wide range of experimental conditions. Also, it would be of interest to examine the effect of various agents, which are known to produce structural changes in the corneum, on the ability of the corneum to hold water and to retain it under a variety of conditions.

Review of Techniques

Before describing the details of the new technique, we shall present a concise review of the most popular methods used to study water vapor sorption and permeability of the stratum corneum.

Permeability of Skin (Stratum Corneum) in vitro. The method most widely used to determine the rate of transfer of water vapor across the skin corneum uses a specially constructed capsule normally called a skin diffusion chamber (5, 6, 7), which is a slight modification of earlier techniques used in the packaging industry (8) for examining the protective properties of coating films. The chamber is depicted schematically in Figure 1. It consists essentially of cup A which contains either water, a saturated salt solution, or a desiccant and a pronged ring (not shown) which holds the stratum corneum against the lip of the cup when compressed by a locking nut. The chamber with the skin is placed in a room, cabinet, or desiccator that has controlled temperature and humidity; the loss or gain in weight during a given time is recorded.

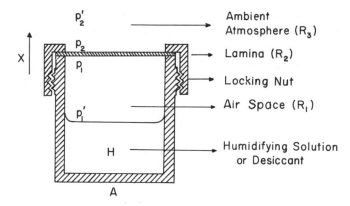

Figure 1.   Schematic of equipment for study of permeability and diffusion in excised skin

Measurements are usually made after stationary state conditions have been established, *i.e.* the rate of loss or gain in weight has become constant. A quantitative estimate of the transport characteristics of the system may be made on the basis of the well known diffusion equations. The equations which form the basis of the quantitative study of linear diffusion processes and from which the definitions of diffusion and permeability are derived are

$$F = -DA \frac{\partial C}{\partial X} \tag{1}$$

$$\frac{\partial C}{\partial t} = \frac{\partial}{\partial X}\left(D\,\frac{\partial C}{\partial X}\right) \tag{2}$$

$$\frac{\partial C}{\partial t} = D\,\frac{\partial^2 C}{\partial X^2} \tag{3}$$

where F is the rate of transfer of diffusate in the X direction through an area A, C is the concentration of diffusate, and D is the diffusion coefficient. For the diffusion of water vapor through the lamina, C is the concentration of water absorbed by or dissolved in the lamina.

Equation 1 applies to stationary states, and Equation 2 to nonstationary states of diffusion. Equation 2 can be derived from Equation 1 by considering the rate of accumulation of diffusate at a given point in the medium; it reduces to Equation 3 when the diffusion coefficient is a constant (9). If the latter condition is satisfied, the diffusion process is said to be ideal, or Fickian, and Equations 1 and 3 represent Fick's first and second laws of diffusion respectively.

When the nature of the diffusion process within the lamina is under consideration, it is advisable to work in terms of the concentrations of water within the lamina, and the foregoing equations are the most appropriate (9). Use of these equations requires knowledge of sorption or solution data since the quantities which are known directly in the experimental design are the vapor pressures or concentrations in the media at the boundaries of the lamina. For practical application of the experimental findings, however, knowledge of the rates of vapor transmission under given vapor pressure differences is required. It is, therefore, convenient in many cases to express the gradient in terms of the vapor pressure at the boundaries of the lamina.

The defining equation for stationary states of diffusion then becomes

$$F = P'\,A\,(p_1 - p_2) = P\,A\,\frac{(p_1 - p_2)}{l} \tag{4}$$

where l is the thickness and p_1 and p_2 are the vapor pressures at the boundaries of the lamina. P' is usually termed the permeability of the lamina and $P(= P' \times l)$ the permeability coefficient or specific permeability of the material.

Equations 1 and 4 are equivalent only when the diffusion and sorption processes are ideal, *i.e.* when the diffusion coefficient is a constant and when Henry's law, which states that there is a linear relation between the external concentration or vapor pressure and the equilibrium concentration in the material, is obeyed (or, in other words, when the sorption isotherm is linear). In this case, Equation 1 may be interpreted to give

$$F = - D A \frac{(C_1 - C_2)}{(x_1 - x_2)} \tag{5}$$

and, since $(x_1 - x_2)$ is necessarily negative,

$$F = D A \frac{(C_1 - C_2)}{l} \tag{6}$$

Henry's law may be written as

$$C = Sp \tag{7}$$

where C is the concentration of water in the lamina when the lamina is in equilibrium with an ambient vapor pressure p, and S is termed the solubility or sorption coefficient. From Equations 4, 6, and 7, the permeability and diffusion coefficients are, therefore, related in this case by the equation

$$P = DS \tag{8}$$

with proper regard to units.

When the diffusion process is not ideal, the diffusion coefficient calculated from Equation 6 for a stationary state process is the mean diffusion coefficient corresponding to a particular concentration difference. Methods of determining the true diffusion coefficient pertaining to a particular concentration were reported by Crank (9). The permeability coefficient calculated from Equation 4 is also a mean value corresponding to a particular vapor pressure difference if the diffusion process is not ideal.

Many investigators have expressed their data in terms of F, the rate of diffusion, and have called this quantity the permeability of the material. O'Neill and Goddard (7) found it more convenient to use the terms resistance and specific resistance. These are defined as the reciprocals of the permeability and the specific permeability respectively. The specific resistance is analogous to electrical specific resistance if the vapor pressure difference and the rate of diffusion are taken to be equivalent to potential difference and current in the electrical case. Diffusion resistances are additive (10), and the effect of each component of a complex system, in which the vapor diffuses through a number of consecutive laminae separated by layers of air, is at once apparent if the resistances of the individual components are known.

With the experimental apparatus (Figure 1), the rate of diffusion F can be measured. When this value is substituted in Equation 4 together with the vapor pressure p'_1 at the surface of H and the vapor pressure

p'_2 in the ambient air, the permeability coefficient can be calculated. Although the method appears simple and straightforward, there are many sources of error which are not obvious. These errors, and the attempts to eliminate them, were reviewed by Newns (*10*). The more important factors include the following:

(a) *Intervening Spaces.* The permeability of the lamina or the permeability coefficient of the material cannot be determined precisely because the vapor has to diffuse through the static layer of air inside the cell and the air at the outer surface of the lamina. In addition to the vapor pressure drop across the lamina, there will be vapor pressure gradients inside and outside the cell. The escape of vapor from the cell is controlled, therefore, not only by the resistance of the lamina R_2 (Figure 1), but also by that of the air space within the cell R_1, and the resistance of the air outside the lamina R_3. The vapor pressures which can be obtained with some degree of certainty are p'_1 at the surface of H and p'_2 in the ambient air, whereas the vapor pressure drop across the lamina $(p_1 - p_2)$ is not known. Only if R_2 is large compared with R_1 and R_3 will the vapor pressures at the surfaces of the lamina approach p'_1 and p'_2. In many cases, because of the high resistance of the lamina or the reduction of R_1 and R_2 by suitable modifications of the apparatus, it is possible to use these values to calculate reasonably accurately the permeability of the lamina. The method is frequently used to compare different materials without obtaining exact values for the permeability coefficients. In an attempt to keep R_1 constant for these measurements, the level of H (Figure 1) is usually adjusted to the same fixed value at the beginning of each experiment.

(b) *Type of Seal.* This factor is especially important when the more resistant types of materials are being examined. The main sources of error are leakage at the edges and inexact delineation of area. Two main types of seal have been used: those made with some type of wax compound and mechanical seals. The advantages of mechanical seals are that the lamina is easily clamped into position and the exposed area is well defined; however, there is a possibility of leakage.

(c) *Wetting of Lamina Surface.* The lamina might be wetted in careless handling, and, when water is used within the cell, there is also a tendency for vapor to condense on the lamina surface as a result of sudden slight decreases in temperature. For this reason, experiments are usually performed at 90–95% relative humidity (RH) on either face of the lamina.

In conclusion, it seems that the principal advantages of this method are that the apparatus and procedure are simple, and the apparatus can be readily duplicated so that a number of measurements can be made simultaneously. A disadvantage is that the cells must be moved during the measurements with the possibility of disturbing the diffusion process. Generally, the method gives better results for materials which do not absorb much water and which do not have low resistance. Although resistance of the stratum corneum may be considered high, calculation of the diffusion coefficient from permeability data may be complicated

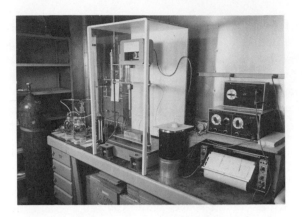

Figure 2.   General view

by the presence of holes or cracks in the corneum (*e.g.,* holes resulting from hair shafts during separation of the corneum from the epidermis). The presence of such holes results in higher D values.

Moisture-Binding Capacity. S in Equation 8 must be evaluated separately from the sorption isotherm in order to calculate D. Such water sorption isotherms have been determined (*5, 11*); they were referred to as moisture-binding capacity curves. The moisture-binding capacity of skin membranes was measured at several relative humidities maintained by the use of saturated aqueous solutions of inorganic salts. The membranes were first dried over Drierite; then they were equilibrated successively at each RH in a desiccator. The amount of water bound by the membranes was estimated gravimetrically. The necessity of removing the sample for weighing and thereby exposing the skin to different ambient conditions may preclude real equilibrium values.

Details of the New Technique

This technique for studying diffusion characteristics of water vapor in skin is based on measurement of sorption and desorption kinetics. The technique features a humidity-generating apparatus to produce an air stream with given RH, a sample chamber, a continuously recording electrobalance (Cahn RG), and a two-pen chart recorder. A time derivative computer (Cahn) connected to the balance in series provides simultaneously the rate of weight change corresponding to the weight *vs.* time trace. A general view of the equipment is presented in Figure 2.

Experimental Equipment. HUMIDITY GENERATING APPARATUS. Figure 3 shows the humidity-generating apparatus. Saturated air and dry air are mixed at predetermined ratios in a gas mixer in order to obtain the desired RH (0–100%). This method of regulating the RH of an air stream was used previously (*12, 13*). The pressure of the dry air from a compressed air cylinder is regulated by a special, low pressure regulator (Matheson model 70, delivery range: 0.5–5 psi), and the pressure of the air stream leaving the regulator (3 psi) is further checked by a mercury manometer. The air stream is then split: one branch passes through two drying columns (Analabs) while the other is bubbled *via* fritted glass discs through the water in a series of saturators. The two air streams enter the two tubes of the gas mixer (proportioner) (Matheson model 665 with two 601 Rotameter tubes calibrated to 1% accuracy) and are mixed in the required ratio in the central mixing tube. Two trap bottles are included in the apparatus to safeguard against back-suction (should the pressure drop suddenly) and carry-over of water droplets from the last saturator. The flow rates in the two mixer tubes were calibrated by the supplier; they are regulated by two metering valves (Matheson NRS High Accuracy Valves 4172-1505). The valve in the wet line is situated before the saturators, and the pressure in the last saturator is maintained very slightly above atmospheric (3 mm Hg). An on-off switch in the wet air line permits quick isolation of the system. The high-precision metering valves of the mixer are used for fine adjustment of flow rate. All connections in the two branches are made with polyethylene tubing (0.25-in. o.d.) and Swagelok couplings. The couplings on the saturators are soldered through the lid.

Weighing Apparatus. The stream of air from the mixer enters a metallic brass chamber ($d = 22$ mm) *via* polyethylene tubing which is

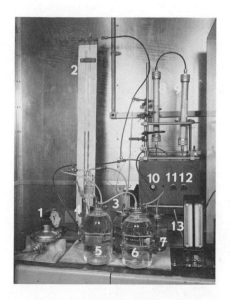

Figure 3.   Humidity-generating apparatus

1, Low pressure regulator; 2, mercury manometer; 3, trap; 4, saturator adapted for thermocouple leads; 5 and 6, saturators; 7, trap; 8 and 9, drier columns; 10, 11, and 12, valves; and 13, mixer

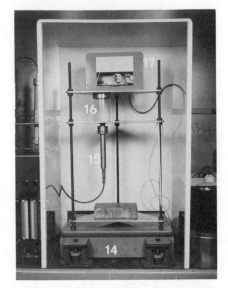

*Figure 4.   Sample chamber and Cahn
electrobalance*

*14, Antivibration base; 15, sample chamber; 16, plexiglas enclosure; and 17, Cahn
balance*

connected to the chamber through a quick-connect coupling (Figure 4).
It was found that the chamber diameter affects the kinetics of establishing
an equilibrium atmosphere in the chamber (*see* below). The sample is
hung on a wire connected to the terminal of the Cahn balance. Care is
taken so that the sample does not touch the walls of the chamber which
is screwed onto a metal collar fixed to a metal block supported on a
tripodal frame. The position of the metal block can be adjusted to ensure
that the wire is centered in the sample chamber. A similar metal block

Figure 5.   Recording equipment

*18, Control unit for Cahn balance; 19,
time derivative computer; 20, recorder;
21, thermos for thermocouple cold junction; and 22, time switch used in conjunction with thermocouple*

with a hole for the wire supports the electrobalance. The tripodal frame rests on an antivibration base. A cylindrical Plexiglas frame between the chamber and the balance shields the sample wire from drafts. The Plexiglas frame and the collar are included with the sedimentation accessories supplied with the balance. Direct connection of the chamber to the base of the balance produces appreciable noise in the output signal.

RECORDING EQUIPMENT. The recording equipment is pictured in Figure 5.

Procedure as Applied to Skin. Human stratum corneum sheets were prepared by the method of Kligman and Christophers (*14*), and the isolated corneum was refrigerated. A small piece of corneum weighing 3–5 mg was normally used. In order to determine the initial dry weight, the skin sample was hung on the wire, the chamber was secured in place, and dry air was then passed into the sample chamber at the rate of 100 ml/min for an average of 16 hrs. Air at a given RH was then admitted to the chamber, and the kinetics of moisture sorption was followed. Changing the RH of the air stream was simply effected by changing the ratio of the volumes of the dry and saturated air streams in the gas proportioner. Similarly, desorption was investigated by passing dry air or air of any given RH into the sample chamber where the skin piece had been equilibrated.

The correlation between the theoretical and actual RH values of the air stream after mixing were examined. Smith (*12*) provided evidence that these values are very similar. He determined directly the equilibrium RH of salt solutions exposed to an air stream of varying RH. The RH of the air stream at which there was no weight loss or gain by the salt solution was very close to the known equilibrium RH of the salt solution. For example, the determined equilibrium RH values for sodium chloride and magnesium chloride were, respectively, 74.4 and 31.5% at 50°C, which are in excellent agreement with the literature values of 74.7 and 31.4% at the same temperature.

We checked the humidity of the air stream in two ways. In the first, the humidity of the moist branch of the air stream was determined by a dew-point method. After leaving the saturators, the moist air branch was indeed saturated; this ensured that the proportioning procedure gave the correct RH. In the other approach, the water vapor isotherm of the same piece of stratum corneum was measured by the dynamic method and a modified static method using a recording microbalance and stirred salt solutions to generate the required RH. The equilibrium sorption values were almost the same for the two isotherms (Figure 6); therefore, the RH of the air stream was the same as calculated. Generally, the flow rates and, therefore, the RH of the air stream remained stable for as long

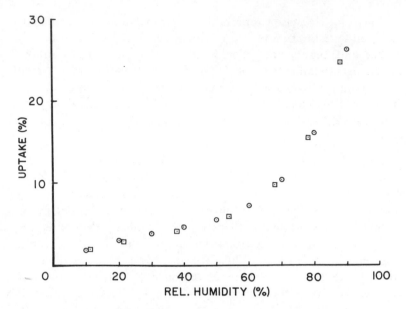

Figure 6. Equilibrium water vapor sorption isotherms obtained on female stratum corneum (age 29) using salt solutions (□) and the dynamic method (○) at 23.3 ± 0.5°C

as 24 hrs with a maximum observed fluctuation of ±2%. Infrequent minor adjustment of the flow rates further reduced these fluctuations.

Calculation of the Diffusion Coefficient from Sorption and Desorption Kinetics in a Plane Sheet

The mathematical treatment of experimental data to calculate the diffusion coefficient of sorbate (diffusate) into sorbents was reviewed by Crank (9). It may be sufficient to illustrate the procedure used to calculate the diffusion coefficient for the simple case where D is considered constant, *i.e.* it does not vary with the concentration of sorbate. More detailed analysis of the characteristics of D will be discussed in future reports.

For the case of sorption and desorption by a plane sheet of thickness l where M_t is the total amount of diffusing substance which has entered the sheet at time t and M_∞ is the corresponding quantity after infinite time, it can be shown (9) that the solution to Equation 3 is given by:

$$\frac{M_t}{M_\infty} = 4 \left(\frac{Dt}{l^2}\right)^{1/2} \left(\frac{1}{\pi^{1/2}} + 2 \sum_{n=0}^{\infty} (-1)^n \operatorname{ierfc} \frac{n\,l}{2\,(Dt)^{1/2}}\right) \tag{9}$$

The uptake is considered a diffusion process controlled by a constant D, and it is assumed that the concentration at each surface of the membrane instananeously attains a constant equilibrium value. With suitable interpretation of M_t and M_∞, Equation 9 also describes desorption from the same sheet which was initially conditioned to a uniform concentration and whose surface concentrations are instantaneously brought to zero at $t = 0$. Equation 9 is particularly useful for small times. Another form of the solution, which is particularly useful for moderate and large times, is given by:

$$\frac{M_t}{M_\infty} = 1 - \frac{8}{\pi^2} \sum_{m=0}^{\infty} \frac{1}{(2m+1)^2} \exp\left[- \frac{D(2m+1)^2}{l^2} \pi^2 t \right] \quad (10)$$

From Equation 10, the value of t/l^2 for which $M_t/M_\infty = \frac{1}{2}$, conveniently written as $(t/l^2)_{\frac{1}{2}}$, is closely approximated by:

$$\left(\frac{t}{l^2} \right)_{1/2} = - \frac{1}{\pi^2 D} \ln \left[\frac{\pi^2}{16} - \frac{1}{9} \left(\frac{\pi^2}{16} \right)^9 \right] \quad (11)$$

with the error being about 0.001%. Thus we find:

$$D = \frac{0.04919}{(t/l^2)_{1/2}} \quad (12)$$

and so, if the half-time of a sorption or desorption process is observed experimentally, the value of the diffusion coefficient, which is assumed to be constant, can be determined. If the diffusion coefficient is not constant, then D represents an average value.

Discussion

The experimental approach for obtaining sorption and desorption isotherms as well as hysteresis effects is well known and will not be discussed further. However, diffusion measurements will be given a somewhat more detailed treatment.

Equations 9 and 10 assume instantaneous establishment of equilibrium surface concentration as a result of sudden change in the ambient environment in the sample chamber. However, it was observed experimentally from the time derivative trace (dM/dt) that a finite response time is required to establish new equilibrium conditions (*see* Figure 7). Under ideal conditions, the rate of sorption or desorption should, to a good approximation, be represented by Curve X. The actual observed curve, Curve R, has a distinct maximum. The position of the maximum

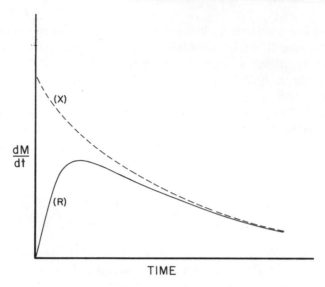

Figure 7.   Schematic depiction of ideal (X) and observed (R) sorption or desorption rates

on the time axis and its height depend on factors that affect the rate of equilibration in the sample chamber, namely chamber size and air stream flow rate.

To demonstrate this effect, five cylindrical chambers were tested. The effective volume of four chambers varied by a factor of about two

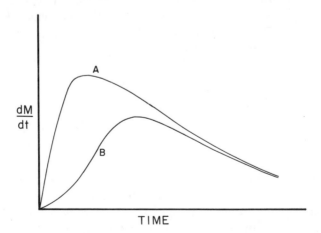

Figure 8.   Schematic presentation of the observed sorption or desorption rate as a function of sample chamber size
Chamber volume: A, 20–50 ml; and B, 400 ml

(20–50 ml), whereas the fifth chamber had an effective volume of about 400 ml. Figure 8 depicts schematically the rate profile obtained with the five chambers. These profiles represent experimental runs with the same skin sample which was initially dry and then subjected to air at 90% RH flowing at 100 ml/min. Whereas the initial sorption rate was not affected by a relatively small change in chamber size (20–50 ml), the effect was very pronounced with the larger chamber. With the smaller chambers, there was a sharper peak after a relatively short time. The maximum is probably related to the gradual mixing of the air stream with the chamber environment. Consequently, the smaller the chamber size, the faster the rate of equilibration. However, for obvious practical reasons, the design of the sample chamber must take into account the use of a large enough sample to increase the precision of the data. The maximum occurred after 1–2 min in the small chamber.

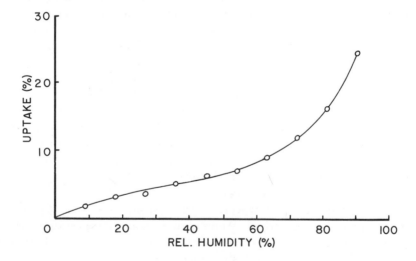

Figure 9.   Equilibrium sorption isotherm of water vapor on female stratum corneum (age 68) at 23.3 ± 0.5°C

The initial deviation of the observed rate from the ideal curve complicates calculation of D from the half-time of sorption or desorption since it would be difficult to locate the start of the ideal diffusion process. However, D values obtained in this manner should still prove useful in assessing the effects of various treatments on the same substrate (*e.g.*, skin) since the relative error attributable to the initial non-equilibrium conditions in the sample chamber will be the same. When D is calculated in this way and the start of the diffusion process is referenced to $t = 0$ (*i.e.* the time required to achieve 50% uptake or loss is referenced to

$t = 0$), the values of D provide useful information on the effect of detergents on excised skin. For example, known aggressive detergents producd a marked increase in D.

In order to obtain more accurate values of D, calculations are performed using Equation 10 which is useful at longer times when the summation can be effectively replaced by its first term, thus:

$$\ln\left[1 - \frac{M_t}{M_\infty}\right] = \ln\left[\frac{8}{\pi^2}\right] - \frac{D\ \pi^2 t}{l^2} \tag{13}$$

so that D is obtained from the slope of the linear portion of the graph of $\ln(1 - M_t/M_\infty)$ vs. time (15). The linear portion of the graph represents sorption or desorption data which are close to equilibrium (for skin, equilibrium sorption and desorption values were attained after approximately 5–20 hrs, depending on the vapor pressure of water in the air stream), and hence it can be safely assumed that the concentration of water vapor at the sample surface is in equilibrium with the surroundings.

A sorption isotherm on excised human abdominal stratum corneum (female, age 68) is presented in Figure 9. The data were obtained on separate pieces of skin which were initially dried and then exposed to an air stream with a given RH; uptake was followed until equilibrium was reached. Values of D calculated from half-time data for each humidity interval were 1.37–5.10×10^{-10} cm^2/sec. Scheuplein and Blank (16) reported a value of 5×10^{-10} cm^2/sec for human stratum corneum. The range of our values would indicate that D is concentration dependent. Full details will be reported in the future.

Materials with different geometric forms were also examined, e.g. hair fibers and powders. However, it was necessary to design appropriate holders for such samples which would allow maximum exposure of sample surface to the air stream. An attempt to use two small magnets to hold hair fibers was not very successful since some fibers fell out during the sorption process causing a sudden decrease in weight. A cylindrical pan (diameter, 10 mm; height, 25 mm; weight, ~ 250 mg) made of very fine stainless steel gauze gave the best results with fibers and powders. One should take care that the fibers or powder in the pan have a low degree of compactness.

An interesting feature of the sorption process was generally observed at the higher humidities (90–100%). Small fluctuations (less than 5% of total uptake) appeared on the weight vs. time curve just as the sample seemed to attain the equilibrium uptake value. It was thought that this was attributable to variations in laboratory temperature that caused condensation. To minimize this effect, the apparatus was moved to a constant temperature and humidity room (22.8°C, 55% RH). Although temperature variation was small under the new conditions, fluctuations in the weight–time curves persisted at higher humidities. An attempt was made to measure the actual temperature in the sample chamber and the temperature of water in the saturaters *via* thermocouples referenced against ice, but no conclusive evidence on the effect of temperature could be obtained. A similar effect was observed by Watt (*17*) during sorption of water vapor in wool under static conditions, but no detailed explanation of the effect was given. In order to resolve this problem and to ascertain its significance, a new setup is being constructed which will allow much more efficient thermostating. The whole unit, except for the balance, will be placed in a water bath.

Summary

The main advantages of the new technique are as follows.

(a) The sorption and desorption processes are followed continuously *in situ* without disturbing the system.

(b) Use of a continuous gas flow ensures good mixing of the air flowing in the chamber, thus minimizing the effects of the stagnant layer around the test sample. In static systems, the effect of the stagnant air layer is usually overcome by working under vacuum. The new technique provides a more realistic dynamic system for the study of sorption, desorption, and diffusion of water in excised skin.

(c) It is easy to change the vapor pressure of water vapor in the air stream at will without disturbing the system. This eliminates inconvenient use of inorganic salts.

Acknowledgment

The authors express their thanks to Lever Brothers Co. for permission to publish this paper.

Literature Cited

1. Idson, B., *J. Soc. Cosmet. Chem.* (1973) **24**, 197.
2. Lynch, L. J., Watt, I. C., Marsden, K. H., *Kolloid Z. Z. Polym.* (1971) **248**, 942.
3. Clifford, J., Sheerd, B., *Biopolymers* (1966) **4**, 1057.
4. Hansen, J., Yellin, W., in "Water Structure at the Water–Polymer Interface," H. Jellineck, Ed., p. 19, Plenum, New York, 1972.
5. Blank, I. H., *J. Invest. Dermatol.* (1952) **18**, 433.
6. Vinson, L., Masurat, T., Singer, E., "Basic Studies in Percutaneous Absorption," U.S. Army Edgewood Arsenal, contract DA 18-108-CML 6573, final report 1965.
7. O'Neill, J., Goddard, E. D., *J. Colloid Interface Sci.* (1967) **25**, 57.
8. Burch, G., Winsor, T., *Arch. Dermatol. Syphilol.* (1946) **53**, 39.
9. Crank, J., "Mathematics of Diffusion," Oxford University, Oxford, 1956.
10. Newns, A., *J. Text. Inst.* (1950) **41T**, 269.

11. Singer, E., Vinson, L., *Proc. Sci. Sect. Toilet Goods Ass.* (1966) **46,** 29.
12. Smith, P., in "Humidity and Moisture," pp. 3, 487, Reinhold, New York, 1965.
13. Rinehart, G., U.S. Army Electronics Command, Res. and Devel. Tech. Rep. (1970) **ECOM-5332.**
14. Kligman, A., Christophers, E., *Arch. Dermatol.* (1963) **88,** 70.
15. Crank, J., Park, G. S., in "Diffusion in Polymers," Academic, New York, 1968.
16. Scheuplein, R., Blank, I., *Physiol. Rev.* (1971) **51,** 702.
17. Watt, I., *Text. Res. J.* (1960) **30,** 443.

RECEIVED May 22, 1974.

Wetting Characteristics of Keratin Substrates

A. EL-SHIMI and E. D. GODDARD

Lever Brothers Research Center, 45 River Road, Edgewater, N. J. 07020

The wettability of living human skin, bovine hoof keratin, and various synthetic polymers has been examined using two liquids, water and methylene iodide, and Wu's empirical approach to obtain $\gamma_s{}^d$ and $\gamma_s{}^p$. The sum of these parameters agreed with reported values of γ_c, the critical surface tension, based on Zisman plots. γ_c values obtained using aqueous ethanol solutions were lower and showed little or no dependence on the type of solid surface. The results show that skin and keratin possess surfaces of low free energy comparable with that of polyethylene. Experimental values of contact angles at the hydrocarbon liquid/water/solid interface showed satisfactory agreement with the Wu equation for Teflon and poly(methyl methacrylate), but those observed for nylon 11 and bovine hoof keratin ($> 170°C$ in oil) were higher than predicted.

An important function of animal skin is to act as a barrier to prevent water loss and entry of foreign materials. In man and other mammals this barrier property is conferred by the outermost layer known as the stratum corneum. Although it is composed essentially of a few layers of cells which undergo extensive keratinization as they are forced upward from the dermis layer, the barrier properties of this stratum corneum layer far exceed those of keratin. The efficiency of stratum corneum as a barrier depends on the presence in it of lipoidal material. This efficiency (1) is severely impaired if the stratum corneum is treated with combinations of polar, water miscible liquids such as methanol and water immiscible lipophilic solvents such as hexane or chloroform.

Much effort is expended cosmetically to improve the condition of skin. This generally involves applying products to restore impaired skin to normal. Our investigation examines the wetting characteristics of normal skin. Clean skin in as natural a state as possible was used. Aggressive solvent treatments were avoided since such treatment would remove a considerable portion of skin lipids and influence the wetting properties. For comparison we have investigated the wetting of a keratin substrate itself, namely bovine hoof. Such a substrate has some similarity to human hair although freshly cleaned human hair will also in time acquire a layer of lipids by migration of sebum constituents from the scalp. To allow more complete evaluation of the properties of skin and keratin surfaces, the wetting behavior of several synthetic polymers were studied. The experiments involve wetting all these materials in air and submerged in water.

Experimental

Substrates and Materials. For studies of contact angles on skin, the area chosen should be as smooth as possible and free from hair. Following earlier practice (2), the area selected was the dorsal side of the index finger with the skin rendered taut by bending of the finger. Prior to contact angle measurements, the female subjects washed their hands with soap and rinsed them under tap water. The experimental surface was allowed to air dry for 15 min; the other areas were lightly padded with paper towels. Two subjects were chosen from a large group as having the smoothest skin in the finger dorsal region. Since the findings for both subjects were very close, the data for only one are reported.

The alcohols used were distilled compounds, and the water was double distilled. A variety of substrates was chosen which includes paraffin wax (Gulf Oil Corp., m.p. 55–56°C); Teflon and nylon 11 (DuPont); poly(methyl methacrylate) (PMMA) (Rohm and Haas); bovine hoof keratin (BHK); and human skin. Bovine hooves were obtained fresh from the slaughter house. The hard keratin was isolated by first boiling the hooves in water and then scraping the unwanted components with a sharp knife. The dried keratin pieces thus obtained were generally irregular in shape, and it was necessary to machine the keratin hoofs into some desirable geometric planar surface. In this work, planar discs 20 mm in diameter and not less than 5 mm thick were used. Discs thinner than 5 mm tended to deform and thus lose their planar geometry. The discs were polished using emery paper of succeedingly finer grade until a highly reflecting surface was obtained. Microscopic examination showed no residual debris on the surface.

Teflon, nylon, and PMMA were supplied in sheet form. These solids were washed in detergent solution, rinsed thoroughly in distilled water, and stored in a vacuum oven at room temperature. Paraffin wax discs were obtained by dipping PMMA discs or glass slides in molten wax, then removing and gently pressing the discs or slides onto warm aluminum foil placed on a planar, black granite dressing plate normally used for gage blocks. The resultant coating was flat and smooth. The surface of the synthetic polymers was carefully protected and used for wetting studies in virgin form.

For the oil/water studies, discs of paraffin wax and the other materials were glued to a glass rod *ca.* 3 in. in length. The junction of the

glass rod and the discs was coated with paraffin wax to ensure no contamination of surrounding liquid with glue (Eastman 910 adhesive).
For convenience Teflon, nylon, and PMMA were also studied in the
form of cylindrical rods of *ca.* 1-in. diameter at the base and 4 in. long.
The upper 3-in. portion was machined to a ~ ½-in. diameter and was
used as a handle to fit in a laboratory clamp. It was secured in a vertical
position by clamping onto a laboratory jack and the broader, lower
portion of the cylinders was used for contact angle measurements. This
set-up provides a convenient way to move the solid cylinder vertically
into the optical cell while the dropping assembly remains stationary. A
smooth reflecting surface was obtained by polishing with a suspension
used for precision optical equipment. Results obtained showed agreement between the disc and rod forms of PMMA and Teflon, but for nylon
the rod surface was evidently scratched and was discarded.

The hydrocarbon liquids were very pure samples: C_7- (spectral grade,
Matheson, Coleman, and Bell), C_8-, C_{10}-, C_{12}-, and C_{16}-*n*-alkanes (Hum-
phrey Chemical Co.), and cyclohexane (spectral grade, Eastman). The
absence of impurities was confirmed by GLC. The mineral oil (Drakeol-
15 U.S.P. grade, Pennsylvania Refining Co.) was further purified by treating with Fluorosil. The methylene iodide was a high purity specimen
(Fisher Scientific Co.) and was used as received. For the oil/water
studies, all the liquids were equilibrated with double distilled water for
24 hrs in glass containers with continuous agitation in a rotating machine.
After equilibration, the liquids were left to stand for a few hours and
separated using a separating funnel. The surface and interfacial tensions
of the systems investigated were measured at room temperature using
the Wilhelmy plate method, and the results are shown in Table I.

Technique. The set-up consisted of an optical bench with two car-
riers. One carrier held the support for the optical cell, the height of
which could be adjusted by a vertically movable mounting shaft. The
second carrier, supporting a low power microscope fitted with a goniometer scale, was capable of vertical and horizontal movements. This design
facilitates microscope movement in three dimensions and eliminates the
need for adjusting the needle position with respect to substrate and
optical path. The liquid drops were dispensed with the help of an ultra
precision micrometer syringe (Manostat) fitted with a stainless steel
needle which was placed in an optical cell containing the solid disc. The
syringe was clamped on a vibration-free stand. In all cases, the sessile

**Table I.   Surface and Interfacial Tensions (dynes/cm)
of Mutually Saturated Liquids**

	$\gamma_{l(water)}$	$\gamma_{l/water}$	$\gamma_{water(l)}$
C_7	20.4	50.2	72
C_8	21.8	50.8	72
C_{10}	23.9	51.2	72
C_{12}	25.4	51.5	72
C_{16}	27.5	52.3	72
cyclohexane	25.5	50.2	72
n-hexanol	26.5	6.9	34.2
mineral oil	31	51.5	72

drop method was used for the measurements which used standard contact angle goniometer equipment. Advancing (θ_A) and receding (θ_R) angles were measured with the needle tip in the liquid phase by changing the volume of the drop until a movement of the contact line was observed. This device allows the measurement of the contact angle at both edges of the drop without any disturbances of the drop configuration. Measurements were carried out at room temperature ($25.5 \pm 1.5°C$) during winter months when the ambient humidity was low ($\sim 30\%$) (2).

For measurements on skin, the subject held her finger around a rubber stopper to generate a skin surface as planar as possible. For measurements of contact angles on solids submerged in water, since all the hydrocarbon liquids investigated were lighter than water, the captive drop method was used. The substrate surface was normally lowered into the aqueous phase with the help of the laboratory jack, and its distance from the tip of the needle was controlled by observation through the microscope. Advancing and receding angles were measured with the needle tip in the oil drop by changing the volume of the drop. All measurements were again carried out at room temperature. Liquid drops were deposited on at least five different locations on the surface of a particular solid as a check on the reproducibility of the measured contact angle. Agreement for the least reproducible surface, hoof keratin, was $\pm 3°$. For attenuated total reflectance (ATR) IR studies, a Perkin-Elmer grating IR spectrophotometer was used with a KRS-5 plate.

Theoretical

The Fowkes (4) equation for the interfacial tension, γ_{12}, between two contacting phases which interact by dispersion forces only, has the form

$$\gamma_{12} = \gamma_1 + \gamma_2 - 2\sqrt{\gamma_1^d \gamma_2^d} \tag{1}$$

where γ_1 = surface tension of phase 1, γ_2 = surface tension of phase 2, γ_1^d, γ_2^d = the respective dispersion force components of γ_1, γ_2.

For the case where polar force interactions are also involved, Wu (5) has proposed, by using reciprocal mean expressions for both the dispersion and the polar interactions, that the interfacial tension γ_{12} between two contacting phases be represented as

$$\gamma_{12} = \gamma_1 + \gamma_2 - \frac{4\gamma_1^d \cdot \gamma_2^d}{\gamma_1^d + \gamma_2^d} - \frac{4\gamma_1^p \cdot \gamma_2^p}{\gamma_1^p + \gamma_2^p} \tag{2}$$

where γ_1^p, γ_2^p = the respective polar force components of γ_1, γ_2.

The Young equation for a solid/liquid system has the form

$$\gamma_l \cos\theta = \gamma_s - \gamma_{sl} \tag{3}$$

and combination with Equation 2 yields

$$\gamma_l \cos\theta = -\gamma_l + \frac{4\gamma_s{}^d \cdot \gamma_l{}^d}{\gamma_s{}^d + \gamma_l{}^d} + \frac{4\gamma_s{}^p \cdot \gamma_l{}^p}{\gamma_s{}^p + \gamma_l{}^p} \qquad (4)$$

where s and l, replacing 1 and 2 respectively, denote solid and liquid.

By Zisman's (6) definition, $\gamma_l = \gamma_c$, the critical surface tension for wetting, when $\cos\theta = 1$, and under these conditions Equation 4 becomes

$$\gamma_c = 2 \left[\frac{\gamma_s{}^d \cdot \gamma_l{}^d}{\gamma_s{}^d + \gamma_l{}^d} + \frac{\gamma_s{}^p \cdot \gamma_l{}^p}{\gamma_s{}^p + \gamma_l{}^p} \right] \qquad (5)$$

It is clear from Equation 5 that if $\gamma_l{}^d = \gamma_s{}^d$ and $\gamma_l{}^p = \gamma_s{}^p$, then

$$\gamma_c = \gamma_s{}^d + \gamma_s{}^p$$

which represents an expression for the total surface free energy of the solid.

In the special case when interaction between solid and liquid at the point of complete spreading is attributable solely to dispersion forces, Equation 5 becomes

$$\gamma_l{}^d = \gamma_c = \frac{2\gamma_s{}^d \cdot \gamma_l{}^d}{\gamma_s{}^d + \gamma_l{}^d} = \gamma_s{}^d \qquad (6)$$

For the same conditions, a similar result emerges from the Fowkes-Young equation

$$\gamma_l \cos\theta = -\gamma_l + 2\sqrt{\gamma_s{}^d \gamma_l{}^d} \qquad (7)$$

On the basis of Equation 4, the values of $\gamma_s{}^d$ and $\gamma_s{}^p$ for a solid polymer surface can be calculated from the contact angles on it of two liquids, the surface tensions of which have been defined in terms of the respective contribution of dispersion and polar force components (5). In this case, Equation 4 is rearranged to two simultaneous equations, and solved for $\gamma_s{}^d$ and $\gamma_s{}^p$.

In the case of a solid in contact with two (mutually saturated) liquids, preferential wetting can be predicted on the basis of the contact angles of the individual liquids on the given solid in air and application of Young's equation. For example, consider the case of an oil (o) and water (w) on a low energy substrate (s), such that we have for the air/water/substrate system

$$\gamma_s - \gamma_{sw} = \gamma_w \cos\theta_w \qquad (8)$$

and for the air/oil/substrate system

$$\gamma_s - \gamma_{so} = \gamma_o \cos\Theta_o \qquad (9)$$

and for the oil/water/substrate system

$$\gamma_{sw} - \gamma_{so} = \gamma_{ow} \cos\Theta_{ow} \qquad (10)$$

where γ_s is the solid surface tension and the other terms have their usual significance.

Substituting Equations 8 and 9 into Equation 10, we obtain the Bartell-Osterhof (7) equation

$$\gamma_{ow} \cos\Theta_{ow} = \gamma_o \cos\Theta_o - \gamma_w \cos\Theta_w \qquad (11)$$

Equation 11 allows contact angle prediction in systems containing two immiscible liquids and a solid. Since Young's equation is only valid in cases where the contact angle is finite, Equation 11 would not be expected to hold in such cases where the contact angle θ equals zero in air as is commonly observed for low energy solids and nonpolar liquids (8). However, Equation 11 can be verified using Teflon substrates since all common nonpolar liquids exhibit a finite contact angle on it. A less direct approach to predict θ_{ow} in liquid/liquid/solid systems, where $\theta = 0$ in air for one of the liquids, is to use an equation like Fowkes' (4) for interfacial tension. Thus we would have for the solid/water system,

$$\gamma_{sw} = \gamma_s + \gamma_w - 2\sqrt{\gamma_s{}^d \cdot \gamma_w{}^d} \qquad (12)$$

and for the solid/oil system,

$$\gamma_{so} = \gamma_s + \gamma_o - 2\sqrt{\gamma_s{}^d \cdot \gamma_o{}^d} \qquad (13)$$

Substituting Equations 12 and 13 into 10 gives

$$\gamma_{ow} \cos\Theta_{ow} = \gamma_w - 2\sqrt{\gamma_s{}^d \gamma_w{}^d} - \gamma_o + 2\sqrt{\gamma_s{}^d \gamma_o{}^d} \qquad (14)$$

where the $\gamma_s{}^d$ terms are the dispersion component of the surface tensions. $\gamma_o{}^d$ is usually taken as equal to γ_o for nonpolar liquids.

Equation 14 allows us to predict θ_{ow} from $\gamma_s{}^d$ and $\gamma_w{}^d$ which has a value of *ca.* 22 erg/cm^2 according to Fowkes (4).

Correspondingly, using the Wu (5) expression we would have

$$\gamma_{ow} \cos\Theta_{ow} = \gamma_w - \frac{4\gamma_s{}^d \cdot \gamma_w{}^d}{\gamma_s{}^d + \gamma_w{}^d} - \frac{4\gamma_s{}^p \cdot \gamma_w{}^p}{\gamma_s{}^p + \gamma_w{}^p}$$
$$- \gamma_o + \frac{4\gamma_s{}^d \cdot \gamma_o{}^d}{\gamma_s{}^d + \gamma_o{}^d} + \frac{4\gamma_s{}^p \cdot \gamma_o{}^p}{\gamma_s{}^p + \gamma_o{}^p} \qquad (15)$$

for the case where polar force interaction across the interface is also involved. Equation 15 allows us to predict contact angles in the oil/water/substrate system taking into account the polar interactions as well as the more universal dispersion interactions. Information on the components of the solid surface free energy is available, and again it is assumed that for a nonpolar liquid $\gamma_o{}^d = \gamma_o$.

Substitution of appropriate values in Equations 14 and 15 thus allows us to predict θ_{ow} values based on two approaches. These values can then be compared with experimental data.

Table II.   Measured Contact Angles

	Water		Methylene Iodide	
	$\theta_A,^\circ$	$\theta_R,^\circ$	$\theta_A,^\circ$	$\theta_R,^\circ$
Paraffin wax	110	108	60	50
Teflon	112	106	85	75
Nylon 11	75	66	37	29
PMMA	74	67	34	28
Bovine hoof keratin	90	65	44	25
Human skin	75	60	56	45
Nylon 66 [a]	66	—	25	—
Poly(vinyl chloride) [a]	83	—	40	—
Polyethylene [a]	95	—	46	—
Poly(ethylene terephthalate) [a]	71	—	23	—

[a] Ref. (*9*).

Results and Discussion

The results of contact angle measurements at the solid/vapor interface are shown in Table II for water and methylene iodide on the surfaces we investigated. Agreement is very good with measurements on the first four matrials by Dann (*9*). Table II includes results on other polymers reported by Dann. Table III contains the computed values, based on θ_A values for the above two liquids, of $\gamma_s{}^d$, $\gamma_s{}^p$ and $\gamma_s = (\gamma_s{}^d + \gamma_s{}^p)$ for the same surfaces along with literature values for γ_c obtained by the Zisman method of extrapolating a plot of γ_l *vs.* $\cos\theta$ to a $\cos\theta$ value of unity. γ_s values obtained by Wu's (*5*) equation are in good agreement with literature values of γ_c.

For low energy surfaces, the dispersion force component is larger than the nondispersion force component which, however, cannot be neglected or assumed to be nonexistent. An exception is paraffin wax. The results of Table III for paraffin confirm that $\gamma_s{}^p$ is zero, consistent with a surface composed completely of hydrocarbon entities with no polar components. By this treatment, Teflon turns out to have a small but definite polar component of γ (*10*).

Table III.   Computed Values of γ_s According to Equation 4[a]

	$\gamma_s{}^d$	$\gamma_s{}^p$	γ_s	γ_c (Ref.)
Paraffin wax	25.5	—	25.5	25.5 (17)
Teflon	17.3	1.7	19.0	18.5 (6)
Nylon 11	31.9	12.2	44.1	43 (9)
PMMA	33.0	12.4	45.4	39 (6), 33–44 (16)
Bovine Hoof Keratin	32.8	5.6	38.4	39
Human skin,	23.1	15.1	38.2	—
Nylon 66[b]	35.0	15.8	50.8	46 (9)
Poly(vinyl chloride)[b]	32.4	8.5	40.9	39 (6)
Polyethylene[b]	34.2	3.4	37.6	31 (6), 36.2 (16)
Poly(ethylene terephthalate)[b]	36.5	13.0	49.5	43 (9), 50 (17)

[a] $\gamma_l{}^d$ and $\gamma_l{}^p$ values for water (22.1, 50) and methylene iodide (44.1, 6.7) were those used by Wu (5) (units are dyne/cm).
[b] Ref. (9).

Two features emerge concerning bovine hoof keratin. The first is its relatively high contact angle with the two liquids. The second is the rather large discrepancy between the advancing and receding contact angles. Keratin is clearly a solvatable material as discussed later. For this reason contact angles were recorded within 1–3 min of placing the drop. The high values of the advancing contact angles, as well as the relatively low value of γ_s and $\gamma_s{}^p$, indicate keratin to be a rather low free energy surface (3) of low polarity.

Values in the literature for the contact angle of water on human skin fall into two groups: one ca. 60°–70° (2, 11) and the other > 100° (2, 8, 12). The former corresponds more to skin in a normal state and the latter to skin that has been subjected to severe cleaning, generally involving exposure to a delipidizing solvent. Severe cleaning may affect the structure of the outer layer of the skin. Since our interest was in the

Table IV.   Comparison of Predicted Θ_{ow} Values (degrees) in Oil/Water/Teflon System ($\gamma_s{}^d = 17$, $\gamma_s{}^p = 1.7$ dynes/cm) with Experimental Data

	Equation 14	Equation 15	Experimental (Θ_A)	Experimental (Θ_R)	Equation 11
C_7	0	29	42	32	25
C_8	0	30	40	31	—
C_{10}	14	33	41	32	—
C_{12}	14	35	40	33	—
C_{16}	19	37	38	34	—
Cyclohexane	9	32	42	33	—
Mineral oil	20	38	37	33	—
n-Hexanol	0	0	48	40	38

normal state of the skin, the skin surface was prepared with bar soap washing followed by thorough rinsing under running tap water.

To check the results, a second procedure was used to eliminate the possibility of lime soap deposits. In this procedure, the hands were vigorously rubbed in a 4% nonionic detergent (Neodol 45-11EO, Shell) solution in distilled water followed by extensive rinsing with distilled water. This second treatment gave contact angles which agreed within experimental error with those obtained by the soap/tap water rinse procedure.

Table V. Comparison of Predicted Θ_{ow} Values (degrees)
in Oil/Water/PMMA System (γ_s^d = 33, γ_s^p = 12
dynes/cm) with Experimental Data

			Experimental	
	Equation 14	Equation 15	(Θ_A)	(Θ_R)
C_7	6	78	94	80
C_8	9	77	92	75
C_{10}	8	76	94	78
C_{12}	9	76	93	77
C_{16}	12	76	92	75
Cyclohexane	0	76	95	79
Mineral oil	9	75	93	78
n-Hexanol	0	180	47	35

The data in Table III reveal that whereas the derived γ_s value of human skin is very close to that of hoof keratin, its polar component is relatively much higher. This reflects the presence of polar lipids in the stratum corneum layer. It is clear, however, that both skin and keratin present low free energy surfaces comparable in wetting characteristics with those of polyethylene.

Literature values of the critical surface tension of skin range from 26 to 27.5 dynes/cm (2, 11, 12). In all these cases, aqueous solutions of materials such as acetone (2, 13) or propylene glycol (11) were used for the γ_c determination to obtain an appropriate range of liquid surface tensions. Dann (9), however, has pointed out that γ_c values obtained in this way will be less than the value obtained with liquids, such as hydrocarbons, which do not possess a γ^p component. Correcting an experimental value of 27.5 dynes/cm for this effect, Rosenberg et al. (12) obtained 37.0 dynes/cm for γ_c. As suggested by Murphy et al. (13) in their work on wetting by aqueous alcohol solutions, the above effects are probably linked to preferential adsorption of the solute onto the solid.

In the work on the wetting behavior of ethanol/water mixtures on nylon 11, PMMA, bovine hoof keratin, and human skin, we found two

Table VI.   Comparison of Predicted Θ_{ow} Values in
Oil/Water/Nylon System ($\gamma_s^d = 32$, $\gamma_s^p = 12$
dynes/cm) with Experimental Data

	Equation 14	Equation 15	Experimental Θ_A
C_7	6	78	>170
C_8	9	77	>170
C_{10}	8	76	>170
C_{12}	9	76	>170
C_{16}	12	76	>170
Cyclohexane	0	76	>170
Mineral oil	9	75	>170
n-Hexanol	0	180	115

important results: the derived γ_c (27 ± 0.5) is independent of the solid surface, and it is considerably lower than that obtained by using nonpolar liquids. These findings thus reinforce those of Murphy et al. (13) who also found that the derived γ_c values were rather insensitive to the water soluble alkanol used (10). They explain this by the adsorption of the alcohol molecules at the solid surfaces which substantially affects their wetting properties (10, 13). Care is obviously needed when using aqueous solutions as wetting liquids to derive values of the critical surface tension of wetting.

Turning now to the solid/water/oil measurements, we compare the predicted θ_{ow} values according to Equations 14 and 15 with experimental values in Tables IV–VII for the substrates investigated. In the case of Teflon, where it is possible to test Equation 11, values are given for heptane and n-hexanol. The dispersion and polar components of the surface tension of water–n-hexanol, i.e. water saturated with n-hexanol, and hexanol–water were obtained by measuring the contact angle of liquid drops on paraffin wax ($\gamma_s^d = 25.5$ dynes/cm), which served as a

Table VII.   Comparison of Predicted Θ_{ow} Values in
Oil/Water/Bovine Hoof Keratin System ($\gamma_s^d = 3.28$,
$\gamma_s^p = 5.6$ dynes/cm) with Experimental Data

	Equation 14	Equation 15	Experimental Θ_A
C_7	6	54	>170
C_8	9	54	>170
$C_{10,}$	8	53	>170
C_{12}	9	53	>170
C_{16}	12	53	>170
Cyclohexane	0	51	>170
Mineral oil	9	51	>170
n-Hexanol	0	136	>170

probe for dispersion interactions, and then applying the Fowkes-Young equation

$$\sqrt{\gamma_l{}^d} = \frac{(\cos\Theta + 1)\gamma_l}{2\sqrt{\gamma_s{}^d}}$$

where $\gamma_s{}^d$ is the surface energy of paraffin wax, and $\gamma_l{}^d$ is the dispersion component of the liquid surface tension.

The values obtained for n-hexanol–water were $\gamma_l{}^d = 24.6$ and $\gamma_l{}^p = 1.9$ dynes/cm and for water–hexanol, $\gamma_l{}^d = 22$ and $\gamma_l{}^p = 12.4$ dynes/cm. As regards the contact angle results obtained with n-hexanol, values of θ_{ow} obtained from Equations 14 and 15 and by experiment show a wide disparity. *See* Tables IV–VII. For paraffin wax, both equations predict zero contact angle for n-hexanol whereas the experimental values are 45° (θ_A) and 37°C (θ_R). As discussed earlier, adsorption of alcohols at the solid/liquid interface may affect the wetting behavior of substrates. These effects are not accounted for by the Fowkes or Wu treatment, and hence it is not unexpected that these equations do not correlate with experimental θ_{ow} values for hexanol/water.

Although Equations 14 and 15 are inadequate in predicting contact angles of hexanol/water systems, Equation 11 based solely on the Young equation leads to much better agreement in the case when it was applied, *viz.*, for Teflon.

Considering the results with nonpolar oils, we find that both Equations 14 and 15 predict zero contact angle of the hydrocarbon liquids on paraffin wax in water. This was confirmed experimentally, with evidence that the hydrocarbon liquids had actually etched the paraffin wax surface. The results for the Teflon surface are interesting in that the experimental values for θ_{ow} are in fair agreement with Wu's equation which, unlike the Fowkes equation, takes into account non-dispersion force interactions. The correlation with θ_R is generally better than with θ_A. This also holds for PMMA (Table V). As an example of the use of Equation 11 to predict θ_{ow} for a hydrocarbon liquid, the values for heptane on Teflon are included in Table IV. Agreement with experiment was satisfactory, but with the predicted value again in better agreement with θ_R.

Tables VI and VII show the results for nylon 11 and bovine hoof keratin. In both cases the predicted values for θ_{ow} according to Equations 14 and 15 vary markedly with the experimental θ_{ow} values; note the disparity in the predicted θ_{ow} values according to Equations 14 and 15. Experimentally, the situation with nylon 11 and hoof keratin was clearly very different from the other polymers as no contact between the hydrocarbon drop and substrate could be established. The observed discrepancy between predicted θ_{ow} values according to both Equations 14 and 15 and experimental values is very likely because of the tendency of nylon and keratin when submerged in water to hydrate, giving rise to stable water films on the surface. The thickness of such aqueous films could conceivably extend to a few molecular layers. These films would tend to increase the attraction between the aqueous phase and the solid substrate at the expense of interaction with the hydrocarbon liquid. It was thought that the film may be slow draining and that if the drop were pressed against the substrate for a sufficient period of time, it would be possible to establish contact between the hydrocarbon liquid and substrate. Even 1 hr was insufficient to obtain a contact angle reading, and the whole drop detached intact upon recession and withdrawal.

We have conducted some qualitative experiments to gain some more information on the behavior of the hydration films on nylon and keratin when discs of the materials are brought into air after being submerged for a period of time under water. The initial aqueous layer present exhibits marked instability and recedes into discrete drops. The discs with these aqueous drops were gently dried either with an ashless filter paper or by blowing the fluorocarbon F-12 (Falcon Safety Products) across the surface for a few seconds. The discs appeared dry after this treatment. When water drops were placed on the surface again, the surfaces were not wetted, and the aqueous drops exhibited a contact angle comparable with those obtained on substrates dried in a vacuum for several days. It seemed inconceivable that such gentle drying procedures of hydrated surfaces could eliminate all the associated surface water and render it hydrophobic again. ATR IR spectroscopy revealed no water band for nylon, but keratin did exhibit a water band at 3000–2800 cm^{-1}. The latter could have been the result of squeezing water out of minute pores when applying the pressure necessary to establish adequate contact between the substrate and the IR plate.

To examine the hydration of bovine hoof keratin further, specimens (supplied by J. Clifford, N. G. Pryce, and G. K. Rennie of Unilever Research, Port Sunlight, England) were studied by two other techniques, viz., pulsed NMR on 6-mm diameter and 6-mm thick BHK discs, and differential scanning calorimetry (DSC) on BHK powder. The keratin discs took up 36% of their weight in water in 25 hrs. The uptake was linear with the square root of time, and the discs suffered less than a 5% change in linear dimensions during this process. After 5 hrs the uptake was $\sim$ 14%, and half of this water, i.e. $\sim$ 7%, had a relaxation time, T_2, characteristic of bound water. DSC on keratin powder exposed to water showed that an uptake of $\sim$ 200% water occurred in 2 days, with 36% being bound. This probably represents the equilibrium value. These results confirm the highly hydratable nature of keratin. While bound water is not readily lost on exposure to laboratory atmosphere, it is not possible to make any deductions about water in the surface layer which would control the wetting behavior of the specimens. Further information is necessary to elucidate the condition of the surface of keratin after its removal from water.

Studies of Blake and Kitchener (*14*) on the stability of aqueous films on methylated silica show that thick aqueous films are essentially meta-stable and sensitive to environmental disturbances. If the thickness is decreased to a critical value, the film becomes unstable and breaks into discrete drops in equilibrium with an adsorbed film, presumably mono-molecular. We have shown that nylon and bovine hoof keratin are essentially hydrophobic in air. It is possible that an analogous mechanism operates in these cases, *i.e.*, water films associated with the surface are inherently unstable after emersion and are easily removed from the surface even by the mild wiping procedure. Another possible mechanism (*15*) to explain the hydrophilic–hydrophobic transition in air of hydratable low energy surfaces is related to the conformation of the polymeric surface molecules which might extend into aqueous phase in the form of loops. The configuration of these polypeptide molecules when totally immersed in water would thus differ from that when in air. In other words, we postulate a degree of flexibility of the molecules in the surface region which can aid their hydration by formation of new hydrogen bonds. The speed of restoration of hydrophobic properties in air emphasizes the reversible nature of any configurational changes.

Another interesting feature of nylon and hoof keratin surfaces is observed when these substrates are treated with mineral oil in air (where $\theta = 0$), then submerged in water. The objective was to study the rate of oil recession as a result of the hydration process described above. In the case of nylon, θ_R attained a maximum value of $> 170°$ in the course of a few hours as expected. However, hoof keratin retained the oil in the form of a wetting drop ($\theta_R \sim 30°$) for a long time (48 hrs); this is about the same θ value obtained on Teflon. It seems that the dispersion interaction forces between oil and keratin are strong enough that non-dispersion interaction between the keratin and water is not sufficient to displace the oil at least within the time limit of the experiment. The particular behavior observed is thus governed by the history of exposure. This may be a specific feature of natural keratin surfaces, *i.e.*, surface molecules are able to adopt and retain a configuration compatible with their immediate environment.

Literature Cited

1. Onken, H. D., Moyer, C. A., *Arch. Dermatol.* (1963) **87**, 584.
2. Ginn, M. E., Noyes, C. M., Jungermann, E., *J. Colloid Interface Sci.* (1968) **26**, 146.

 3. Alter, H., Cook, H., *J. Colloid Interface Sci.* (1969) **29,** 439.
 4. Fowkes, F. M., *J. Phys. Chem.* (1962) **66,** 382.
 5. Wu, S., *J. Polym. Sci. Part C* (1971) **34,** 19.
 6. Zisman, W. A., ADVAN. CHEM. SER. (1964) **43,** 1.
 7. Bartell, F. E., Osterhof, H. J., *Colloid Symp. Monogr.* (1927) **5,** 113.
 8. Adamson, A. W., Knichika, K., Shirley, F., Orem, M. J., *J. Chem. Ed.* (1968) **45,** 702.
 9. Dann, J. R., *J. Colloid Interface Sci.* (1970) **32,** 302.
10. El-Shimi, A., Goddard, E. D., *J. Colloid Interface Sci.* (1974) **48,** 242.
11. Schott, H., *J. Pharm. Sci.* (1971) **60,** 1893.
12. Rosenberg, A., Williams, R., Cohen, G., *J. Pharm. Sci.* (1973) **62,** 920.
13. Murphy, W. J., Roberts, M. W., Ross, R. H., *J. Chem. Soc. Faraday I.* (1972) **68,** 1190.
14. Blake, T. D., Kitchener, J. A., *J. Chem. Soc. Faraday I.* (1972) 1435.
15. Van den Tempel, M., personal communication.
16. Schonhorn, H., *Encyclopedia of Polymer Science and Technology* (1969) **13,** 543.
17. Wu, S., *J. Phys. Chem.* (1968) **72,** 3332.

RECEIVED December 4, 1973.

Wetting Properties of Collagen and Gelatin Surfaces

R. E. BAIER[1] and W. A. ZISMAN

U.S. Naval Research Laboratory, Washington, D.C. 20375

Collagenous proteins dominate the organic matrix of connective tissue and teeth, so their wetting properties are important for surgical and dental adhesives. Gelatin, a randomized collagen product, is used extensively in interfacial applications. Collagen maintains a bound layer of water under all practical conditions which prevents complete spreading of low surface tension, organic liquids. Higher surface tension liquids wet collagen surfaces as they would simpler organic materials with critical surface tensions of about 40 dynes/cm. Gelatin exhibits even stronger interfacial interactions with hydrogen-bonding liquids as a distinct class. The combination of a high critical surface tension and a separate contact angle pattern for hydrogen-bonding liquids provides evidence of accessible amide groups at the surfaces of polyamide materials.

Although proteins are extraordinarily complex, high molecular weight polymers, a common feature is their polyamide backbone chain. The surface properties of specific proteins and their potential range of interfacial behavior are important commercially and are also of biological interest. Work at the Naval Research Laboratory during the last 30 years has demonstrated the value of contact angle measurements in assessing the surface characteristics of simpler polymer films; therefore, this series of experiments was initiated in order to lay a foundation for eventual reliable interpretation of contact angle data for protein surfaces.

Our earlier reports (*1, 2*) considered the nature of the contribution to wetting behavior of the common polyamide backbone chains. It was

---

[1]Present address: Calspan Corp., P.O. Box 235, Buffalo, N.Y. 14221.

demonstrated that the simplest protein analogue, polyglycine, was wet differently by hydrogen-bonding and nonhydrogen-bonding liquids (2). Investigation of polyamides of the nylon family (2, 3) established that such wettability differences reflect a special interaction between hydrogen-bonding liquids and the surface-accessible amide links of the polymer chain. The diagnostic value of this feature was explored with the more complex polypeptide poly(methyl glutamate) (PMG) (1, 4); it was discovered that the accessibility of the hydrogen-bonding polyamide backbone at the surface of the solid polymer was a function of the molecular configuration as well as the chemical constitution of the material. The polyamide backbone was effectively masked from the interface by the array of methyl ester side chains thrust uppermost when PMG was forced by solvent changes into the extended chain, beta conformation. When PMG was cast into films with a coiled alpha helical structure or a random tangle of chains, the contact angles signaled the availability of the polyamide backbone to H-bonding at the solid–gas interface.

These investigations were aided substantially by use of the multiple attenuated internal reflection (MAIR) spectroscopic technique to record surface specific, infrared (IR) spectra in order to determine structure and monitor preparation and cleaning methods.

In this report, these concepts are applied to real proteins: to collagen, an important structural material in tendons, bones, teeth, and skin, and to gelatin, the denatured product of collagen that is so important industrially. These materials are complex because of their 18 different, component amino acid side chains; in addition, they present experimental difficulties because of their water solubility—they cannot be washed (e.g., with an aqueous detergent) to assure surface cleanliness. Furthermore, they are often of unknown purity. They do have the common polyamide backbone, and it is possible to transform the molecular configuration. The data are indicative of the potential utility of contact angle measurements of important, natural materials. No claim is made for adequate attention to the complex biochemistry of these materials.

Our tentative conclusions are as follows. Purified collagen preparations present no evidence for the availability of H-bonding polyamide links at the surface of solid films, but increasingly denatured (i.e. randomized) gelatins do reflect the stronger interaction of the polymer with the H-bonding wetting liquids. In addition, collagen and gelatin evidence anomalous, nonspreading behavior with low surface tension, organic liquids on their surfaces; this is attributed to the presence of strongly adsorbed water at their surfaces. This attribution is based on experiments with water-swollen material and with synthetic polyacrylamide, and on work (5, 6) on other intrinsically high surface energy materials like glass and metals.

Collagen and Gelatin

Wilkie summarized the important interfacial properties of gelatin (7), and Stenzel and co-workers prepared a similar review of the features of collagen which underlie that protein's fundamental importance as a biomaterial (8, 9).

Collagen is the most abundant protein in animals. In fact, one-third of all the protein in mammals is collagen. Collagen's physical properties of inertness and strength enable it to function as a protective and supporting structure in skin, tendon, and cartilage. Collagen is also the organic component of teeth and bone.

Collagen, like all proteins, is made up of α-amino acids connected in peptide sequence. The structure of collagen, as determined by its x-ray diffraction pattern, consists of three polypeptide chains each of which has a helical structure. The three helices coil around one another in a gradual, right-handed helix to form molecular collagen known as tropocollagen. A tropocollagen molecule is about 14 A in diameter and 2800 A long with a molecular weight of about 300,000 (10). Collagen represents a particular aggregation state of constituent tropocollagen molecules. Non-helical appendages determine the molecular interactions and immunologic properties of collagen (11).

The amino acid content of collagen is one-third glycine, one-third cyclic, and one-fifth dibasic or dicarboxylic amino acids; a small proportion of oligosaccharide groups is also present. This unusual blend no doubt accounts for the extraordinary physical properties of collagen. The glycine content is unusually high, and that of the aromatic and sulfur-containing amino acids is low. Collagen is unusual among proteins in having a high content of the amino acids proline and hydroxyproline. In fact, collagen has the highest concentration of proline of any known protein, and it is the only protein which contains both hydroxyproline and hydroxyglycine. The compositions of various collagen and gelatin specimens are summarized in Table I (12–33).

Collagen is classified as a fibrous or scleroprotein. The fibrous structure of solid collagen can be divided into very small fibrils with a periodicity of about 640 A depending on the degree of hydration (24). Native collagen fiber is built up by a highly ordered process of linear and lateral aggregation of the thin, highly elongated tropocollagen molecules. Three types of fibrous collagen constitute the protein components of connective tissue. Collagen is insoluble in water, but 0.5–3% of connective tissue collagen is either neutral salt soluble or acid soluble. The remainder, insoluble collagen, can be almost completely solubilized by treatment with proteolytic enzymes without destroying its native molecular structure (25). Each collagen molecule contains three subunits or chains.

Table I.   Amino Acid Composition of Collagens and Gelatins[a]

Amino Acid	Content, wt %
Alanine	5.9–11.6
Arginine	4.6 – 9.3
Aspartic acid	4.4 – 7.0
Cystine	0 – 0.1
Glutamic acid	7.0–12.4
Glycine	20.2–33.5
Histidine	0.4 – 1.0
Hydroxylysine	0.4 – 2.7
Hydroxyproline	9.3–15.7
Isoleucine	1.1 – 1.9
Leucine	2.3 – 4.0
Lysine	2.5 – 5.6
Methionine	0.4 – 1.0
Phenylalanine	1.2 – 3.6
Proline	12.4–18.0
Serine	2.9 – 4.2
Threonine	1.5 – 2.4
Tryptophan	0.0
Tyrosine	0.1 – 1.1
Valine	2.0 – 3.3

[a] From References 12–23.

Newly formed collagen extracted with cold, aqueous NaCl solutions consists of three equal-sized chains (α-components) of two different composition types (α-1 and α-2). The two chains of similar composition are the α-1 chains. The α-2 chain differs from the α-1 in a number of amino acids, particularly hydroxyproline, proline, lysine, and histidine (26). As the collagen molecule matures, the α-chains crosslink intramolecularly in pairs; this older protein can be readily extracted with acidic solutions such as dilute acetate and citrate buffer, but not with salt solutions. The crosslinked chains are called β components; the crosslinks are probably covalent bonds (26) that arise by condensation of the side chains of strategic lysyl residues after enzymatic oxidative deamination. Older collagen also forms intermolecular bonds, but the nature of this crosslink has not yet been determined (27).

One of the most important and useful products derived from collagen is gelatin. Gelatin is a collective name covering a wide range of materials that are distinguished by good clarity and the ability to form tough gels at relatively low concentrations (18). Gelatin is derived from collagen by irreversible hydrolytic procedures. The properties of a gelatin sample depend on pH, electrolytes present, the collagen source, the method of manufacture, and its thermal history, aging, and concentration. Pure, dry commercial gelatin is tasteless, odorless, transparent, brittle, glasslike, and very faint yellow to amber in color. Yet gelatin is extremely hetero-geneous—composed of many-sized polypeptides (18). Its molecular weight ranges from 15,000 to several hundred thousand, reflecting the degree of hydrolysis of the higher molecular weight collagen source material. Source collagen and product gelatin differ little since gelatin has the same amino acid composition as the collagen from which it was derived. Differences among gelatins of diverse origins appear to be related to the degree of degradation of the parent collagen.

Crystallites of collagenlike structure may form after the gelation of gelatin. When gelatin is dried down from a gel (gel-dried), its wide angle x-ray diffraction pattern is similar to that of collagen whereas, when gelatin is dried from a warm solution (sol-dried), there is no evidence of crystallinity (24). As a result, the tensile strength of gel-dried layers is twice that of sol-dried layers. Furthermore, the IR absorption of gel-dried layers is similar to that of collagen (24).

In the process of producing gelatin from collagen, the collagen coils are split apart both laterally and longitudinally into separate strands and groups of strands (21). Tangled protein strands are ravelled or snipped short, or are present as a diverse combination of ravelled structures; gelatin is therefore described as being randomly coiled. There are two principal methods for manufacturing gelatin. In one, gelatin is derived from an acid-processed collagen stock, primarily pigskin. In the other, collagen stock (such as calf, cattle, and water buffalo hides and demin-eralized bone) is treated with lime.

The lime treatment method, used especially to manufacture photo-graphic gelatin, involves the following steps (24): (a) collagen stock is treated with a lime slurry at 10°–20°C for weeks or months; (b) the stock is washed and neutralized with acid; and then (c) the gelatin is extracted with warm water at neutral or slightly acid pH using successively higher temperatures in a series of cooks. The lime keeps the pH at about 12 and hydrolyzes some of the peptide bonds. It also hydrolyzes the side chain amide groups of glutamine and asparagine yielding glutamine and aspartic acid residues (24).

The acid process of gelatin preparation is as follows (21): (a) the hides, usually pig, are washed with water and soaked in dilute sulfuric acid (pH ~2; (b) they are washed free of acid and soluble proteins; (c) they are placed in extraction kettles and hydrolyzed with successive portions of hot water at an acid pH; (d) the dilute solution is filtered and evaporated; (e) concentrated solutions are chilled to a gel; and (f) the gel is dried with filtered and conditioned air in drying tunnels or in continuous driers.

The acid degradation method is much faster than lime treatment since the acid-soaking requires only a few hours. There are, however, a number of important physical differences between acid- and alkaline-derived gelatin. The reactive group sites vary in number, producing different properties (21). Important differences in structure are indicated by variations in the isoelectric point (where the net charge of the gelatin is zero). In the acid process, amide groups of glutamine and asparagine are not hydrolyzed so the gelatin has fewer free carboxyl groups than lime-processed gelatin and, therefore, a higher isoelectric point (24). The isoelectric point of acid-produced gelatin is between pH 7.0 and 9.0 whereas that of lime-treated gelatin is between 4.7 and 5.1 (21).

No two gelatin preparations will have the same properties unless the stock and method of preparation are identical. The gel prepared from each successive extract is less rigid than the previous batch; the viscosity may increase or decrease from extract to extract depending on the stock used.

Eight companies in the United States annually produce a total of approximately 60 million pounds of gelatin. The photographic industry exploits almost every property of gelatin including its gelation, protective colloid, viscous flow, and film-forming functions as well as the poorly understood chemical properties leading to sensitization (28). The gelatin layer coats individual microscopic particles of silver bromide and prevents their agglutination and flocculation; it also helps to regulate the size and growth of the silver halide particles (21). In the paper industry, the tensile strength of thin gelatin films increases the bursting strength of paper. Paper can accept ink without smudging because of gelatin's hydrophilic nature and ink can be removed from ledger paper with a penknife because of gelatin's brittleness (28). In the textile industry, the adhesive film-forming properties of gelatin strengthen warp fibers and prevent the shedding of minute filaments during weaving. Gelatin glue forms a strong tacky gel on cooling since its polar groups wet many surfaces and its long chain molecular structure provides a tough joint.

Materials and Methods

The primary experimental material was a sample of water-soluble collagen acid-extracted from rat skin (29) that was donated by Karl A. Piez, National Institute of Dentistry. The original sample was a dry, pure white, fibrous, cottonlike mass that had been dehydrated by freeze-drying. The collagen sample was ultracentrifugally homogeneous (Figure 1A); it sedimented with a single, hypersharp peak and had a sedimentation coefficient of 2.82×10^{-13}. Solution of the sample in water warmer than $40°C$ led to the formation of a mildly denatured collagen form, described here as gelatin, which also sedimented with a single peak (Figure 1B).

Figure 1.   Ultracentrifuge Schlieren photographs of sedimentation behavior of highly purified rat skin collagen and its mildly gelatinized product at corresponding times

Top, *hypersharp sedimentation peak for 0.15% solution of rat skin collagen in water (sedimentation coefficient = 2.82 × 10⁻¹³); and* bottom, *broader sedimentation peak for 0.16% solution of rat skin collagen in water after heating to 40°C (sedimentation coefficient = 2.35 × 10⁻¹³)*

The gelatin peak was considerably broader, and the sedimentation coefficient was lower—2.35×10^{-13}.

A second collagen sample was bovine Achilles tendon collagen (Schwartz Mann, Inc.). This material resisted solvation in all simple aqueous solutions and even in strong acids, but it could be solubilized with time by dichloroacetic acid (DCA). Thus, films formed from this sample had to be cast from DCA.

Except for a commercial sample of Knox gelatin, all gelatin samples were obtained from Fisher Scientific Corp.: 1099, gelatin (purified calfskin); 1099-P, gelatin (practical); 5247, gelatin (purified pigskin); G-5, white gelatin (silver label); and bacteriological gelatin.

All films prepared for contact angle studies were cast from an excess of dry polymer swollen and fluidized by a small quantity of added liquid (usually triple-distilled water, the last two distillations being in an all-quartz apparatus). The pH of these weak gels at the initiation of film formation was always 5–6. All our surface films were gel-dried rather than sol-dried. Although swelling may not have reached equilibrium prior to film drying, the films were only about 1 mm thick and their surface properties, as determined by contact angle measurements, were always reproducible within 5°.

Thin films of the collagen and gelatin samples were formed on freshly flamed platinum sheets by the drop spreading method (*1*). A few grains

Table II.   Wettability of Purified Rat Skin Collagen

Average Contact Angle,[a] degrees

Wetting Liquid	Surface Tension at 20°C, dynes/cm	Film Formed at 20°C[b]	Water-Swollen 20°C Film[c]	Film Formed at 80°C[d]
Water	72.8	92	90	90
Glycerol	63.4	79	89	70
Formamide	58.2	81	85	65
Thiodigylcol	54.0	54	97	48
Methylene iodide	50.8	48	68	48
sym-Tetrabromoethane	47.5	43	69	46
1-Bromonaphthalene	44.6	35	58	38
o-Dibromobenzene	42.0	34	50	35
1-Methylnaphthalene	38.7	30	33	29
Dicyclohexyl	33.0	16	19	12
n-Hexadecane	27.7	0	11	0
n-Decane	23.9	0	0	0

[a] Averages of at least 10 readings on at least two independently prepared films of each type.
[b] Drop spread from distilled water on platinum sheet and air dried in grease free container at 20°C.
[c] Drop-spread film swollen with distilled water.
[d] Drop spread from distilled water at 80°C (*i.e.* gelatinized) and air dried at 80°C in greasefree container.

of dry polymer are placed in the center of a freshly flamed but cool platinum plate and an appropriate solvent is transferred to this sample dropwise with a freshly flamed, but cool platinum wire. The solvent drops and their dissolved polymer burden spread spontaneously over the high energy metal plate, and, after slow air drying in a covered, grease-free container, a specularly smooth film is left over most of the flat plate surface and remnants of the solid polymer in the plate center. This technique has the obvious benefits of simplicity and general applicability, and it requires only small amounts of the sometimes scarce polymeric material. In addition, the thin films equilibrate most quickly with the room environment, various test humidities, and swelling liquids such as water.

MAIR spectroscopy at IR wavelengths was used to identify the polymers, monitor any induced structural transformations after the various treatments, and verify the freedom of the thin film from residual trapped solvent or adventitious contaminants (*1, 2*). A model 9 internal reflection accessory (Wilks Scientific Corp.) was used in conjunction with Beckman IR-12 and IR-7 and Perkin-Elmer 21, 257, and 457 IR spectrometers to record reflection spectra of the immediate interface of the sample which was clamped against the multiple internal reflection prisms made from the thallium bromide salt KRS-5.

Slowly advancing contact angles were determined with pure reference liquids placed dropwise on each specimen surface (*1, 2, 3*); these liquids included hydrogen-bonding and nonhydrogen-bonding compounds with a wide range of surface tension and structural type. Each cited value for contact angle is the average value recorded reproducibly within the first 10–20 sec after the drop was slowly advanced over a fresh surface region. With water, formamide, and ethylene glycol, the contact angles sometimes changed rapidly after this interval as the result of gradual penetration of the liquid into the plastic solid. With glycerol, the contact angles also changed in some experiments, but much more slowly. With thiodiglycol, the contact angles were often constant for many minutes before they also began to diminish. The contact angles measured with the nonhydrogen-bonding organic liquids were generally constant for many minutes, and usually there was almost no hysteresis when receding angles were measured. All data were recorded with samples and liquids equilibrated in a clean room maintained at 20°C and 50% relative humidity unless specifically stated otherwise.

Results

Average contact angle values for various highly purified diagnostic liquids on films prepared from purified rat skin collagen are presented in Table II. Contact angles were measured with the films (a) drop spread from distilled water on a platinum sheet and air dried to a specularly smooth continuous film in a greasefree container at a temperature that never exceeded 20°C; (b) prepared as in Method a but then reswollen with distilled water for at least 1 hr and measured while still completely water swollen in equilibrium with relative humidities of 50% and >99%

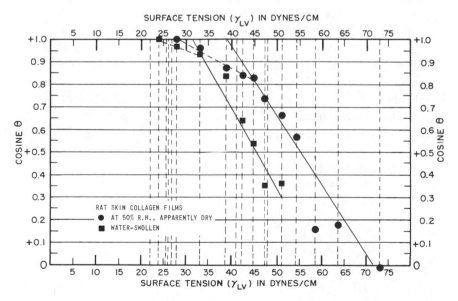

Figure 2.   Plot of contact angle data for rat skin collagen films at 50% relative humidity and when completely water swollen

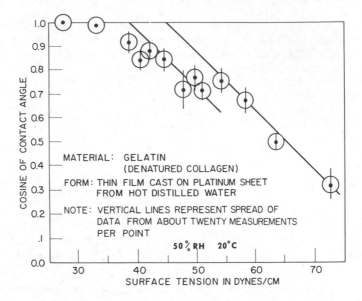

Figure 3.   Plot of contact angle data for mildly gelatinized
rat skin collagen

(no significant differences were observed at these two humidities); and (c) drop spread from distilled water at 80°C and air dried at 80°C in a grease-free container. The data for the 20°C dry and wet collagen films are plotted in Figure 2. The data for films gelatinized by preparation method c, which involved heating the gel above 40°C, are plotted in Figure 3; the data spread was more than for replicate films of the native collagen preparation. There were significant differences among these three types of films formed from identical starting material (*see* Table II and Figures 2 and 3). The apparently dry collagen cast at low temperature had a critical surface tension of about 39 dynes/cm which decreased significantly to about 32 dynes/cm when the film was totally water swollen. In contrast, when the collagen was mildly randomized in the hot water casting technique, all liquids capable of reacting across the interface by hydrogen bond formation, in addition to the more universal dispersion force, did so react. Data for the H-bonding liquids fall on a separate critical surface tension line indicating that the surface free energy sensed by such liquids was increased by the randomization of the collagen structure. In all preparations, anomalous nonspreading of low surface tension, dispersion force only liquids was noted; we interpret this anomaly —as compared with all simpler polymers we have examined so far—as reflecting the presence of adsorbed and organized water on proteinaceous surfaces (*see* below).

We suggested (2) that thiodiglycol and methylene iodide be used as indicators of the availability of hydrogen bond forming groups at the solid surface. In the absence of hydrogen bond formation across the interface, the higher surface tension liquid should produce the larger contact angle; consequently, equality of the contact angle values with thiodiglycol and methylene iodide, or a reversal of their normal order, provides a quick indication of hydrogen bond accessibility across a solid–liquid interface. Both these liquids have large enough molecular sizes and poor enough solvency powers to minimize potential complications such as penetration into the polymer surface and solution of the polymer in the liquid droplet. The contact angle values measured with hydrogen-bonding liquids on the completely water-swollen specimens have been omitted from the plot in Figure 2; all such data were nonequilibrium measurements of angles around 90° that resulted from a marked inter-action with the swollen polymer immediately after placement of certain droplets on the test specimen surfaces.

Typical IR reflectance spectra of films formed by cold water casting and by hot water casting are presented in Figure 4. The only discernable difference between the native and the denatured or gelatinized samples was that the latter material was characterized by slightly broader ab-sorption bands indicative of the presence of a wider variety of molecular chain configurations.

Contact angles were measured with the same series of wetting liquids on the surfaces of DCA-spread films of the bovine Achilles tendon collagen sample (Table III). The films were of three types: (a) those formed

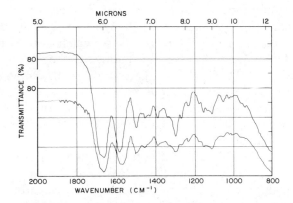

Figure 4.   Internal reflection IR spectra of the surface zones of cold water cast rat skin collagen films (top) and hot water cast (i.e., gelatinized) films from the same original sample (bottom)

Table III.　Wettability of Bovine Achilles Tendon Collagen

Wetting Liquid	Surface Tension at 20°C, dynes/cm	Film Formed at 20°C from DCA[b]	20°C Film after Water Wash[c]	20°C Film Heat Denatured at 80°C[d]
		Average Contact Angle,[a] degrees		
Water	72.8	87	79	56
Glycerol	63.4	80	76	70
Formamide	58.2	72	70	60
Thiodiglycol	54.0	65	64	53
Methylene iodide	50.8	60	51	51
sym-Tetrabromoethane	47.5	50	45	47
1-Bromonaphthalene	44.6	41	37	40
o-Dibromobenzene	42.0	39	35	34
1-Methylnaphthalene	38.7	33	33	30
Dicyclohexyl	33.0	25	21	15
n-Hexadecane	27.7	14	11	0
n-Tetradecane	26.7	13	—	—
n-Tridecane	25.9	10	—	—
n-Dodecane	25.4	0	8	—
n-Decane	23.9	0	0	—

[a] Averages of at least 10 readings on at least two independently prepared films of each type
[b] Drop spread from dichloracetic acid on platinum sheet and air dried at 20°C in covered, greasefree container.
[c] After extensive water wash and drying.
[d] Heat denatured at 80°C in water and redried.

at 20°C from DCA and air dried for extended periods of time for solvent removal; (b) those formed like Type a, then washed with distilled water (which did not solubilize them even though they had apparently been acidified) and again air dried; and (c) those which were heat denatured or gelatinized while water swollen by conducting the water-swelling and drying processes at 80°C. These data confirm the findings with the better-characterized collagen sample (see Table I and Figures 1, 2, 3, and 4). These data particularly demonstrate the increased randomization, as judged by accessibility of hydrogen bond interactions across the interface, in the heated sample.

IR spectra were made of the thin films of the rat skin and the bovine Achilles tendon collagen samples actually used for the contact angle measurements (Figure 5). The considerably greater breadth of the bands for the bovine tendon sample belies the suggestion, which might have derived from Figure 4, that simple breadth of diagnostic protein absorption bands might be an independently reliable indicator of the randomization or hydrogen bond interaction potential of these protein materials.

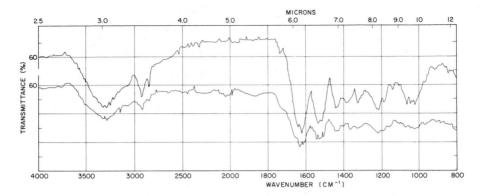

Figure 5.   Internal reflection IR spectra of the surface zones of films of a highly purified rat skin collagen sample (top) *and a more heterogeneous collagen preparation from bovine Achilles tendon collagen* (bottom)

The contact angles measured with the series of purified wetting liquids on cold water cast, thin films of gelatins of various derivation are listed in Table IV. The table has been arranged to provide in the sequence of the columns a sequence of materials with progressively less apparent hydrogen bond capability across their interfaces, and, concomitantly, apparently lower degrees of randomization from the native structure which masks such hydrogen-bonding potential. In Figure 6 are plotted the contact angle values for purified calfskin gelatin; as with

Table IV.   Wettability of Gelatin Films

Average Contact Angle,[a] degrees

Wetting Liquid	Surface Tension at 20°C, dynes/ cm	Calf- skin	Silver Label	Knox	Pig- skin	Bac- terio- logi- cal	Prac- tical
Water	72.8	72	67	69	70	62	63
Glycerol	63.4	59	62	61	64	55	73
Formamide	58.2	51	49	47	60	44	54
Thiodiglycol	54.0	37	42	43	42	47	62
Methylene iodide	50.8	41	44	45	42	45	49
sym-Tetrabromoethane	47.5	36	35	41	39	38	45
1-Bromonaphthalene	44.6	24	30	34	33	34	38
o-Dibromobenzene	42.0	22	28	30	29	25	32
1-Methylnaphthalene	38.7	15	23	26	23	16	27
Dicyclohexyl	33.0	0	11	15	11	5	16
n-Hexadecane	27.7	0	0	0	0	0	0

[a] Averages of at least 10 readings on at least two independently prepared films of each type.

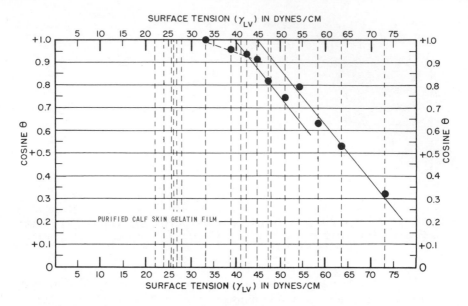

Figure 6.   Plot of contact angle data illustrating the duality of wetting behavior of a purified calfskin gelatin film when tested with hydrogen-bonding and non-hydrogen-bonding liquids

the randomized pure collagen sample, there were separate critical surface tension intercepts for hydrogen-bonding and nonhydrogen-bonding liquids. In Figure 7 are the internal reflection IR spectra for these same six samples in the identical form that was prepared and used for the contact angle measurements; there was great diversity in the IR absorption patterns which reflect the molecular species and their conformations. There was no obvious correlation between particular features of individual IR spectra and the wetting data. Thus the contact angle data reflect the actual outermost atomic constitution of the various samples much more sensitively than even this remarkably surface sensitive spectral technique which characterizes no more than a micron or so of the sample surface phase.

Interpretation and Discussion of Results

There are three major complications in contact angle studies of proteins and other water sensitive polymers. First, no simple or safe method of guaranteeing the cleanliness or uniformity of specimen surfaces is available; the initial purity of the sample must be relied upon as the primary criterion of surface uniformity. Second, there is a strong effect from adsorbed water molecules which remain on the surface and usually also within the bulk of hydrophilic materials such as polyamides. Third, water sensitive materials may be rapidly swollen or solubilized by many liquids applied to their surfaces whereas a more simple wetting and spreading context is desired for a number of biological and/or industrial purposes. In our previous work, such complications were overcome in other instances when it would have been destructive to attempt to dehydrate thin film specimens completely. Drastic water removal procedures almost certainly induce other undesirable alterations in the basic polymer chemistry and structure.

The diagnostic criteria presented (*1, 2*) as indicators of the accessibility of hydrogen-bonding sites at polymer interfaces are applicable even for highly water soluble polyacrylamide. The plot of contact angle data in Figure 8 includes values published earlier (*20*) and also the less reproducible data for water, glycerol, and formamide—three more important hydrogen bond forming liquids—recorded on these surfaces at the same time (*21*). Our diagnostic criterion of a split in the data plots for hydrogen-bonding and nonhydrogen-bonding liquids is met in this instance, reflecting the marked degree of interaction between the hydrogen-bonding liquids and the surface accessible amides which in this polymer are present in the impossible-to-mask side chains. The contact angle value

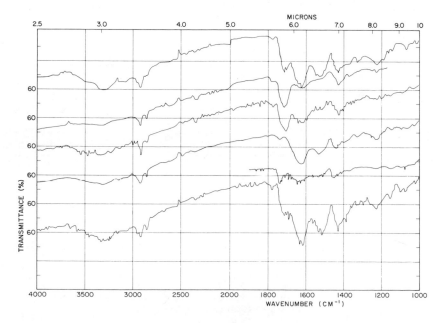

Figure 7.   Internal reflection IR spectra revealing the variety of structures dominating the surface zones of six different gelatin preparations cast from hot water into thin, smooth films

obtained with the hydrogen-bonding liquid of largest molecular size, thiodiglycol, was both stable and reproducible; this suggests that its inability to penetrate the sample surface eliminates the complications (scatter in data of about 5–10 degrees) caused by simple penetration of the smaller, hydrogen-bonding molecules.

There are three possible critical surface tension intercepts which might independently characterize water sensitive polymers, depending on the image forces and special interactions that can be manifested across the liquid–solid interface. An intercept above 40 dynes/cm is obtained by extrapolating data for hydrogen-bonding liquids; a value at about or slightly below 40 dynes/cm is obtained by extrapolating data for non-interacting organic liquids, and a much lower intercept—from a plot with a markedly different slope—at about 30 dynes/cm is obtained by extrapolating data for dispersion force only liquids with very low surface tensions (which give anomalous nonzero contact angle values). As was demonstrated earlier with simpler polyamides and here with the collagen–gelatin transformation, the separate critical surface tension intercept for the hydrogen-bonding liquids is not observed when the polymer structure is such that amide groups which could enter into such special interactions are masked from the surface. With exceptionally water sensitive polymers such as the proteins and polyacrylamide, however, the apparent

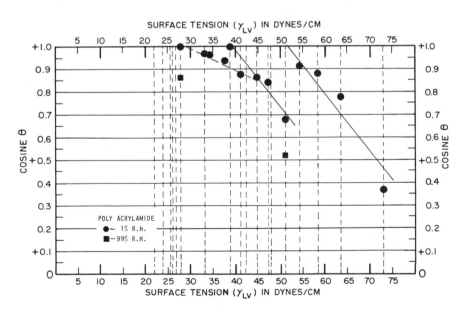

Figure 8. Plot of contact angle data illustrating great disparity in the wetting of a polyacrylamide surface by liquids capable of sensing different surface chemical groupings

lowest critical surface tension intercept for the low surface tension, organic liquids remains constant even when a large variety of configurations is present, as illustrated with our series of gelatins. We attribute this apparent anomaly to the presence on all such surfaces of a strongly adsorbed and organized layer of water.

The difficulty of swelling and solubilization of the polymers by certain test liquids arises in any system where the liquids and the polymers are potentially or known to be miscible. The effect of organic liquid interactions with polystyrene was noted many years ago (22). In the current experiments, when hydrogen-bonding liquids were applied to the various collagen and gelatin films, swelling and partial solubilization did occur noticeably in some instances. Reproducible values could always be obtained, however, by advancing the interacting droplets over fresh surface areas and recording only the initial advancing contact angles. When drops which had been on the surface for a minute or more were carefully incremented, the droplet perimeter remained constant rather than advancing, and the apparent contact angle values increased to well over 100° and remained at such anomalously high values for many minutes. Retraction of these incremented drops of interacting liquids from the protein surface, for example by simply touching them with a wick of filter paper, left behind an obviously liquid-swollen region which was capped by a surface-spread protein film which collapsed in the shape of a truncated liquid droplet of the original size.

Because of these many complications, it is most encouraging that the data derived from this study were remarkably orderly and that they agreed with predictions based on our studies of simpler model polymers (1, 2, 3). The following can be concluded about the wetting properties of collagen and gelatin:

(a) collagen maintains a tenaciously bound layer of adsorbed moisture under the experimental conditions (50% relative humidity at 20°C), and it is unlikely that this adsorbed aqueous component could be completely eliminated without destroying the polymer;

(b) collagen has an estimated critical surface tension of about 40 dynes/cm, and, in its native fibrous structure, it does not expose accessible hydrogen-bonding, backbone amide groups at the solid–gas interface; and

(c) gelatin, a randomized and destructured product of collagen, can be induced by simple temperature increases, and it can exhibit marked changes in surface interactions with hydrogen-bonding liquids even though no large changes in the IR spectra or other parameters of the bulk polymer are observed.

The combination of a high critical surface tension intercept and a split in the contact angle data for the hydrogen-bonding and nonhydrogen-bonding liquids is the best indicator of the presence of accessible amide groups at the surface of polyamide materials, even when such materials are as complicated as natural fibrous proteins. The occurrence of anomalous nonzero contact angles for simple organic liquids with very low surface tensions suggests the presence of a strongly adsorbed layer of water on such surfaces. Depending on the practical use to be made of contact angle data, any or all of the three different critical surface tension values might be the most important. It would be unwise to attempt to deduce indicators of the surface free energy of such biological macromolecules by relying upon contact angles with only one type of liquid. A common error of biologists and even some polymer chemists is to place complete faith in measurements of water contact angles alone.

Although considerable additional work with more model systems, a greater variety of wetting liquids, and proteins of the greatest possible purity and uniformity is necessary, the orderly pattern of the data obtained so far deserves attention. This article illustrates only the first successful application of these methods to the protein collagen and its variously modified structural forms in common gelatin preparations. Collagen-derived proteins constitute the bulk of the organic material in connective tissue and in teeth, so the wetting and spreading of a variety of liquids on each of these materials is of great practical importance. Surgical adhesives, for example, must efficiently wet and spread upon the protein of cut tissue and form strong bonds with it. Similarly, dental adhesives, especially those for use within the prepared tooth cavity, must wet the proteinaceous component (coating that surface) which was derived from the collagen-based dentine if the restorative material is to make a firm, void-free bond.

The preliminary wetting studies reported here describe and delineate the types of spreading behavior which are to be expected with liquid compositions of various types. Further, our work demonstrates the practical utility which induction of modest configurational changes of proteins might have in predisposing a protein-dominated surface to either favor or reject spontaneous wetting by a given liquid or class of liquids. The potential range of applicability of these findings, after they are extended and confirmed, is exceptionally large when one recalls that the surface properties of proteins have already been implicated in the catalysis or control of most biochemical events; in the permselectivity of biological membranes; in the acceptability of organ grafts and grafts of engineering or structural materials; in the irreversible damage occurring in extracorporeal devices for blood handling; in the acceptance of medications, cosmetic agents, and other treatments by the skin; and in the resistance of skin surfaces to adverse penetration by contaminating agents.

Summary and Conclusions

Previously, we examined the wetting properties of seven model poly-amides, including the polypeptides poly(methyl glutamate) and poly-(benzyl glutamate); polyglycine; nylons 66, 6, and 11; and polyacrylamide. All were successfully characterized by contact angle criteria as amide-containing materials. When structural transitions were induced in these experimental materials which alternately exposed hydrogen-bonding amide groups to the surface and masked them from it, the diagnostic wetting criteria were validated. These model materials had the inherent advantage of being synthetic products of defined constitution that could be freed from contamination by simple techniques. When the study was extended to the more complicated natural protein system of collagen–gelatin, it was noted that anomalous contact angle values for these water sensitive specimens were in a characteristic range of their own that did not interfere with the contact angle diagnosis of accessible amide groups at the polymer interface. The anomalous nonzero contact angles for low surface tension, organic liquids are indicative of the retention of a strongly bound water layer on the protein surfaces. Despite this adsorbed mois-ture, collagen demonstrates a surface composition with a critical surface tension of about 40 dynes/cm when in its native form, and a second, higher critical surface tension intercept between 40 and 50 dynes/cm with hydrogen-bonding liquids only when randomized to its gelatin form.

Acknowledgments

Helpful suggestions by members of the Surface Chemistry Branch of the Naval Research Laboratory, especially George Loeb, Elaine G. Shafrin, and Marianne K. Bernett, are gratefully acknowledged. D. S. Cain of the Physical Chemistry Branch rendered valuable assistance in recording the IR spectra. Albert J. Fryar provided the data on sedi-mentation coefficients.

Literature Cited

1. Baier, R. E., Zisman, W. A., *Macromolecules* (1970) **3**, 70.
2. *Ibid.* (1970) **3**, 462.
3. Ellison, A. H., Zisman, W. A., *J. Phys. Chem.* (1954) **58**, 503.
4. Baier, R. E., Loeb, G. I., in "Polymer Characterization: Interdisciplinary Approaches," C. D. Craver, Ed., p. 79, Plenum, New York, 1971.
5. Shafrin, E. G., Zisman, W. A., *J. Amer. Ceram. Soc.* (1967) **50**, 478.
6. Bernett, M. K., Zisman, W. A., *J. Colloid Interface Sci.* (1968) **28**, 243.
7. "Gelatin," Gelatin Manuf. Inst. Amer., New York, 1973.
8. Stenzel, K. H., Miyata, T., Kohno, I., Schlear, S., Rubin, A. L., ADVAN. CHEM SER. (1975) **145**, 26.
9. Stenzel K. H., Miyata, T., Rubin, A. L., *Ann. Biophys. Bioeng.*, in press.

10. Rubin, A. L., Drake, M. P., Davison, P. F., Pfahl, D., Speakman, P. T., Schmitt, F. O., *Biochemistry* (1965) **4**, 181.
11. Stenzel, K. H., Dunn, M. W., Rubin, A. L., Miyata, T., *Science* (1969) **164**, 1282.
12. Grahan, C. E., *J. Biol. Chem.* (1949) **177**, 529.
13. Neuman, R. E., *Arch. Biochem.* (1949) **24**, 289.
14. Block, R. J., Bolling, D., "The Amino Acid Composition of Proteins and Foods," Charles C. Thomas, Springfield, 1951.
15. Hess, W. C., Lee, C., Neidig, B. A., *Proc. Soc. Exper. Biol. Med.* (1951) **76**, 783.
16. Tristram, G. R., in "The Proteins," H. Neurath, Ed., vol. 1A, p. 181, Academic, New York, 1953.
17. Eastoe, J. E., *Biochem. J.* (1955) **61**, 589.
18. Idson, B., Braswell, E., *Advan. Food Res.* (1957) **7**, 235.
19. Hughston, H. H., Earle, L. S., Binkley, F., *J. Dent. Res.* (1959) **38**, 323.
20. Hess, W. C., Dhariwal, A., Chambliss, J. F., Alba, Z. C., *J. Dent. Res.* (1961) **40**, 87.
21. "Gelatin," Gelatin Manuf. Inst. Amer., New York, 1962.
22. Rubin, A. L., Drake, M. P., Davison, P. F., Pfahl, D., Speakman, P. T., Schmitt, F. O., *Biochemistry* (1965) **4**, 181.
23. Knox Gelatine, Inc., unpublished data.
24. Curme, H. G., in "The Theory of the Photographic Process," 3rd ed., C. E. K. Mees and T. H. James, Eds., chap. 3, Macmillan, New York, 1966.
25. Nishihara, T., Rubin, A. L., Stenzel, K. H., *Trans. Amer. Soc. Artif. Intern. Organs* (1967) **13**, 243.
26. Piez, K. A., Lewis, M. S., Martin, G. R., Gross, J., *Biochim. Biophys. Acta* (1961) **53**, 596.
27. Page, R. C., Benditt, E. P., *Science* (1969) **163**, 578.
28. Blake, J. N., in "Recent Advances in Gelatin and Glue Research," G. Stainsby, Ed., p. 219, Pergamon, New York, 1958.
29. Piez, K. A., in "Treatise on Collagen," Vol. I: "Chemistry of Collagen," G. N. Ramachandran, Ed., p. 207, Academic, New York, 1967.
30. Jarvis, N. L., Fox, R. B., Zisman, W. A., ADVAN. CHEM. SER. (1964) **43**, 317.
31. Jarvis, N. L., private communication.
32. Ellison, A. H., Zisman, W. A., *J. Phys. Chem.* (1954) **58**, 260.

RECEIVED June 19, 1974. Robert E. Baier was supported as a Post-Doctoral Research Associate of the National Research Council, National Academy of Sciences in 1966–1968; during this time, most of the experimental work discussed here was completed.

Modification of Collagenous Surfaces by Grafting Polymeric Side Chains to Collagen and Soft and Hard Tissues

G. M. BRAUER and D. J. TERMINI

Dental Research Section, National Bureau of Standards, Washington, D. C. 20234

Collagen, soft tissue, and bone can be modified at 37°C by allowing them to react with acrylic, methacrylic, or vinyl monomers using ceric ions, persulfate–bisulfite or comonomers forming donor–acceptor complexes as initiators. The polymeric methacrylate side chain is chemically attached to collagen; similar bonding may occur on reaction with other monomers. With rat skin, the reaction takes place mainly at the surface whereas a higher yield of more homogeneous product is formed on grafting onto collagen. Grafting onto bone is best accomplished with persulfate–bisulfite initiator. Modification of the collagenous surface is indicated by changes in wettability, decreased water sorption, and improved resistance to mold growth; e.g., hydrophobic, oleophobic surfaces are obtained with fluorinated monomers. The modified surfaces could be useful as adhesion-promoting liners for restorative materials.

Grafting of polymeric side chains offers an attractive technique for altering the properties, especially the surface characteristics of the substrate. Many studies have dealt with grafting onto cellulosic (*1, 2*) and proteinaceous materials such as wool, silk, and collagen (*3*). Comparatively little information has been reported regarding the chemically initiated grafting of monomers onto collagen. Grafting onto collagenous surfaces, especially *in vivo* grafting onto soft and hard tissues to form covalent bonds between the collagenous material and the polymeric side chain, is an effective means to obtain chemical adhesion and to improve the physico–chemical properties of the substrate. Such techniques, if successful under clinical conditions, might find applications as soft tissue or bone cements in surgical procedures or in improving adhesion of dental restoratives to dentinal surfaces. Other characteristics that could presumably be improved by such treatments are greater resistance of teeth to caries, increased resistance of bone and skin to fungal and bacterial diseases, and increased protection in photochemical reactions. Furthermore, grafting procedures have indicated upgrading certain properties of leather such as water penetration and abrasion resistance (3–13).

This paper summarizes chemical grafting techniques explored in this laboratory that have potential biomedical application. These reactions, initiated by ceric ions, persulfate–bisulfite redox systems, or the presence of comonomers forming donor–acceptor complexes, were carried out in an aqueous environment under conditions which, with suitable modifications, might be tolerated *in vivo*. Grafting onto tissue surfaces by means of ionizing radiation will not be discussed since techniques for avoiding undesirable side reactions have not yet been developed.

Grafting onto Collagen

During the past few years, grafting to collagen using a variety of initiators has been reported. Rao and co-workers grafted methyl methacrylate, acrylonitrile, and acrylamide to collagen using ceric ammonium nitrate (CAN) as initiator (4–9). Russian investigators treated collagen powder with ferrous ammonium sulfate and aqueous solutions or emulsions of various monomers and hydrogen peroxides usually at 60°– 80°C. By this procedure they were able to graft acrylonitrile (14), acrylamide (15), vinyl acetate (16), methacrylic acid (17), methyl and glycidyl methacrylate (18, 19, 20), and phosphorus-containing polymer (21, 22, 23, 24) side chains onto collagen. Because of the elevated temperatures used in these reactions, denaturation of the collagen may occur. The same investigators used metavanadic acid (25), manganese pyrophosphate (26), ozonized collagen (27, 28), and hydrogen peroxide (19, 29, 30) to graft methyl methacrylate onto collagen. Reducing groups present in collagen form a redox system on leather powder (29). Acrylonitrile can also be grafted onto collagen with potassium persulfate as initiator in the presence of air, using tetrakishydroxymethyl phosphonium chloride as oxygen scavenger (3). However, this method proved unsuitable with other vinyl monomers. Grafting ethyl and butyl acrylate onto chrometanned sheepskins and kangaroo skins has been more successful using the potassium persulfate–sodium bisulfite redox system as initiator in a carbon dioxide atmosphere (10, 11, 12, 13).

A graft copolymer of collagen (as well as other proteinaceous materials) with methyl methacrylate has been made by polymerization initiated by tri-*n*-butylborane (*31*). Kudaba and co-workers also modified collagen by treating it with epichlorohydrin (*32, 33*) or epoxy resins (*34*) with $BF_3 \cdot Et_2O$ as catalyst. Such treatments could be valuable for improving certain properties of leather, but the experimental conditions preclude their use for clinical applications.

UV-induced grafting onto collagen is the most advantageous method for covering the substrate with a polymeric surface coating. The low energy radiation does not degrade the collagen to any appreciable extent. Generally, a-UV sensitive dye or other photosensitizer is added to the reaction mixture to improve the yield of the copolymerization. This proceeds *via* a free radical abstraction of a hydrogen followed by subsequent polymer chain growth from the free radical site on the substrate. The extent of grafting depends on the concentration of the sensitizer, its light absorption characteristics, and the radiant energy corresponding to the wavelength at which the sensitizer absorbs light. Activation energy of grafting depends on the type of monomer used. Photosensitizers which have been used to graft methyl methacrylate to collagen include benzene (*35, 36*), benzil (*36, 37*), benzoin (*38*), benzophenone (*37, 38, 39*), eosin (*37, 40*), and iodoeosin (*37, 41, 42*). Benzil and iodoeosin, which absorb strongly around 260 nm and 546 nm, are more effective than eosin and benzophenone. Riboflavin or fluorescein also can be used to graft acrylamide, *N,N*-dimethylacrylamide, *N*-vinylpyrrolidone, acrylic acid, and acrylonitrile onto collagen (*43, 44*). The quantity of grafted polymer introduced into the substrate depends on the chemical nature and concentration of monomer, photosensitizer, and fibrous substrate. Some homopolymerization of monomer also takes place. In the presence of oxygen, an induction period occurs, but oxygen removal retards the photopolymerizations.

In our studies with collagen, CAN was used as an initiator since free radicals are formed on the side chains of the substrate, and thus a high grafting efficiency compared with other redox systems can be expected. The formation of homopolymer is kept to a minimum and there is little degradation of the backbone substrate. Furthermore, this system makes it possible to conduct the reaction under relatively mild conditions at 37°C. The initial objective of this investigation was to determine which acrylic or vinyl monomers containing a variety of functional groups would react with the collagen substrate to yield a graft copolymer (*45*). A second objective was to determine the optimum conditions for grafting onto soft and hard tissues. In most of the experiments using collagen, rat skin, or bone, the following procedure was used: approximately 1 g of material was stirred for 1 hr in 50 ml of a 2% aqueous dioctyl sodium sulfosuccinate solution. One ml $0.05M$ CAN in $1N$ HNO_3 was added; the mixture was deaerated with nitrogen for 15 min and 2.25 ml of monomer was added. The reaction was continued in an inert atmosphere at 37°C. Homopolymer was removed by thorough extraction with a suitable solvent.

Efficient removal of oxygen is required to obtain optimum yields. Exposure to swelling agents such as 5% aqueous zinc chloride or potassium thiocyanate prior to grafting onto collagen did not increase the yield. However, the presence of a wetting agent proved beneficial.

Results of grafting various monomers onto powdered steerhide collagen are given in Table I. As was established in the preliminary experiments, results depend on the physical state (powder or film) and pretreatment of the specimen, length of its storage in water prior to the reaction, presence and concentration of wetting agent, and concentration and purity of the monomer. A considerable weight increase (21%) was obtained when only collagen was allowed to react with nitric acid or CAN solution in the presence of wetting agent. The anionic dioctyl so-

Table I.   Grafting of Polymeric Side Chains to Collagenous Surfaces
Reaction Times: 3 hours; Temperature: 37°C

	Avg. Weight Increase,[a] g/100 g substrate				
	Initiator:			Initiator:	
	CAN			$K_2S_2O_8$-$NaHSO_3$	
Monomer	Col-lagen	Rat-skin	Bone	Rat-skin	Bone
Control (no monomer—CAN[b])	21	2	− 9	—	− 9
Control (methyl methacrylate, no CAN)	2	0.3	− 5	—	− 1
Acrylic acid	43	11	−48	—	−52
Acrylates					
ethyl	106	7	−10	—	4
butyl	110	—	—	26	6
butyl + 10% acrylic acid	—	—	—	40	—
isodecyl	153	24	− 3	—	7
isodecyl (CO_2 atmosphere)	—	22	—	—	—
2-ethylhexyl	134	9	1	—	6
2,2,2-trifluoroethyl	124	13	5	—	− 3
hexafluoroisopropyl	120	16	—	—	2
$1H$,$1H$,$5H$-octafluoropentyl	117	17	—	—	—
pentadecafluorooctyl	—	130	—	—	—
cyanoethyl	129	16	− 5	—	201(212)[c]
Cellosolve	200	4	—	—	8
2% aq. calcium	11	—	—	—	48(45)[c]
2% aq. zinc	—	—	—	—	78
Methacrylic acid	56	11	−38	—	14

Table I.   Continued

Avg. Weight Increase,[a] g/100 g substrate

	Initiator: CAN			Initiator: $K_2S_2O_8$-$NaHSO_3$	
Monomer	Collagen	Rat-skin	Bone	Rat-skin	Bone
Methacrylates					
methyl	187	14	17	28	45
methyl (no wetting agent)	51	—	—	—	—
ethyl	182	—	− 3	—	12
isobutyl	91	—	− 1	—	− 1
lauryl	148	16	− 1	—	10
2-chloroethyl	96	—	—	—	—
2,2,2-trifluoroethyl	118	14	—	—	—
hexafluoroisopropyl	134	11	—	—	—
1H,1H,5H-octafluoropentyl	30	12	—	—	—
hydroxyethyl	226	4	−11	—	85(38)[c]
glycidyl	239	20	61	—	138
t-butylaminoethyl	2	—	—	—	—
dimethylaminoethyl	12	4	− 1	—	− 3
dimethylaminoethyl (acidified to pH 2.5)	34	4	—	−5	−14
Ethylene dimethacrylate– methyl methacrylate	—	24	—	—	—
Ethylene dimethacrylate	104	23	−10	−4	142(16)[c]
1,3-Butylene dimethacrylate	58	—	—	—	—
Acrylonitrile	74	11	—	—	49
α-Chloroacrylonitrile	77	—	—	—	—
Vinyl acetate	28	13	− 1	—	− 5
Styrene	33	15	−10	—	37
Vinyltoluene	25	—	—	—	—
Divinylbenzene	—	9	—	—	63
N-Vinyl-2-pyrrolidone	27	—	—	—	− 9
4-Vinylpyridine	0	—	—	—	—
Diallyl phosphite	29	16	—	—	−14
Triallyl phosphate	25	16	− 7	—	—
Butenediol	18	13	—	—	—

[a] After extraction of homopolymer with appropriate solvent.
[b] Ceric ammonium nitrate.
[c] Reaction time: 15 min.

dium sulfosuccinate wetting agent which is present in a fairly high concentration (2%) is strongly sorbed by the positively charged collagen molecules in the acidic reaction mixture. It is probably insolubilized in the presence of the nitric acid electrolyte and thus is not removed by washing the substrate with water.

Increase in weight (after extraction of soluble homopolymer) in excess of the increase in weight obtained in the absence of monomer was used as the criterion for successful grafting. Yield (based on monomer added and assuming that the weight increase after extraction results solely from graft polymer formation) varied widely, but it was highest for acrylates and methacrylates. Yields did not change greatly with the higher homologues. Yields for these two homologous series were generally in the 40–90% range (90–239% increase in weight). Monomers containing hydroxyl or glycidyl groups such as hydroxyethyl or glycidyl methacrylate were converted nearly quantitatively to the polymeric form. With acrylic and methacrylic acid, the yields were considerably lower than those obtained with their esters. With monomers containing basic groups, no apparent grafting took place. However, some grafting occurred when di(methylaminoethyl methacrylate) hydrochloride was allowed to react in an acid environment. Fluorinated acrylates and methacrylates also grafted onto the collagen substrate, generally with good yields.

Graft copolymerization with monomers containing 2-chloroethyl, hydroxyethyl, or glycidyl groups leaves residual potential reaction centers for further chemical modification of the product. Graft copolymers with ethylene or butylene dimethacrylate are probably crosslinked and should show reduced solubility and increased chemical resistance.

Most vinyl monomers other than those containing acrylic or methacrylic groups were not as readily grafted onto collagen in an aqueous environment. Little or no weight increase compared with the CAN blank took place with 4-vinylpyridine, vinyltoluene, triallyl phosphate, or butenediol. Grafting conditions for these monomers are more suitable when the reaction is conducted in appropriate non-aqueous solvents.

The presence of polymer on the collagen was confirmed from IR spectra obtained after removal of soluble homopolymer. Appearance of noncollagenous bands was most pronounced for collagen treated with the lower homologues of the acrylic or methacrylic esters. In general, collagenous products that increased more than 25% in weight had more distinctive spectra possessing sharper peaks and additional absorption bands when compared with the original collagen powder. For example, all esters (acrylates, methacrylates, acetate) gave the $C{=}O$ stretching band around 1740 cm^{-1} and many had another strong ester —C—O stretching band between 1140 and 1160 cm^{-1}.

In one series of experiments, collagen films (0.0013 cm and 0.005 cm thick) made from collagen fibrils of about 99% purity were substituted for the collagen powders in the standard grafting experiment. Although it was not possible to measure weight increase in the thinner film specimens, the IR spectra of the treated and solvent-extracted films indicated that polymer was probably grafted onto the substrate.

Under the experimental conditions, surface grafting occurs less than 30 min after addition of the monomer. Thus, after a 30-min reaction time, weight increased (after removal of homopolymers) 100% when methyl methacrylate and isodecyl acrylate were grafted onto collagen powder. Similarly, the increase in weight on grafting glycidyl methacrylate onto 0.005-cm collagen films demonstrated conclusively that measurable amounts of surface graft were formed within 30 mins.

Spontaneous grafting under high swelling conditions has been achieved with the wool–ethyl acrylate system without using any of the normal means of initiating grafting reactions (46). This reaction is presumed to be initiated by the free radical formation as a result of the strong anisotropic swelling brought about initially by water. With this mechano–chemical technique, polystyrene and PMMA could be grafted to collagen (47) in states that were not extracted by boiling toluene, chloroform, or acetone. When the reaction was conducted under nitrogen for relatively short periods, that is, up to 24 hrs, grafting did not take place. A more drastic swelling reaction appears to be required which would limit the usefulness of this technique for many applications.

Grafting onto Soft Tissues

Since many monomers of varying polarity could be grafted onto collagen, it is desirable to experiment with grafting onto soft tissue. If successful, it might be possible to vary widely and thus control surface properties of such a substrate. Defatted rat skins were lyophilized and then grafted in the presence of CAN (48). Control runs indicated that addition of wetting agent and CAN initiator or monomer results in no appreciable increase in weight of the treated rat skin (Table I).

A weight increase after extraction of soluble homopolymer was obtained on reaction of the rat skins with most monomers investigated (Table I). Yield of side chain polymer grafted onto the substrate was much lower for rat skin than for collagen with most weight increases being 15–25% (average of two or more runs). The highly fluorinated pentadecafluoroctyl acrylate yielded by far the largest amount of insoluble polymer (130% weight increase). On the other hand, monomers containing hydroxyl or ethoxy groups did not form any appreciable amount of graft polymer with rat skin despite the fact that these monomers gave nearly quantitative yields of polymer with steerhide collagen. With monomers containing basic groups, only a small weight increase was observed even when these monomers were allowed to react after complete neutralization of the basic groups. Apparent grafting took place not only with acrylate or methacrylate monomers but also with other vinyl monomers such as vinyl acetate and styrene. Reaction of diallyl phosphite or triallyl phosphate with skin produced modified surfaces containing phosphite or phosphate groups. All fluorinated acrylates and methacrylates could be grafted onto the rat skin substrate. Treatment of rat skin with ethylene dimethacrylate also resulted in an increase in weight. No appreciable change in yield occurred when carbon dioxide was substituted for nitrogen to deaerate the reaction mixture.

Reducing the reaction time below 3 hrs decreased the amount of polymer formed on rat skin. However, with a reaction time as low as 20 min, significant weight increases were recorded with such diverse monomers as methyl methacrylate, isodecyl acrylate, and glycidyl methacrylate. Hence, relatively short reaction times may be sufficient to modify significantly chemical and biological characteristics of soft tissue surfaces.

As may be expected for polymerization reactions of this type, the use of rat skins as substrates decreased the reproducibility of the polymer yield. The coefficients of variation for the weight increases averaged 28% and were above 50% for a few monomers, generally when the yield of polymer formed was low. The lower graft polymer yield obtained with rat skin compared with steerhide collagen may be attributed to (a) the different chemical composition (such as keratin) of the epidermis, and (b) the coherent nature of the rat skin providing a much smaller surface area than the collagen powder. Thus, the number of sites accessible for grafting is greatly reduced. Furthermore, wetting agent, initiator, and monomer diffuse only slowly into the interior of the substrate. With the thicker rat skin specimens, equilibrium conditions may not be reached readily. With rat skin, the grafting takes place mainly at the surface, whereas a more homogenous product is formed by grafting throughout the collagen powder.

Persulfate–bisulfite-initiated graft polymerization was accomplished (Table I) using the procedure of Feairheller et al. (11). Lyophilized rat skins were soaked in 100 ml of a 0.4% aqueous solution of an octylphenyl ether of polyethylene glycol containing 9–10 ethylene oxide groups. After 1 hr, 0.4 g potassium persulfate and 0.135 g sodium bisulfite were added, and CO_2 was passed through the solution before 5 ml of monomer were added.

acceptor monomer (A) + Lewis Acid ($ZnCl_2$) $\rightleftharpoons$ adduct (A---$ZnCl_2$)

(methacrylate,   +
maleic anhydride)
                                                              donor monomer
                                                              (styrene)

                                                       +   .   _ $\Updownarrow$
$nZnCl_2$ + [donor-acceptor]$_x$ $\longleftarrow$ [donor - - - acceptor · $ZnCl_2$]
           usually 1:1   complex
           alternating copolymer

Yields with methyl methacrylate were somewhat larger than those obtained in the comparable CAN-initiated polymerization. Even greater increases in weight were obtained with butyl acrylate, especially on addition of 10% acrylic acid both for 3-hr and 20-min reaction times.

Electron-acceptor monomers such as acrylates, methacrylates, and acrylonitrile become stronger acceptors on complexing with Lewis acids such as metal halides. Interaction of this complex with a strong electron-donor monomer, *e.g.*, styrene, leads to the formation of a charge-transfer (donor–acceptor) complex. This complex undergoes spontaneous or radical-initiated polymerization, propagating as a monomeric unit to yield equimolar alternating copolymers, irrespective of the initial monomer composition (*49, 50, 51, 52*). A radical initiator such as CAN causes a competing radical-initiated homopolymerization of the charge-transfer complex concurrently with the spontaneous reaction.

Donor–acceptor polymerization takes place on reacting rat skin with equimolar amounts of methacrylate (or isodecylacrylate), zinc chloride, and styrene even for as short a period as 20 mins. With a large excess of zinc chloride, polymer formation on collagen takes place with the maleic anhydride–styrene comonomer system. In the reactions where donor–acceptor polymerization occurred, yields were of the same order as those obtained solely with CAN initiator.

Thus the grafting reaction onto soft tissues is highly versatile. It takes place with many monomers and different initiator systems such as CAN, persulfate–bisulfite, and donor–acceptor monomers. Although the polymerization is inhibited by oxygen and thus air, it is conducted readily in a nitrogen or carbon dioxide atmosphere.

Grafting onto Hard Tissue

Bone. Monomers were also grafted onto bone. Powdered bone marrow was pulverized in a Wiley mill in the presence of dry ice and was subjected to the standard grafting procedure using CAN as initiator (53). Significant weight increases of the substrate, before extraction of homopolymer, ranging up to 80% (average for two or more runs) were obtained with a few selected monomers. However, a considerable amount of the added polymer was removed as homopolymer on solvent extraction. Consistent grafting was evidenced only with methyl methacrylate, glycidyl methacrylate, and 2,2,3-trifluoroethyl acrylate (Table I).

Much more successful was the reaction of various monomers with bone, using the persulfate–bisulfite redox system as initiator in a carbon dioxide atmosphere. Over 10 acrylate and methacrylate and a few vinyl monomers polymerized on the substrate, some in nearly quantitative yield. Reaction times of 15 min were sufficient for grafting. The amount

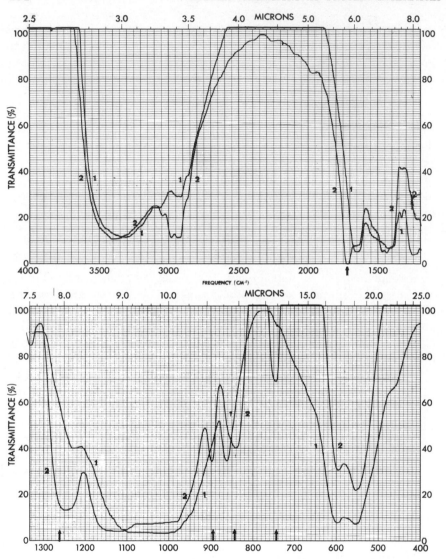

Figure 1. IR absorption spectra (KBr pellets): Curve 1, bone; Curve 2, bone to which glycidyl methacrylate had been grafted after extraction of the product with acetone

of soluble homopolymer was quite large for some products. However, weight increases of over 138% were obtained (after extraction) in the reaction with cyanoethyl acrylate, glycidyl methacrylate, and ethylene dimethacrylate. Since the chains obtained on polymerization with the latter two monomers are likely to be highly crosslinked, homopolymer extraction is less efficient, and it becomes increasingly difficult to ascertain if covalent bonding to hard tissue occurred. Nevertheless, the surface properties of the resulting products were modified.

An interesting reaction is that of calcium or zinc acrylate with bone. The resulting products after acetone and water extraction had a 40% and 76% increase in weight, respectively. A potential graft of this calcium-containing monomer onto highly mineralized bone or dentin may incorporate calcium in defective tissues and thereby improve bone healing. Monomers which could not be grafted onto bone were those containing basic or acidic side groups. The large loss in weight on reaction of bone with acrylic on methacrylic acid is indicative of dissolution of the bone under the experimental conditions.

IR spectra of the products indicated the presence of polymerized material on the bone (Figure 1). Thus, for the glycidyl methacrylate copolymer initiated with CAN, IR spectra showed the oxirane ring of the glycidyl group (899 cm^{-1} and 843 cm^{-1}) as well as the C$=$O stretching band at 1730 cm^{-1}, a C—O stretching band at 1270 cm^{-1}, and an absorption band at 747 cm^{-1} which is characteristic of the methacrylate group.

Dentin. Preliminary studies have been conducted to determine the feasibility of grafting onto dentin (53). With CAN as initiator, some modification of dentin occurred on treatment with methyl methacrylate. With other monomers, no increase in weight was found. Since dentin is the most highly mineralized collagenous substrate that has been studied, its lack of reactivity towards grafting is expected.

Thus, the relative ease of grafting monomers onto water-insoluble collagenous substrates using CAN as initiator decreases in the following order:

powdered collagen > collagen film > epidermal rat skin,
bone > powdered dentin

Apparent grafting to dentin and other proteinaceous materials has been reported through polymerization of methyl methacrylate with tri-*n*-butylborane (31, 54, 55, 56, 57). The mechanism on the top of p. 186 has been suggested.

Complexes of tributylborane with ammonia or primary and secondary amino groups are more stable in air and may also be used. They are easily activated by isocyanates and acid chlorides. Bond formation of the borane-cured resin to dentin is enhanced by moisture, and bond strength is retained fairly well after water immersion. Commercial dental restoratives containing borane initiators have become available. Compared with other restorative filling materials, they showed improved adhesion to dentin (58, 59, 60, 61). Bonding is only to the dentinal collagen whereas any retention to enamel is of a mechanical nature. Therefore, cavity preparations using conventional undercuts to retain

$$R_3B \xrightarrow{\text{O}_2} R_2BOOR \xrightarrow{2R_3'B} 2R'\cdot + R_2BOBR'_2 + \text{other products.}$$

$$R'\cdot + -NH-\underset{\underset{CH_3}{|}}{CH}-\underset{\underset{O}{||}}{C}-NH-\underset{\underset{R'}{|}}{CH}-\underset{\underset{O}{||}}{C}- \rightarrow -NH-\underset{\underset{CH_3}{|}}{C}-\underset{\underset{O}{||}}{C}-NH-\underset{\underset{R''}{|}}{CH}-\underset{\underset{O}{||}}{C}-$$

methyl | methacrylate (MMA)

$$|(MMA)_n$$

$$-NH-\underset{\underset{CH_3}{|}}{C}-\underset{\underset{O}{||}}{C}-NH-\underset{\underset{R''}{|}}{CH}-\underset{\underset{O}{||}}{C}-$$

the dental restorations must be used. Furthermore, certain pretreatments and cleansing processes of the cavity, as well as applying too viscous a liquid, diminish adhesion.

Changes in Properties of Modified Surfaces

Surface behavior of solid materials and biological tissues depends almost solely on the nature and the packing density of the outermost or exposed atoms and functional groups. Even a unimolecular layer grafted to active sites may significantly change such properties as water sorption, wettability, and critical surface tension, thus affecting the ease of adhesion to the surface. The relatively short reaction times found to yield substantial weight increases in the various substrates should be sufficient to modify greatly the physical, chemical, and biological characteristics of the treated surfaces.

Table II. Water Sorption of Graft Copolymers

Monomer	Water Sorption at 50% Relative Humidity, g/100g substrate		
	Collagen[a]	Ratskin[a]	Bone[b]
Substrate	19.2	16.8	8.1
Control (no monomer—CAN)	11.6	7.1	—
Control (methyl methacrylate, no CAN)	17.4	8.0	—
Acrylic acid	9.4	8.9	7.9
Acrylates			
ethyl	4.6	12.1	7.4
butyl	4.4	—	6.9
isodecyl	4.7	9.4	7.5
2-ethylhexyl	3.8	9.4	6.3
2,2,2-trifluoroethyl	3.8	7.4	—
hexafluoroisopropyl	4.3	—	—

Table II.   Continued

Water Sorption at 50% Relative Humidity,
g/100g substrate

Monomer	Collagen[a]	Ratskin[a]	Bone[b]
Acrylates (continued)			
1H,1H,5H-octafluoropentyl	4.1	8.6	—
pentadecafluorooctyl	—	3.1	—
cyanoethyl	4.2	7.4	2.8
Cellosolve	5.5	12.5	7.0
2% aq. calcium	14.2	—	9.1
Methacrylic acid	10.8	14.0	9.2
Methacrylates			
methyl	5.8	7.9	6.0
methyl (no wetting agent)	8.7	—	—
ethyl	4.4	—	—
isobutyl	6.1	—	—
lauryl	4.2	8.7	6.7
2-chloroethyl	5.4	—	—
2,2,2-trifluoroethyl	4.3	7.5	—
hexafluoroisopropyl	6.0	8.0	—
1H,1H,5H-octafluoropentyl	7.8	7.8	—
hydroxyethyl	7.4	10.4	4.3
glycidyl	4.4	—	3.0
t-butylaminoethyl	18.4	—	—
dimethylaminoethyl	15.9	—	7.9
dimethylaminoethyl			
(acidified to pH 2.5)	8.1	8.6	—
Ethylene dimethacrylate	6.9	9.7	4.5
1,3-Butylene dimethacrylate	9.0	—	—
Acrylonitrile	9.3	7.2	5.5
α-Chloroacrylonitrile	5.3	—	—
Vinyl acetate	7.7	8.5	—
Styrene	7.8	7.4	5.7
Vinyltoluene	9.4	—	—
Divinylbenzene	—	7.2	3.8
N-Vinyl-2-pyrrolidone	7.2	8.5	—
4-Vinylpyridine	16.4	—	—
Diallyl phosphite	8.3	7.4	—
Triallyl phosphate	10.1	7.2	—
Butenediol	9.3	9.8	—

[a] Initiator: Ceric ammonium nitrate.
[b] Initiator: Potassium persulfate–sodium bisulfite.

The appearance of the products of the reaction of collagen and various monomers often varied considerably from the original collagen powder (45, 48). With some monomers, the graft copolymer consisted of powders while with others mats or films were formed. The modification of surface properties could often be detected by visual inspection.

Thus with isodecyl, 2-ethylhexyl acrylate, or lauryl methacrylate, rubbery mats were formed. Similarly, rat skins treated with isodecyl or pentadecafluorooctyl acrylate possessed a very tacky, rubbery surface differing significantly from the original substrate.

Water sorption of collagen powder, which at 50% relative humidity was 19.2 g H_2O/100 g collagen, was always lowered on treatment of the collagen (Table II). This even applied for products containing hydrophilic groups in the side chain. Inspection of Tables I and II indicates that water sorption depends not only on the presence of hydrophilic groups in the side chains but also to a very marked extent on the amount of copolymer incorporated in the product. Lowest water uptake of about 4% was obtained for products containing fluorinated acrylates or the higher molecular weight esters of acrylic or methacrylic acid.

Water sorption at 50% relative humidity changed from 17 g H_2O/100 g substrate for untreated rat skin to 7–10 g H_2O/100 g substrate for the treated specimen. The only exception was the pentadecafluorooctyl acrylate-modified rat skin which had by far the largest weight increase and a greatly reduced water uptake (3%). It is not surprising that water sorption, which is predominantly a function of the water absorption throughout the specimen, is not altered as much for the rat skins as for collagen powder. The high yields for grafting onto collagen indicate that the reaction takes place throughout the specimen. Even considering the reduced amount of collagen present per unit weight of skin substrate, the much lower weight increase on grafting onto skin is apparently caused by restriction of the graft process to the exterior surface.

The wetting behavior of the collagenous surface is also changed greatly on reacting with most monomers. Collagen is wetted by water slowly (Figure 2). However, the reaction product of collagen with acrylic and methacrylic acid was hydrophilic and gave a zero contact angle within 15 secs. A majority of the materials, such as those containing grafts of fluorinated acrylates or methacrylates or the higher acrylate or methacrylate homologues, became completely hydrophobic (Figure 2). Exact contact angles were difficult to determine because of the porous and uneven surface of the specimens. The collagen–lauryl methacrylate and collagen–2-ethylhexyl acrylate substrates had approximate contact angles of 85° whereas those of the fluorinated acrylates and methacrylates had contact angle values over 90°. On introduction of sufficient fluorine content, the graft polymers also became oil repellent. After removal of drops of water and oil that contained dyes, the mats showed no sign of stains (Figure 2), thus establishing not only the water and oil repellency of the materials but also their stain resistance.

Water-repellent rat skin surfaces were also obtained with fluorinated acrylates or methacrylates whereas polar groups in the side chain im

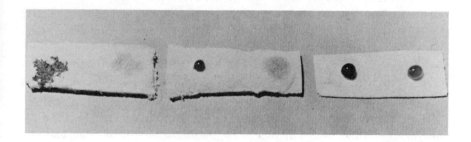

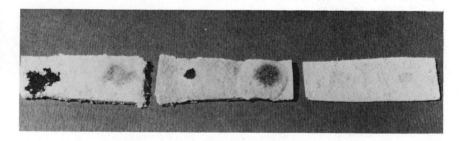

Figure 2. Water and oil repellency of graft polymers: left, acrylic acid–collagen; center, collagen; right, hexafluoroisopropyl methacrylate–collagen mat. Top and center, spreading of water and oil drops placed respectively on left and right of each mat; bottom, mats after blotting off drops with filter paper.

parted hydrophilic characteristics. Oil repellency was incorporated in the subdermal surface on polymerization with highly fluorinated monomers, but efforts to obtain oleophobic epidermal surfaces were unsuccessful. This different behavior of the skin toward non-polar liquids results from the keratinous nature of the epidermal surface and its rough texture even after removal of hair. Generally, modification of wetting characteristics was more easily accomplished with collagen powder than with rat skin.

Under mold-growing conditions, growth was observed on the original rat skin substrate, but not on skins treated with a variety of monomers. Although no detailed mycological studies have been undertaken, these results agree with the resistance to microbial and fungal attack reported for graft products of other naturally occurring materials. Some growth took place at the edges where the rat skins had been cut after treatment, indicating that grafting occurred predominantly at the surfaces.

Modification of bone brings about some decrease in its water sorption behavior. The decrease in sorption at 50% relative humidity was less pronounced than that found for collagen or rat skin. However, when high yields of polymer were formed on the substrate such as in the reaction of bone with cyanoethyl acrylate, hydroxyethyl and glycidyl methacrylate, ethylene, dimethacrylate, or divinylbenzene, water sorption at 50% relative humidity was reduced to 3.0–4.5 g H_2O/g bone.

Grafting was considered successful if the weight increase after extraction of soluble homopolymer was greater than the weight increase in the absence of monomer. It is recognized that some homopolymer, especially homopolymer of the bifunctional methacrylates that crosslink during the reaction, is not removed from the substrate by this treatment. A separate, distinct interpenetrating polymer network entwined with macromolecules of the collagen substrate may be formed in the reaction. Removal of such a phase by solvent extraction may not be possible. Even if the substrate and polymer chains are not covalently bonded, the surface properties of the resulting product should differ from those of the original substrate. Such modification may be used advantageously to develop products with improved properties.

Characterization of Graft Copolymers

A number of studies to determine the mechanisms of grafting onto collagen initiated by CAN, including characterization of the reaction products, have been reported (5, 8, 9, 53, 62). Ceric ammonium nitrate forms an effective redox system with alcohols, aldehydes, amines, and thiols. Alcohols form a ceric ion–alcohol complex, and the dissociation of this complex is the rate-determining step (63):

$$Ce^{IV} + RCH_2OH \rightleftarrows ceric\ ion–alcohol\ complex \rightarrow Ce^{III} + H^+ + RCHOH$$

Collagen contains alcoholic groups in the hydroxyproline, serine, threonine, and hydroxylysine moieties. Free radicals are probably formed at such sites which, in the presence of a vinyl monomer, serve to initiate grafted side chains. Since the free radicals are formed on the side chains of the substrate, a high grafting efficiency and a minimum amount of homopolymer formation compared with other redox systems can be expected.

To determine the probable location where grafting is initiated by CAN on the collagen molecule, the rate of polymerization of methyl methacrylate in the presence of amino acids found in collagen was studied (*62*). Polymer formation (homo- and graft polymerization) took place in the presence of all amino acids studied with the exception of tyrosine and methionine. The relative yields after the 3-hr reaction period were of the following order: tyrosine, methionine < control < proline < glycine, alanine, hydroxylysine < glutamic acid, threonine < *n*-butylamine < aspartic acid << serine, hydroxyproline, 1-butanol. These results indicate that the phenolic hydroxyl group, and possibly the sulfide group, act as inhibitors. Even in the absence of the carboxylic acid group, such as in butylamine, polymerization takes place; that is, the amino group is sufficient to take part with CAN in the redox initiation. Highest yields, however, are obtained in the presence of alcoholic groups such as in *n*-butyl alcohol.

Polymerization of methyl methacrylate initiated by CAN in the presence of polyglycine, polyhydroxyproline, and polyserine was also studied. Somewhat lower yield of polymer was formed in the presence of polyglycine. However, the poor reproducibility of the results did not allow a statistically supported conclusion.

These studies do not prove conclusively the exact site of graft formation on the collagen molecule. A preferred site for initiating grafting appears to be the serine residue. However, other amino acids having accessible hydroxyl and/or amino groups may also act as grafting sites.

Several lines of evidence have been sought to establish unequivocally the formation of a true graft copolymer as opposed to an intimate mixture containing no primary bonds between collagen and the added polymer. Despite the large number of apparent graft products, formation of a true graft has been established only for the collagen–methyl methacrylate and collagen–acrylamide copolymer. Indications that these products are true grafts are provided by the appearance on electron microscopic examination, swelling behavior and the solution properties of the polymeric product, the IR spectra, and presence of amino acid end groups in the polymer chain isolated by acid and enzymatic hydrolysis from the collagen substrate.

Electron microscopy of collagen and ultrathin sections of goat skin grafted with PMMA, PMA, and PBA did not have cross striations when the product contained a large amount of added polymer (9). These observations indicate that the polymer formed on the collagen has penetrated into the fibrils, resulting in the masking of the cross striations, and that the polymer may be bound chemically to the collagen molecules.

The general behavior of equilibrium swelling of collagen–methyl methacrylate and collagen–methyl acrylate copolymers in a series of solvents comprising chlorinated hydrocarbons, ketones, and esters as well as the change of intrinsic viscosity of these copolymers in dichlorobenzene on addition of benzene, provides evidence that indeed a collagen graft copolymer has been formed (8).

Furthermore, soluble collagen grafted with polyacrylamide formed fibrils on heating to 37°C at neutral pH, but unlike the native collagen, these fibrils did not redissolve on cooling to 2°C (8). These results indicate that the redispersion property of soluble collagen is impaired, probably by attachment of the polyacrylamide side chains to the collagen molecule.

Turbidimetric titrations of collagen, PMMA, a mixture of the two, and graft copolymer of these two components dissolved in dichloroacetic acid and precipitated with diisopropyl ether, show rapid discontinuous precipitation of the physical mixture of the collagen and PMMA fractions (8). In the case of the collagen graft copolymer, the titration curve is more or less continuous and no well-marked inflection was observed. The solubility of the collagen graft copolymer is intermediate between those of the corresponding polymers, a behavior which is similar to the solubility characteristics displayed by other graft copolymers.

On hydrolysis of the collagen–methyl methacrylate graft copolymer, the water-insoluble component of the hydrolyzate had solubility characteristics similar to PMMA and a viscosity average molecular weight of 1,400,000 (62). It gave a positive ninhydrin test for protein which was not obtained from a residue after enzymatic digestion of a physical mixture of PMMA and collagen (8). In the PMMA isolated by acid hydrolysis, the characteristic amide IR absorption bands were not seen prominently. However on enzymatic hydrolysis of the grafted substrate, the absorption bands for the amide group are very obvious (9). This should be expected since proteolytic enzymes give rise to longer fragments attached to the grafted polymer.

Grafts isolated by pepsin and trypsin digestion had stronger amide absorption bands than after pronase digestion. Pronase, because of its broad specificity, is capable of hydrolyzing the collagen trunks more extensively than the other two enzymes. Thus, the grafts isolated by pepsin and trypsin have longer fragments of the collagen trunk attached to the ends of grafted polymer chains. The nitrogen contents in the different isolated grafts were in the following order: pepsin > trypsin > pronase > acid hydrolysis. The failure of the grafts isolated by acid hydrolysis to show the characteristic amide bands should be attributed to the fact that the molecular weight of PMMA branches were too high and the attached amino acid residues were too few to be detected by IR.

Dinitrophenylation of the amino end groups of the isolated grafts confirmed the presence of amino acid residues in the water-insoluble hydrolyzate. Spectrophotometric analysis of the dinitrophenylated hydrolyzates dissolved in ethyl acetate indicates that, depending on the conditions used in the original graft polymerization, the method of hydrolysis of the copolymer (acid or enzymatic), and the molecular weight of the isolated grafts, the number of dinitrophenylated amino acid end groups varied from trace amount to one per PMMA side chain (*8, 62*).

Literature Cited

1. Arthur, J. C., "Cellulose Graft Copolymers," ADVAN. CHEM. SER. (1969) **91**, 574.
2. Arthur, J. C., "Graft Polymerization onto Polysaccharides," in *Advan. Macromol. Chem.*, W. Pasika, Ed., **2**, 1, Academic, London, 1970.
3. Rao, K. P., Joseph, K. T., Nayudamma, Y., *J. Sci. Ind. Res. (India)* (1970) **29**, 1.
4. Rao, K. P., Joseph, K. T., Nayudamma, Y., *Leather Sci. (Madras)* (1967) **14**, 73.
5. Rao, K. P., Joseph, K. T., Nayudamma, Y., *Leder* (1968) **19**, 77.
6. Rao, K. P., Joseph, K. T., Nayudamma, Y., *Leather Sci. (Madras)* (1969) **16**, 401.
7. Rao, K. P., Joseph, K. T., Nayudamma, Y., *J. Sci. Ind. Res.* (1970) **29**, 559.
8. Rao, K. P., Joseph, K. T., Nayudamma, Y., *J. Polym. Sci. Part A-1* (1971) **9**, 3199.
9. Rao, K. P., Joseph, K. T., Nayudamma, Y., *J. Appl. Polymer Sci.* (1972) **16**, 975.
10. Korn, A. H., Feairheller, S. H., Filachione, E. M., *J. Amer. Leather Chem. Ass.* (1972) **67**, 111.
11. Feairheller, S. H., Taylor, M. M., Harris, E. H., Jr., Korn, A. H., Filachione, E. M., *Amer. Chem. Soc., Div. Polym. Chem., Prepr.* **13** (2), 736 (New York, August, 1972).
12. Korn, A. H., Taylor, M. M., Feairheller, S. H., *J. Amer. Leather Chem. Ass.* (1973) **68**, 224.
13. Harris, E. M., Taylor, M. M., Feairheller, S. H., *J. Amer. Leather Chem. Ass.* (1974) **69**, 182.
14. Kudaba, J., Ciziunaite, E., *Mater Yubilenoi Respub. Nauch.-Tekh. Konf. Vop. Issled. Primen. Polim. Mater., 8th, Vilnyus* 1966, 289; *Chem. Abstr.* (1968) **68**, 78969.
15. Kudaba, J., Ciziunaite, E., *Liet. TSR Aukst. Mokyklu Mokslo Darb., Chem. Chem. Technologija* (1967) **8**, 157; *Chem. Abstr.* (1969) **70**, 12225.
16. Kudaba, J., Ciziunaite, E., *Liet. TSR Aukst. Mokyklu Mokslo Darb., Chem. Chem. Technol.* (1969) **9**, 167; *Chem. Abstr.* (1970) **72**, 32428.

17. Kudaba, J., Ciziunaite, E., Mikelionyte, A., *Liet. TSR Aukst. Mokyklu Mokslo Darb., Chem. Chem. Technol.* (1969) **10**, 153; *Chem. Abstr.* (1970) **73**, 4980.
18. Kudaba, J., Ciziunaite, E., Gribauskiene, E., Martinkiene, A., *Polim. Mater. Ikh. Issled.* (1969) 7; *Chem. Abstr.* (1971) **75**, 78166.
19. Kudaba, J., Ciziunaite, E., Tsimmperman-Meskauskite, I., *Polim. Mater. Ikh. Issled.* (1969) 3; *Chem. Abstr.* (1971) **75**, 89087.
20. Kudaba, J., Ciziunaite, E., Gribauskiene, E., *Liet. TRS Aukst. Mokyklu Mokslo Darb., Chem. Chem. Technol.* (1971) **13**, 215; *Chem. Abstr.* (1973) **78**, 17654.
21. Martinkiene, A., Kudaba, J., Ciziunaite, E., *Polim. Mater. Ikh. Issled., Mater. Respub. Nauch.-Tekh. Konf.,* 12th (1971) 118; *Chem. Abstr.* (1973) **78**, 99075.
22. Kudaba, J., Ciziunaite, E., Martinkiene, A., *Polim. Mater. Ikh. Issled.* (1969) 11; *Chem. Abstr.* (1971) **75**, 78165.
23. Kudaba, J., Ciziunaite, E., Martinkiene, A., *Liet. TSR Aukst. Mokyklu Mokslo Darb., Chem. Chem. Technol.* (1971) **13**, 221; *Chem. Abstr.* (1973) **78**, 73685.
24. Kudaba, J., Ciziunaite, E., Martinkiene, A., *Liet. TSR Aukst. Mokyklu Mokslo Darb., Chem. Chem. Technol.* (1970) **12**, 267; *Chem. Abstr.* (1972) **76**, 100458.
25. Kudaba, J., Ciziunaite, E., Cimmperman-Meskauskaite, I., *Polim. Mater. Ikh. Issled. Mater. Respub. Nauch.-Tekh. Konf.,* 12th (1971) 123; *Chem. Abstr.* (1973) **78**, 85958.
26. Meskauskaite, I., Kudaba, J., Ciziunaite, E., *Liet. TSR Aukst. Mokyklu Mokslo Darb., Chem. Chem. Technol.* (1969) **10**, 165; *Chem. Abstr.* (1970) **73**, 4979.
27. Kudaba, J.,Ciziunaite, E., Jonutiene, D., *Liet. TSR Aukst. Mokyklu Mokslo Darb., Chem. Chem. Technol.* (1969) **10**, 147; *Chem. Abstr.* (1970) **73**, 4982.
28. Kudaba, J., Ciziunaite, E., Lukaitis, I., *Polim. Mater. Ikh. Issled. Mater. Respub. Nauch.-Tekh. Konf., 12th* (1971) 134; *Chem. Abstr.* (1973) **78**, 85960.
29. Kudaba, J., Ciziunaite, E., Cimmperman-Meskauskaite, I., *Liet. TSR Aukst. Mokyklu Mokslo Darb., Chem. Chem. Technol.* (1971) **13**, 229; *Chem. Abstr.* (1973) **78**, 73686.
30. Okamoto, K., Yamamoto, T., *Kogyo Kagaku Zasshi* (1971) **74**, 527; *Chem. Abstr.* (1971) **75**, 22524.
31. Masuhara, E., *Deut. Zahnärzt. Z.* (1969) **24**, 620.
32. Kudaba, J., Ciziunaite, E., Grybauskiene, E., *Liet. TSR Aukst. Mokyklu Mokslo Darb., Chem. Chem. Technol.* (1969) **10**, 159; *Chem. Abstr.* (1970) **73**, 4981.
33. Kudaba, J., Ciziunaite, E., Grybauskiene, E., *Liet. TSR Aukst. Mokyklu Mokslo Darb., Chem. Chem. Technol.* (1970) **12**, 261; *Chem. Abstr.* (1972) **76**, 87198.
34. Kudaba, J., Ciziunaite, E., Grybauskiene, E., Baronaite, E., *Liet. TSR Aukst. Mokyklu Mokslo Darb., Chem. Chem. Technol.* (1971) 13; *Chem. Abstr.* (1973) **78**, 73687.
35. Kudaba, J., Ciziunaite, E., Jakubenaite, V., *Liet. TSR Aukst. Mokyklu Mokslo Darb., Chem. Chem. Technol.* (1971) **13**, 203; *Chem. Abstr.* (1973) **78**, 73674.
36. Kudaba, J., Ciziunaite, E., Jakubenaite, V., *Zh. Vses Khim. Obshchest* (1969) **14**, 596; *Chem. Abstr.* (1970) **72**, 45046.
37. Kudaba, J., Ciziunaite, E., Jakubenaite, V., *Liet. TSR Aukst. Mokyklu Mokslo Darb., Chem. Chem. Technol.* (1971) **13**, 209; *Chem. Abstr.* (1973) **78**, 73684.

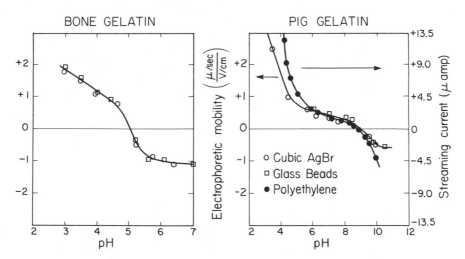

Figure 1.   Charge vs. pH profiles at 24° ± 1°C for gelatins adsorbed to the substrates indicated

IEP also signifies other physical characteristics such as the pH of minimum swelling, osmotic pressure, viscosity, etc. (5). In these studies several techniques were used to show the variations in the IEP as a function of gelatin type and its derivatives as well as the IEP dependence on hydrolysis and ionic strength.

Experimental. The gelatin solutions were prepared as 0.75–1.0 wt % in distilled water. The samples were made from the dry stock daily by heating the gelatin–water mixture to 40°C for complete dissolution and then cooling to 24 ± 1°C for the IEP determinations.

Measurements were carried out using three methods: (a) A Zeta meter (Zeta Meter, Inc., New York) was used for determining electrophoretic mobilities. The gelatin was first coated on glass beads or AgBr grains, and the mobility of these particles was measured as a function of pH according to established procedures (6).

(b) A streaming current detector (SCD) (Waters Associates, Framingham, Mass.) was used to measure the alternating current flow resulting from the reciprocating movement of the piston. The streaming current originates from the shearing of the double layer associated with charged particles or films adsorbed to the polyethylene walls of the sample container and is detected across two platinum electrodes placed at opposite ends of the sample cavity. The solution pH was continuously monitored in the SCD cup with a miniature pH combination electrode (Sargent-Welch). The experimental error of the IEP is ~ ±0.2 pH unit. For a more complete description of the SCD, introductory articles by the developers of the instrument are available (7, 8).

(c) Using the technique of Janus, Kenchington, and Ward (9), isoelectric points were also determined by passing gelatin solutions through a mixed-bed, ion-exchange resin and measuring the resultant pH.

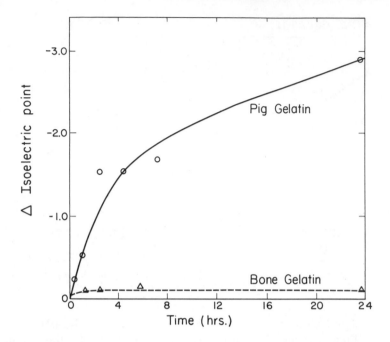

Figure 2.   Change in isoelectric point vs. time of alkaline hydrolysis (pH 10.4 at 24° ± 1°C) of 0.75% solution of indicated gelatin

Results and Discussion. The pH-charge profiles of two types of gelatin at constant temperature are shown in Figure 1. The electrophoretic mobility data for bone gelatin adsorbed on AgBr and on glass beads can be represented by a single curve. This observation agrees with the general conclusion that adsorbed gelatin, independent of the substrate composition, determines the electrokinetic behavior (10). The electrokinetic results for pig gelatin show a similar independence of substrate. Streaming-current data for pig gelatin on polyethylene substrate are also plotted on Figure 1. The same IEP was observed for pig gelatin regardless of the technique or substrate used. Table I lists the pH of the IEP for several gelatin types. Acylation will change gelatin properties (11), and the IEP of these derivatives ranged from pH = 3.9 for a phthalated gelatin (12) to 10.6 for a lysylated gelatin (13). Good agreement was observed between the electrokinetically determined IEP and those obtained by measuring the pH of gelatin solutions after treatment with mixed-bed ion-exchange resins (9). However, streaming current determinations were convenient for measuring changes in IEP as a function of hydrolysis time and ionic strength.

The IEP of acid-treated pig gelatin (8.9) rapidly decreased in alkaline solution (Figure 2) whereas no such change was observed for the limed bone gelatin (4.8) under the same conditions. This IEP decrease was probably caused by base hydrolysis of the acid-derived gelatin. Such hydrolysis (*14, 15*) involves deamination of the amide groups to ionizable carboxyl groups with evolution of amines, thus shifting the IEP to a lower pH. The stability of the limed bone gelatin is attributed to the presence of fewer hydrolyzable amide groups which are prevalent in acid-treated gelatins (*1*).

In contrast to the effect of alkaline hydrolysis of pig gelatin, enzymatic hydrolysis (HT Proteolytic Takamine) of a 1% bone gelatin solution did not alter the IEP. The viscosity of the enzyme-treated gelatin was also observed with time at 25°C using an Ostwald viscometer (Figure 3). Since the viscosity of gelatin may be related to its molecular weight (*14, 15*), the observed decrease suggested the rapid breaking of the gelatin macromolecule into smaller fragments but without changing the IEP. This tentative conclusion assumes there is no selective adsorption of gelatin fragments to the instrument surfaces during the streaming

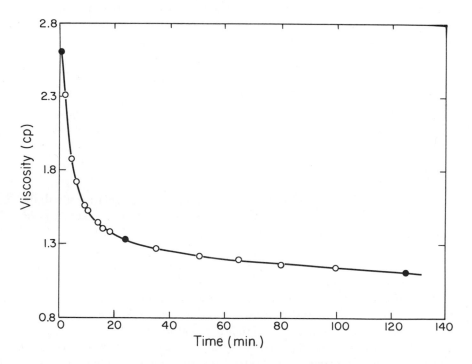

Figure 3.   Viscosity vs. *time for 1% limed bone gelatin solution treated at 24° ± 1°C with 0.032% HT Proteolytic Takamine enzyme*
The solid circles indicate measured isoelectric points of pH 4.8

Table I.   pH of Isoelectric Point, 24° ± 1°C

Gelatin Type	Electrophoretic Mobility	Streaming Current	Ion Exchange
Polylysyl	10.6	—	—
Pig	9.1	8.9	8.8
Bone	5.0	4.8	4.8–5.0
Phthalyl	3.9	4.1	4.1

current determinations. This qualification applies to all electrokinetic measurements including the electrophoresis data in Table I.

Variations in ionic strength are reported to exert a particularly large effect on the IEP of acid-treated gelatins (14, 16). The bone gelatin was insensitive to changes of ionic strength (addition of KNO_3) and exhibited a decrease in IEP of 0.4 pH unit in $0.1M$ KNO_3. However, under identical conditions the pigskin gelatin gave IEP decreases of 3 pH units and no longer exhibited the weakly positive plateau shown in Figure 1. In view of the indicated lability of this gelatin towards hydrolysis, this reaction may be promoted by inert electrolytes which would screen the few ionic charges associated with this gelatin in the approximate pH range of 5.5–9 (Figure 1). However, an irreversible, salt-promoted reaction was ruled out when, after salt removal by dialysis, the initial IEP of the gelatin was again observed (Table II).

Table II.   Effect of Salt on the Isoelectric Point at 24° ± 1°C

Gelatin Type	Original Sample	$0.1M$ KNO_3	After Dialysis
Bone	4.9	4.5	4.9
Pig	8.7	5.9	8.7

Many of the studies of ionic strength dependence of gelatins attribute the decrease in IEP to specific-ion adsorption (14). More recent data of Donnelly et al. (16) showed the IEP to be linearly dependent on the square root of ionic strength. They suggested that the binding of ions is related to the interaction of the total ionic atmosphere with sites of proton binding in the protein molecule rather than to a semistoichiometric binding. Donnelly et al. (16) chose the square root function on the basis of the effect of ionic strength on dissociation according to the Henderson–Hasselbach equation and Debye–Hückel variation in activity coefficients. The authors also pointed out that the straight lines may be mere artifacts. Their data for dialyzed, acid-treated gelatin gave a slope of -13.7 (IEP/$\sqrt{\mu}$) over the ionic-strength range of zero to 0.14 (NaCl). Hörmann and Ananthanarayanan (17), using higher NaCl concentrations of

0.1–0.6*M,* also found a linear decrease but with the much smaller slope of −2.5 (IEP/$\sqrt{\mu}$). To verify these differences, the IEP of deionized pigskin gelatin was determined over the concentration 0–0.79*M* KNO$_3$. KNO$_3$ was chosen rather than NaCl because of the application to AgBr dispersions which frequently contain KNO$_3$ as a byproduct. The data (Figure 4) show two distinct slopes with a break point at approximately 0.35 $\sqrt{\mu}$, almost identical with the combined data of Donnelly *et al.* (*16*) and Hörmann and Ananthanarayanan (*17*). No explanation is offered for this behavior. A similar test for bone gelatin gave a linear decrease in IEP from 4.8 to 4.0 with ionic strength changing from *ca.* 0.0001μ to 1μ (KNO$_3$). Both bone and pigskin gelatins extrapolated to 1*M* KNO$_3$ gave the same IEP of 4.0.

pH Dependence of AgBr Grain Growth

Although the influence of gelatin charge on the crystal habit, size, and dispersity of AgBr can be evaluated at fixed pH by varying the gelatin type, in this investigation we used a single type of gelatin and varied the charge by changing the pH of the solution. The gelatin used was a de-ashed, lime-processed bone with IEP at pH 4.8.

An important function of gelatin is to peptize AgBr. A schematic drawing of the well known, double-jet technique (*18, 19, 20*) is presented

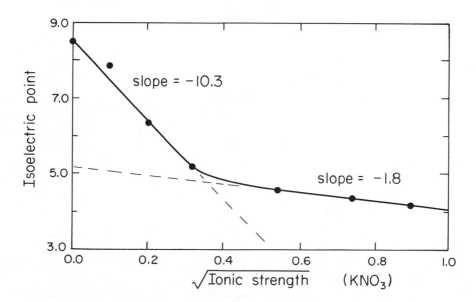

Figure 4.   Isoelectric point of deionized pigskin gelatin in KNO$_3$ solution at
24° ± 1°C

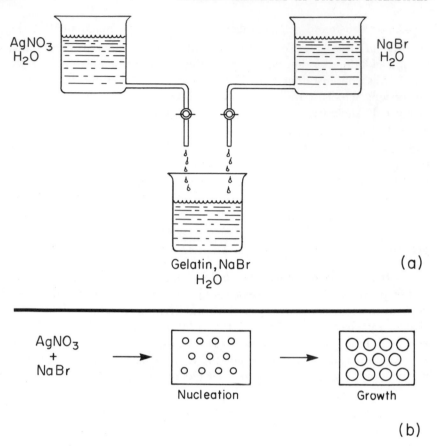

Figure 5.   Schematic representation of double-jet precipitation of AgBr–
gelatin dispersion
(a) addition of reactants, (b) nucleation and growth stages of AgBr formation

in Figure 5. The gelatin is contained in the precipitation kettle along with a fixed concentration of NaBr. The importance of the bromide ion concentration, $[Br^-]$, is well established (18, 19, 20, 21). Figure 6 provides a summary in the form of electron micrographs of AgBr microcrystals: regular cubes are obtained at $[Br]^- < 10^{-3}M$, octahedra between $10^{-2}M$ and $10^{-3}M$ and a variety of hexagonal and triangular tablets at concentrations above $10^{-2}M$. Some AgBr tablets may exhibit unusually high growth rates in certain directions unlike the regular cubes and octahedra which grow at the same rate in all directions (22).

Results of some double-jet precipitations (Figure 7) which led to the present investigation show that at a given pBr ($-\log a_{Br^-}$) both the morphology and size of the AgBr grains are influenced by pH. From

Cubes   Octahedra   Tablets

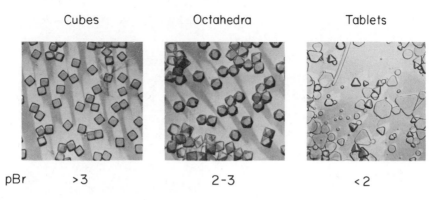

pBr   >3   2-3   <2

Figure 6. Electron micrographs (preshadowed carbon replicas) showing the morphologies of AgBr crystals obtained from double-jet precipitation in bone gelatin at the indicated bromide concentration

other work, the morphology differences may be attributed to changes in peptizer charge (23). At pBr 3.3 ripening activity increased monotonically with pH while at pBr 2 ripening action was at a minimum at pH 6.

Turbidimetric Methods. Established turbidimetric methods (24–29) were used to measure the ripening rates of fine-grain AgBr–gelatin dispersions (Figure 8). The dispersion was prepared by controlled pAg (pBr ~3 at 35°C), double-jet, constant-flow-rate addition of 0.1–0.2

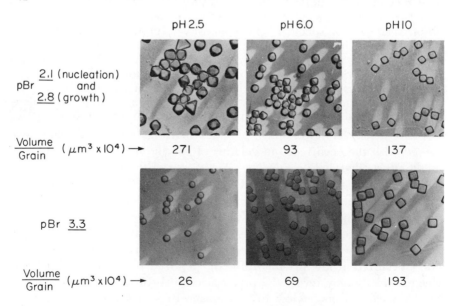

pH 2.5   pH 6.0   pH 10

pBr $\frac{2.1 \text{ (nucleation)}}{2.8 \text{ (growth)}}$ and

$\frac{\text{Volume}}{\text{Grain}}$ ($\mu m^3 \times 10^4$) →   271   93   137

pBr 3.3

$\frac{\text{Volume}}{\text{Grain}}$ ($\mu m^3 \times 10^4$) →   26   69   193

Figure 7. Electron micrographs of AgBr–gelatin dispersions prepared at 70°C in double-jet precipitations under the conditions indicated

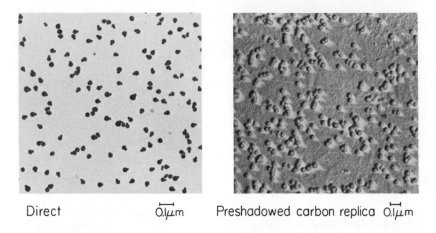

Direct   0.1μm   Preshadowed carbon replica 0.1μm

Figure 8.   Electron micrographs of 0.03 μm AgBr–gelatin dispersion used for turbidimetric studies

mole of $AgNO_3$ and NaBr into 1 l. of 1 or 1.5% de-ashed, lime-processed bone gelatin. This dispersion was stored at pH 6, pBr 3, 23.3mM AgBr, 0.12 or 0.36% gelatin at ∼5°C without significant change in turbidity for at least six weeks. For each kinetic experiment, the dispersion was diluted 1:3 and the pH and pAg were adjusted potentiometrically at 25°C. Turbidities were monitored with a Beckman model B or DK2A in a 1-cm cell jacketed to maintain temperature.

The basis of the light-scatter method is shown in Figure 9. The absorbance (A) or specular optical density is proportional through the turbidity (τ) to the molarity (M) and average volume ($\overline{V}$) of the scattering particles; μ_D is the index of refraction of the medium; m is the index of refraction ratio of the particles to the surrounding medium. Figure 9b shows a trace of absorbance (at a given wavelength, λ, generally 436 nm in the present work) *vs.* time. A ripening rate was calculated from the slope. Rate measurements were reproducible to *ca.* 10%.

Literature Review. Many previous investigations have been concerned with the occurrence in gelatins of specific microcomponents that might affect silver halide ripening (*30*). Some of the more recently published data specifically relate to pH effects on silver halide grain growth. Using turbidimetric methods, Jones (*31*) observed that at the equivalence point of AgBr, ripening rates were faster at high pH both for an acid- and a lime-treated gelatin. In each case, ripening appeared to cease at a pH *ca.* 2 units below the IEP. Hirata (*32*) found that rate acceleration generally occurred at low pH for AgBr/Br⁻ peptized by an inactive gelatin; electron microscopy revealed a preference for AgBr tablets when nucleation was carried out at low pH. Ammann-Brass (*33*) showed that for AgCl/Cl⁻ systems and several types of lime-processed bone gelatins ripening rates were faster at high and/or low pH depending on the gela-

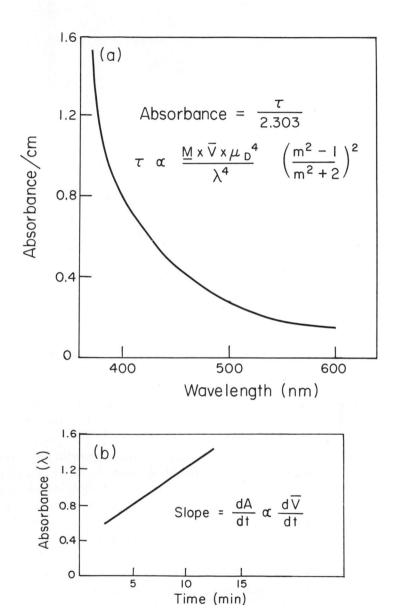

Figure 9.   Use of Rayleigh scatter to measure growth rates of fine-grain AgBr dispersions
(a) absorbance vs. wavelength for 1-cm path of 7.8mM AgBr in 0.12% gelatin solution at room temperature, (b) absorbance at a fixed wavelength (e.g. 436 nm) for a growing AgBr dispersion

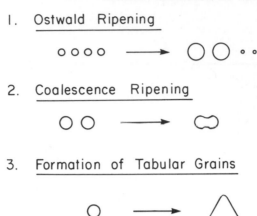

Figure 10.   Schematic representation of three possible mechanisms for AgBr grain growth

tin. The pH dependence could be generated by adding nucleic-acid-growth restrainer to an inert gelatin which alone showed little or no pH dependence of grain growth. Similar studies were carried out by Ohyama (*34, 35*) and by Irie and Ishida (*36*). Klein, Moisar, and Roche (*37*) observed that the rate of ripening of AgBr in an inert bone gelatin increased monotonically with pH at pBr ~2 (that is, for AgBr with adsorbed Br⁻ (*38*)) and decreased with increasing pH at pAg ~3. Contrary to the interpretations of Ammann-Brass (*33*) and Ohyama (*34, 35*), the latter workers attributed the observed pH dependence to electrostatic interactions between the peptizer and the silver halide surface.

Figure 10 summarizes schematically the three growth mechanisms that have been considered: (a) *Ostwald ripening*, in which large particles grow at the expense of smaller ones. Factors which lead to increases in AgBr solubility are expected to favor this process (*39, 40*); (b) *Coalescence ripening*, in which two or more particles coagulate and grow as a single unit. Such processes are generally unimportant in the presence of sufficient quantities of good peptizers such as gelatin; (c) *Formation of tabular grains*, which may lead to accelerated growth rates (*22*).

Results and Discussion. Figure 11 shows the dependence of turbidimetrically determined ripening rates on pH at three Br⁻ levels. These data show the same trends as the precipitation results of Figure 7 despite the fact that the latter were carried out at somewhat variable ionic strength, whereas these ripening rates were measured at constant ionic strength ($0.18M$). Therefore, the rate–pH profile at the highest [Br⁻] can not be attributed to increasing ionic strength.

The monotonic increase in rates with pH at low [Br⁻] suggested ripening action by amines. Indeed, the pH dependence of Ag⁺ binding (*41*) by lime-processed calfskin gelatin (Figure 12) bears a striking resemblance to the rate–pH profile at pBr 3.3 (Figure 11). Furthermore, amino acid analysis (*42*) of the gelatin indicates that there is sufficient amine of the arginine, lysine, and histidine types to account for the observed ripening rates—at least for the corresponding low-molecular-weight, unbound amines (*cf.* Figure 12). Additional support for ripening action by amines was obtained by measuring the ripening rate as a function of gelatin concentration. As Figure 13 shows, the rate increased monotonically with gelatin concentration at pH 2, 6, and 10. The effects probably do not arise, as Klein and Moisar have suggested (*37*), from gelatin charge since saturation coverage occurs at or slightly below the lowest level plotted (*11*).

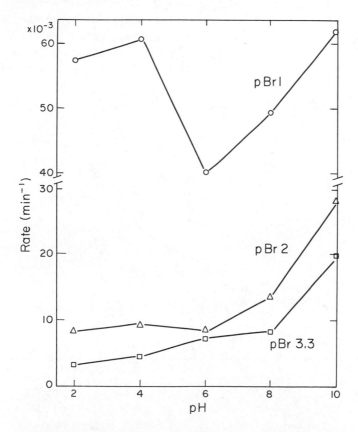

Figure 11. Turbidimetric ripening rates vs. pH at the indicated bromide concentration; all rates measured at 48°C 0.18M ionic strength, 7.8mM AgBr, 0.12% gelatin

A study of the effects of gelatin phthalation on ripening rate also suggests ripening action by amines. Table III shows the rate decreased monotonically with increasing phthalation, *i.e.*, as the amines of the gelatin were converted to the very weakly basic, weakly Ag^+-complexing amides. The fact that the rate decreased only slightly with 3% gelatin phthalation (compared with 5 or 5.9%) suggests that phthalation occurred first with the more basic amines (*e.g.*, arginine and lysine) which at pH 6 should be protonated and inactive in the nonphthalated gelatin. The IEP data further suggest that gelatin charge (*37*) has little bearing on these results.

The specific source of ripening action by gelatin appears to be intimately tied up with the gelatin itself. Thus, prolonged dialysis or washing failed to decrease the ripening rate at pH 2, 6, or 10 by more than ~10%. Neither of these procedures should have resulted in significant loss of the non-gelling component of bone gelatin (*42, 43*). Other experiments verify that rates at high pH do not arise from gelatin degradation (*i.e.*, to give low-molecular-weight amines) (*14, 15*), since preheating the gelatin alone under all but the most extreme conditions of temperature and pH (*e.g.* 70°C, pH 12, 0.18M $NaNO_3$, 1–2 hrs) failed to affect the ripening rate at pH 10.

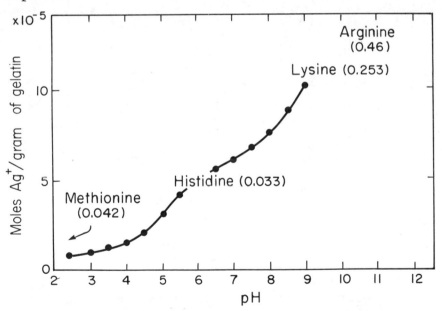

Figure 12.   Dependence on pH of Ag^+ binding by lime-processed calfskin gelatin.   Curve from Ref. 41.

Numbers in parentheses are values for mmoles amino acid/gram bone gelatin.   Data from Ref. 42.

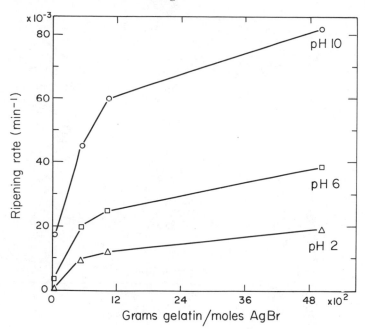

Figures 13. Turbidimetric growth rates vs. gelatin concentration at the indicated pH

All rates measured at 48°C, pBr 3.3, 0.18M NaNO₃, 7.8mM AgBr (1200 g gelatin/mol AgBr corresponds to ~1% gelatin by wt)

The gelatin itself, by virtue of its amine functional groups, is apparently responsible for the observed dependence of ripening rate on pH at low [Br⁻].

To pursue further the nature of the rate increases at low pH and high [Br⁻] (Figure 11), the temperature dependence of ripening at pBr 1 was measured as a function of pH (Figure 14). The Arrhenius plots of Figure 14b indicate the same activation energy, E_a, within experimental error, at all pH values from 2 to 10. Table IV lists E_a for AgBr ripening under a variety of experimental conditions including those of the

Table III. Effect of Gelatin Phthalation on AgBr Ripening Rates
(48°C, pBr 3.3, pH 6, 0.18M NaNO₃)

	Control	Gelatin 1	Gelatin 2	Gelatin 3
Rate (min⁻¹ × 10³)	26	17	8	7
CPA (%)[a]	0	3.01	4.98	5.86
IEP (pH)[b]	4.80	4.30	4.14	4.10

[a] % combined phthalic anhydride (47).
[b] Measured with the Waters streaming current detector (cf. text).

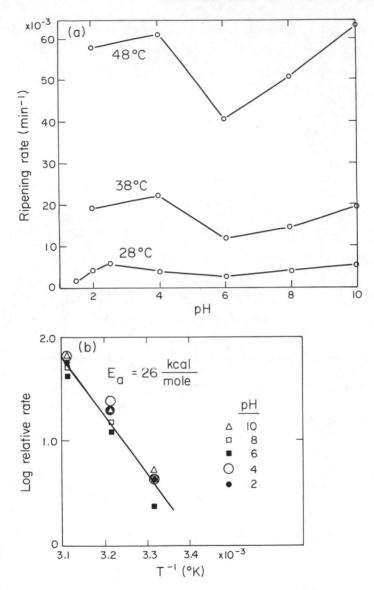

Figure 14.   Dependence of AgBr growth rate at pBr 1
on temperature
(a) rate vs. *pH at indicated temperature, (b) correspond-*
ing Arrhenius plots

present study. These data provide no further insight into the origin
of the ripening processes at low pH and high Br⁻ levels.

One of the earliest suggestions for the rate acceleration at low pH
was that the gelatin might be functioning as a cationic surfactant to in-

Table IV. Activation Energies for AgBr Ripening (kcal/mole)

$pBr1$[a]	$pBr3.3$ $(+0.18M\ NaNO_3)$[a]			$pBr2$[a]
$(pH2-10)$	$(pH2)$	$(pH6)$	$(pH10)$	$(pH6,10)$
26	21	29	26	25

$pBr3.3$ $(sols)$[b]	NH_3[c]	*Ethylene*[c] *Diamine*	KBr[c]	NH_3[d]	
				$(cubes)$	$(oct.)$
24	25	13	8	66	8

[a] This work ($\pm \sim 10-15\%$).
[b] Ref. 25.
[c] Ref. 29.
[d] Ref. 44.

crease the [Br⁻] within the diffuse double layer (45). Although this mechanism might be favored at low [Br⁻], at least for lower-molecular-weight surfactants such as N-dodecylpyridinium ion, the situation could differ in the case of gelatin as surfactant. There are at least two other objections to this mechanism: (a) like the results at lower [Br⁻] (Figure 13), the ripening rates at pBr 1 also increased with gelatin concentration well beyond saturation coverage; (b) AgBr sols prepared and ripened in the absence of any peptizer also showed rate enhancements at low pH and high [Br⁻] (Table V). Compared with gelatin dispersions, the AgBr sols were larger, more polydisperse, and less well defined (Figure 15) and had lower, less precise ripening rates (Table V); nevertheless, it is tentatively concluded that rate acceleration at low pH and high [Br⁻] is independent of gelatin. Experiments with nonionic polymers may support this conclusion under more favorable experimental conditions.

One ripening mechanism favored at high [Br⁻] is related to the formation of tabular grains with their unusually high growth rates (22). Figure 16, in fact, suggests that at least the singly twinned variety of

Table V. Ripening Rates[a] of AgBr Sols at pBr2
(43 °C, 3 × 10⁻⁴M AgBr ± gelatin)

Gelatin, %	pH		
	3	*6*	*10*
0	4.0	2.3	2.0
0.004	4.0	2.9	3.2

	pH		
	3.8	*6*	*10*
0	3.6	2.3	2.3
0.12	5.7	5.1	5.4

[a] Measured turbidimetrically at 350 nm; units are min⁻¹ × 10³; uncertainty $\pm \sim 10\%$.

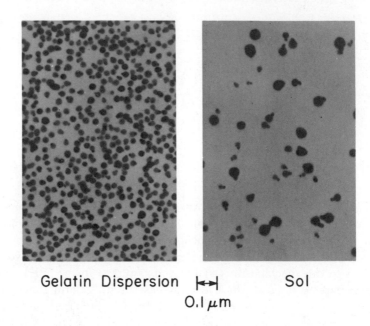

Gelatin Dispersion |←→| Sol
0.1 μm

Figure 15.   Direct electron micrographs of AgBr dispersions
used in turbidimetric studies

(a)   (b)

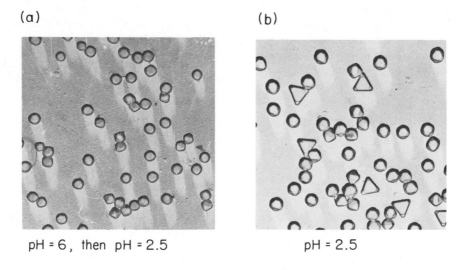

pH = 6, then pH = 2.5   pH = 2.5

Figure 16.   Electron micrographs of AgBr gelatin dispersions grown at 70°C,
pBr 2.8, pH 2.5
(a) nucleation at pBr 2.1, pH 6; (b) nucleation at pBr 2.1, pH 2.5

such crystals may be favored at low pH. Hirata previously observed a preference for such flat crystals at low pH (32). Two lines of evidence, however, argue against the involvement of tabular grain growth: (a) electron micrographs of AgBr–gelatin dispersions ripened at pBr 1 showed no more tabularlike grains at pH 3 than 6; (b) recrystallization at pBr 1.2 of initially well-defined cubic grains showed enhanced activity (rounding) at low as well as high pH despite the obvious absence of any twin planes (Figure 17).

E. P. Przybylowicz of these laboratories suggested a possible pH-dependent formation of the soluble complexes $AgBr_n^{-(n-1)}$ ($n > 1$). If such complexes were sufficiently basic, lowering the pH to the vicinity of the pK_a of the conjugate acid might increase AgBr solubility and thereby the ripening rate. However, solubility measurements (with radioactive tracer) by R. Chapas and D. Shiao of these laboratories failed to show any significant changes in β_2 or β_3 [stability constants for $AgBr_n^{-(n-1)}$ formation] in the pH range 5–1.7.

Although ripening rates of AgBr gelatin dispersions are pH dependent, the charge of the gelatin is probably not the determining factor (37). At low Br$^-$ concentrations, the pH dependence is attributed to the amine moieties of the gelatin; at higher Br$^-$ levels, the rate acceleration at low pH appears to be independent of gelatin.

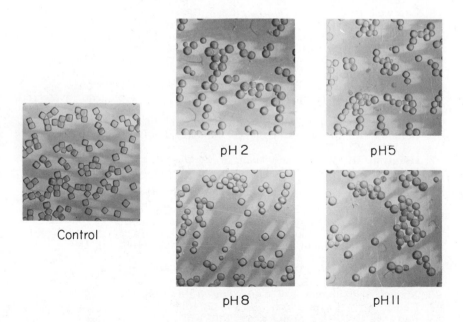

Figure 17. Electron micrographs of AgBr gelatin dispersions recrystallized at 70°C and pBr 1.2 at the indicated pH

Addendum. Although changes in configuration of adsorbed gelatin may account for the pH dependence of fixation time observed by Kragh and Peacock (*46*) such configurational changes alone cannot account for the pH profile of ripening rates, the dependence of ripening rate on gelatin concentration, or the effects of gelatin phthalation.

Acknowledgments

We are indebted to C. R. Berry, D. D. F. Shiao, and P. Bagchi for many helpful discussions; to C. Oster for all the electron micrographs; and especially to R. V. Brady, Jr., for his very competent experimental assistance.

Literature Cited

1. Mees, C. E. K., James, T. H., "The Theory of the Photographic Process," 3rd ed., chap. 3, Macmillan, New York, 1966.
2. Breeuwama, A., Lyklema, J., *J. Colloid Interface Sci.* (1973) **43**, 437.
3. Hardy, W., *J. Physiol.* (1899) **24**, 285.
4. *Ibid.* (1905) **33**, 251.
5. Gustavson, K., "The Chemistry and Reactivity of Collagen," Academic, New York, 1956.
6. Weiss, G., Ericson, R., Herz, A. H., *J. Colloid Interface Sci.* (1967) **23**, 277.
7. Gerdes, W., "12th National ISA Analysis Instrumentation Symposium," Houston, May, 1966.
8. Cardwell, P., *J. Colloid Interface Sci.* (1966) **22**, 430.
9. Janus, J., Kenchington, A., Ward, A., *Research* (London) (1951) **4**, 247.
10. Abramson, H., Moyer, L., Govin, M., "Electrophoresis of Proteins," p. 14, Reinhold, New York, 1942.
11. Curme, H., Natale, C., *J. Phys. Chem.* (1964) **68**, 3009.
12. Yutzy, H., Frame, G., U.S. Patent **2,614,928** (1952).
13. Wilkins, D., Meyers, P., *Brit. J. Exptl. Pathol.* (1966) **47**, 568.
14. Veis, A., "The Macromolecular Chemistry of Gelatin," Academic, New York, 1964.
15. Groome, R., Clegg, F., "Photographic Gelatin," Focal, London, 1965.
16. Rehfeldt, K., McGinnes, Jr., R., Donnelly, T., *J. Colloid Interface Sci.* (1968) **27**, 667.
17. Hörmann, H., Ananthanarayanan, S., *Z. Physiol. Chem.* (1967) **348**, 995.
18. Berry, C. R., Marino, S. J., Oster, Jr., C. F., *Photogr. Sci. Eng.* (1961) **5**, 332.
19. Berriman, R. W., *J. Photogr. Sci.* (1964) **12**, 121.
20. Klein, E., Moisar, E., *Mitt. Forschungslab. Agfa-Gevaert* (1964) **4**, 38.
21. Moisar, E., "Proceedings 6th International Conference on Corpuscular Photography," p. 17, Florence, Italy, July, 1966.
22. Berry, C. R., Skillman, D. C., *Photogr. Sci. Eng.* (1962) **6**, 159.
23. Cohen, J. I., unpublished data.
24. Meehan, E., Beattie, W., *J. Phys. Chem.* (1960) **64**, 1006.
25. *Ibid.* (1961) **65**, 1522.
26. Maillet, A., Pouradier, J., *J. Chim. Phys.* (1961) **58**, 710.
27. Heller, W., *J. Chem. Phys.* (1965) **42**, 1609.
28. Berry, C. R., *J. Opt. Soc. Amer.* (1962) **52**, 888.
29. Berry, C. R., Skillman, D. C., *Photogr. Sci. Eng.* (1969) **13**, 69.
31. Jones, V. V., *Z. Wiss. Photogr.* (1955) **50**, 138.

30. Breslav, Y. A., *Zh. Nauch. Prikl. Fotogr. Kinematogr.* (1970) **15**, 458.
32. Hirata, A., *Nippon Shashin Gakkai Kaishi* (1960) **23**, 71.
33. Ammann-Brass, H., "Proceedings 2nd Symposium on Photographic Gelatin," p. 251, Cambridge, August–September, 1970.
34. Ohyama, Y., "Proceedings 2nd Symposium on Photographic Gelatin," p. 251, Cambridge, August–September, 1970.
35. Ohyama, Y., "Proceedings International Conference on Science and Applications of Photography," p. 37, London, September, 1953.
36. Irie, H., Ishida, T., "Proceedings 2nd Symposium on Photographic Gelatin," p. 293, Cambridge, August–September, 1970.
37. Klein, E., Moisar, E., Roche, E., *J. Photogr. Sci.* (1971) **19**, 55.
38. Herz, A. H., Helling, J. O., *J. Colloid Sci.* (1962) **17**, 293.
39. Wagner, C., *Zh. Elektrochem.* (1961) **65**, 581.
40. Berry, C. R., Skillman, D. C., *J. Photogr. Sci.* (1968) **16**, 137.
41. Carroll, B. H., Hubbard, D., *J. Res. Nat. Bur. Stand.* (1931) **7**, 811.
42. Rose, P. I., Research Laboratories, Eastman Kodak Co., personal communication.
43. Johnson, P., Metcalfe, J. C., *Eur. Polym. J.* (1967) **3**, 423.
44. Claes, F. H., Borginon, H., *J. Photogr. Sci.* (1973) **21**, 155.
45. Oppenheimer, L. E., James, T. H., Herz, A. H., "Symposium on Particle Growth in Suspensions," p. 159, Brunel University, April, 1972.
46. Kragh, A. M., Peacock, R., *J. Photogr. Sci.* (1967) **15**, 220.
47. Judd, M., Eastman Kodak Co., private communication.

RECEIVED June 7, 1974.

10

Adsorption of Blood Proteins onto Polymer Surfaces

SUNG WAN KIM and RANDY G. LEE

Department of Applied Pharmaceutical Sciences and Division of Materials Science and Engineering, University of Utah, Salt Lake City, Utah 84112

Adsorption of albumin, γ-globulin, and fibrinogen from single solutions onto several hydrophobic polymers was studied using internal reflection IR spectroscopy. The adsorption isotherms have a Langmuir-type form. The calculated rate and amount of protein adsorbed was dependent on the polymer substrate and the flow rate of the solution. Competitive adsorption experiments were also investigated to determine the specific adsorption of each ^{125}I-labelled protein from a mixture of proteins. Platelet adhesion to these proteinated surfaces is discussed in relation to a model previously proposed.

Although the adsorption of polymers onto solid surfaces has been thoroughly studied (*1*), relatively few studies can be found in the literature on the adsorption of proteins onto polymer surfaces. In 1905, Landsteiner and Uhliz (*2*) discussed the interaction of serum proteins with synthetic surfaces. Blitz and Steiner (*3*) showed that albumin adsorption onto solid surfaces increased with increasing albumin concentration and that adsorption was nearly irreversible. Hitchcock reported (*4*) that adsorption of egg albumin onto collodion membranes followed a Langmuir isotherm with maximum adsorption occurring near the isoelectric point. Later, Kemp and Rideal (*5*) reported that protein adsorption onto solids conforms with Langmuir adsorption.

There is good evidence that proteins adsorbed from solutions onto some solid surfaces form films in which the molecules remain predominantly in the compact configuration, are unoriented at the interface (*6, 7*), or may not unfold at liquid/solid interfaces (*8*). Protein adsorption onto synthetic polymer surfaces is important because of its possible involvement in the initial stages of the interaction of blood with implantable polymeric materials. Baier and Dutton (9) showed that proteins were rapidly adsorbed from blood after exposure; they contended that the protein layer transmits the particular characteristics of a given surface to the blood. Ellipsometry has proved to be a highly sensitive technique for studying the kinetics of protein adsorption onto solid surfaces exposed to blood plasma. Vroman and Adams have used a recording ellipsometer to monitor the adsorbed amount and rate of protein adsorption at a solid–plasma interface (10). Recently, Smith *et al.* (11) measured the amount of adsorbed protein *vs.* time on metal and polyethylene surfaces by ellipsometric techniques. Electron microscopy (12) has been used to count directly individual fibrinogen molecules adsorbed onto mica surfaces. Both methods are difficult to apply to polymer substrates because proteins and polymers have similar refractive indexes, and many polymer surfaces are too rough for accurate measurement. Hoffman and his group (13) have been studying adsorption onto hydrogel substrates by using radiolabelled proteins. Hydrophobic polymers such as polypropylene and poly(dimethyl siloxane) adsorb greater amounts of protein than do grafted surfaces of poly(hydroxyethyl methacrylate) gel (14).

Studies of adsorption of albumin, γ-globulin, and fibrinogen onto several polymer membranes including cation exchangers have been carried out by Dillman and Miller (15), and two types of adsorption were characterized. One is relatively hydrophilic, exothermic, and easily reversible with a heat of adsorption of *ca.* −10 kcal/mole. The other is apparently tightly bound, hydrophobic, and endothermic with heats of adsorption ranging from 5 to 20 kcal/mole.

Mattson and Smith (16) observed that the adsorption depends on the surface charge and that negative charges enhance protein adsorption at the solid–solution interface. IR internal reflection spectroelectrochemistry was applied for this experiment. A theoretical approach has been attempted by Levine (17), who explored the thermodynamics of adsorbed protein films in terms of surface charge and surface energy. The conformation of adsorbed proteins was measured by the IR bound fraction method, and the results indicated that the internal bonding of these globular proteins is sufficient to prevent changes in the structure while adsorbed, even at low surface populations (18).

We obtained adsorption kinetic data and isotherm curves by determining the amount of protein adsorbed onto polymers by internal reflection IR spectroscopy which gave good reproducibility (19). The adsorption character was consistent with the Langmuir adsorption type and adsorption *vs.* time curves showed the expected plateau usually found in macromolecular adsorption. A competitive adsorption study is being carried out using radiolabelled proteins; the surface composition of each adsorbed protein and its respective equilibrium time (which are significantly different from the adsorption times obtained for single protein solutions) on various polymer surfaces have been obtained from this experiment (*20*).

Several investigators have found that foreign surfaces, when exposed to blood, adsorb plasma proteins (*21, 22*). Platelet adhesion to this proteinated surface is the first observable event in thrombosis on the foreign surface. If we consider the proteinated surface as an acceptor in a platelet adhesion mechanism (*23*), the significance of the nature and composition of adsorbed protein is its role in platelet adhesion (*24*).

Experimental

Proteins used in this study are albumin (bovine, crystalline, Nutritional Biochemical Co.), γ-globulin (bovine, Fraction II, Nutritional Biochemical Co.), prothrombin (bovine, Fraction III-2, Nutritional Biochemical Co.), ^{125}I-γ-globulin and ^{125}I-fibrinogen (New England Nuclear Co.), and ^{125}I-albumin (Squibb Co.).

Substrate materials included poly(dimethyl siloxane)(SR) (Silastic Rubber, medical grade, Dow Corning), fluorinated ethylene/propylene copolymer (FEP) (Teflon FEP, Dupont), and a segmented copolyether–urethane–urea (PEUU) based on poly(propylene glycol), methylene bis-4-phenylisocyanate, and ethylenediamine. This PEUU was provided by D. J. Lyman.

Static Kinetic Adsorption. This study was carried out in 60 mg % solutions which had been degassed by aspiration before the protein was

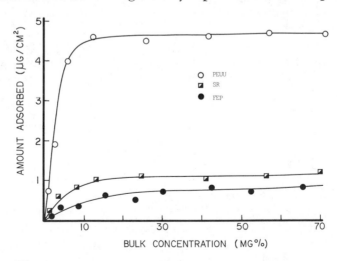

Figure 1.   Adsorption isotherm of albumin onto polymer surfaces

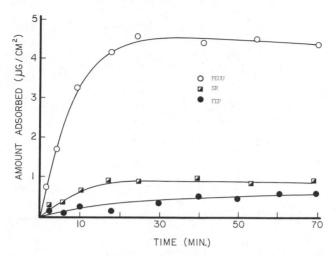

Figure 2.   Adsorption of albumin from a 60 mg % solution onto polymer surface

introduced. Following adsorption (1 hr for the isotherm samples), polymer samples were clamped into designed cells where they were rinsed at a flow rate of 10 ml/sec for 1 min. This procedure was necessary to remove all excess clinging solution. They were then air dried in a dust-free hood. Flow adsorption was conducted the same way except that the protein solution was passed across the sample in the cell by a previously calibrated roller pump. The amount of protein adsorbed onto the polymer surfaces was determined by internal reflection spectroscopy (IRS) using a Perkin-Elmer 521 Grating IR spectrometer and a Wilks IRS attachment with germanium crystals ($1 \times 20 \times 50$ mm).

A net adsorption value (corrected for any IRS prism contamination) which could be related to calibration curves for surface concentration was obtained from the ratio of the amide I ($C=O$ stretching) band at 1640 cm^{-1} to a standard band for each substrate polymer. The standard bands used were the CH_3 bending vibration at 1400 cm^{-1} for poly(dimethyl siloxane) and the CF_2 scissoring deformation at 1450 cm^{-1} for fluorinated ethylene/propylene copolymer. Because of the amide I band in polyurethanes, a $4\times$ expanded abscissa was used; for the other two polymers, the $1\times$ scale was sufficient.

Competitive Adsorption. For these studies mixtures of albumin, γ-globulin, and fibrinogen at a physiological concentration ratio (with one of the proteins labelled with ^{125}I) were studied to determine the surface composition of each adsorbed protein and their respective equilibrium times on selected polymer surfaces. The solution consisted of 50 mg of albumin, 30 mg of γ-globulin, and 15 mg of fibrinogen in 200 ml of phosphate-buffered saline (pH 7.4). An adsorption cell with a washing compartment was used in this experiment (25). This design avoids an air–water interface, controls the washing time to remove excess clinging solution, is convenient in manipulation, and allows one to obtain kinetic

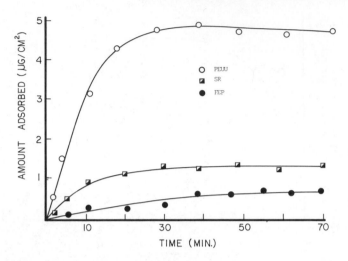

Figure 3.   Adsorption of γ-globulin from a 60 mg % solution onto polymer surfaces

adsorption data.   A Nuclear Chicago 2-channel liquid scintillation spectrometer was used to determine radioactivity on polymer surfaces.

Results

Adsorption of Single Protein System. One minute of rinsing (at a flow rate of 10 ml/sec) was sufficient to reach a constant surface con-

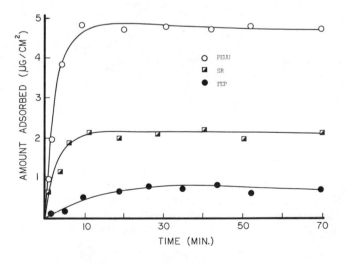

Figure 4.   Adsorption of prothrombin from a 60 mg % solution onto polymer surfaces

Table I.   Protein Adsorption onto Various Polymer Surfaces

Surface: Protein	Isotherm Plateau Bulk Conc., mg %	Plateau Time, min	Plateau Concentration, μg/cm²	Rate Constant, min⁻¹
SR:				
albumin	12	25	1.0	0.13
γ-globulin	25	30	1.3	0.15
prothrombin	15	10	2.3	0.67
FEP:				
albumin	30	60	0.55	0.044
γ-globulin	30	60	0.80	0.083
prothrombin	30	25	0.85	0.19
PEUU:				
albumin	15	25	4.5	0.11
γ-globulin	15	30	4.7	0.23
prothrombin	15	10	4.7	0.38

centration value except for FEP where 2 min were needed (*19*). Static protein adsorption isotherms were determined for PEUU and for the two control surfaces, SR and FEP. The results with albumin are shown in Figure 1. The amount of adsorbed protein at saturation is considerably higher on the PEUU surface than on the other surfaces. The adsorption isotherms were verified as nearly Langmuir types (*19*). The experiments determining protein adsorption kinetics were carried out under static conditions, and the kinetic adsorption curves are shown in Figures 2, 3, and 4. Plateau concentrations, plateau times, and rates of adsorption calculated from the first order reaction equation are given in Table I. FEP has a much longer plateau time and lower surface concentration with albumin than does PEUU with SR intermediate.

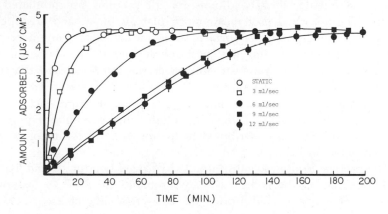

Figure 5.   Albumin adsorption onto PEUU at various flow rates

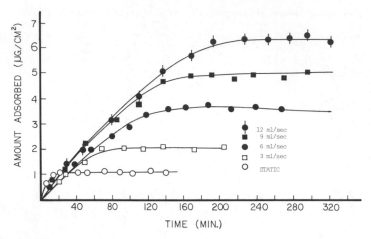

Figure 6.   Albumin adsorption onto SR at various flow rates

Similar experiments were also conducted under varying flow conditions. The albumin adsorptions onto SR and PEUU under flow conditions are shown in Figures 5 and 6, respectively. The plateau concentration depends on the flow rate when SR is the substrate, but not with PEUU as substrate. Scanning electron micrographs showed that SR surfaces are very rough, and those of PEUU are the smoothest of the three. FEP, intermediate in surface roughness, was intermediate in dependence of plateau concentration on flow.

A rough surface can provide a greater surface area and thus more anchoring sites for the greater numbers of protein molecules in the vicinity with the increased flow rate. The plateau concentrations and plateau times under flow conditions for all proteins are tabulated in Ref. *19*. The adsorbed protein amounts, plateau times, and adsorption rates depend on the polymer surface, and limiting factors might be the ability of hydrogen bonding and hydrophobic interaction between adsorbate and adsorbent, water structuring at the interface, and the configurational entropy of the proteins at the adsorbed sites. An overall theory is difficult to formulate since any of these multiple factors can explain the data. In general, the higher rates and amounts of proteins adsorbed onto PEUU are attributed to hydrogen bond formation which is not available on FEP.

The protein binding to substrate *via* hydrogen bonding or *via* hydrophobic bonding has been discussed by Vroman (*26, 27*). In his study on the effect of protein adsorption on the wettability of a surface, he noted that adsorption decreased the wettability of glass but increased the wettability of Lucite. This difference in the behavior is dependent

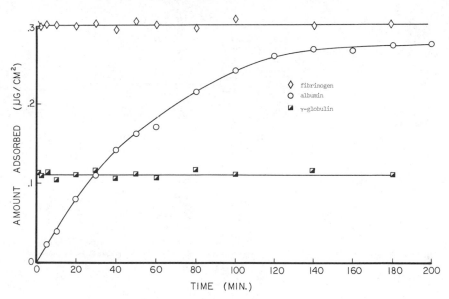

Figure 7. Competitive adsorption of plasma proteins onto FEP

on the capability of the protein to form hydrogen bonds or hydrophobic bonds at a surface. We found (28) that a protein adsorbed onto a hydrophobic polymer surface may expose its more hydrophilic sites if the hydrophobic bonding occurs between the protein and the surface.

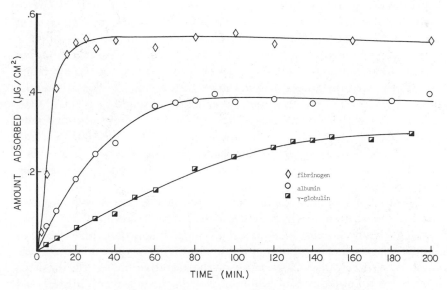

Figure 8. Competitive adsorption of plasma proteins onto SR

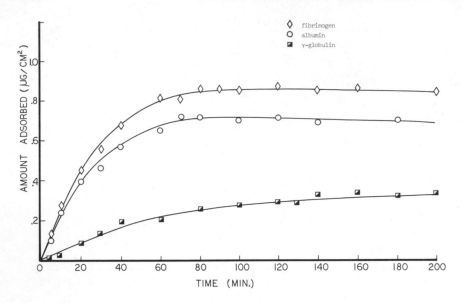

Figure 9.   Competitive adsorption of plasma proteins onto PEUU

The contact angle of water on untreated FEP was 103°; on albuminated FEP the angle was 81°. The PEUU water contact angle was 55°, and it was not changed after coating with albumin, indicating hydrogen bonding between PEUU and albumin.

Adsorption of Protein Mixtures. The adsorption of each protein in single systems had a different pattern from that in the mixture system (20). Protein adsorption onto the FEP surface had significantly different plateau times. The plateau time of albumin adsorption was delayed in the mixture system although γ-globulin and fibrinogen were adsorbed much faster in the mixture system than in the single system. The PEUU and SR showed similar albumin adsorption times with the single system and with each protein in the mixture except for the larger amount adsorbed onto the PEUU. The adsorption of each protein singly onto PEUU, SR, and FEP was studied by using ^{125}I-labelled proteins. Figures 7, 8, and 9 show the adsorption of proteins onto PEUU, SR, and FEP from the mixture solution.

Fibrinogen generally reaches the surface plateau concentration faster than the other proteins which agrees with Vroman's results (29). However, its adsorption onto FEP was much faster than onto PEUU or SR. Both fibrinogen and γ-globulin in the mixture system were adsorbed very rapidly onto FEP, perhaps because of the negative charge on the FEP surface and a unique role of albumin as a dispersing agent. PEUU and SR adsorption patterns are similar except that PEUU adsorbs more albumin at a faster rate. For the mixture solution, we used the three proteins which are the most abundant in blood. We determined the competitive adsorption kinetics. The adsorption pattern would likely change upon adding proteins. A technique that measures simultaneously the adsorption of each protein from plasma or whole blood would aid blood–foreign surface interaction studies.

Discussion

Blood compatibility with foreign surfaces has been discussed in terms of a number of surface-dependent parameters (*30*). Surface hydrophilicity, hydrophobicity, charge, critical surface tension, interfacial free energy, and surface roughness have all beeen introduced as controlling factors. Unfortunately, no experiment or theory has explained exactly the mechanism of blood clotting on foreign surfaces; consequently trial-and-error approaches are being used to develop non-thrombogenic surfaces.

Several investigators (*9, 21, 22*) have found that foreign surfaces, when exposed to blood, adsorb plasma proteins. Since platelet adhesion to the surface is the first observable event occurring in clotting on foreign surfaces, and since platelets are known to participate in hemostasis and coagulation, the indication is that platelet adhesion onto the plasma protein-coated surface plays a major role in the *in vivo* initiation of thrombus formation on foreign surfaces.

Packham *et al.* (*31*) demonstrated that an albumin-coated surface showed little platelet adhesion, but on the contrary, fibrinogen- or γ-globulin-coated surfaces showed more platelets adhering. Fibrinogen is known to be important for platelet adhesion to glass (*32*). We proposed (*23, 24*) that platelet adhesion to foreign surfaces is mediated by formation of enzyme–substrate complex bridges between platelet glycosyl transferases and incomplete heterosaccharides present in surface-adsorbed glycoproteins. The number and stability of these complexes would determine the strength and duration of adhesion. To test this model, an *in vitro* series of experiments assayed two glycosyl transferases using adsorbed plasma proteins as acceptors. Sugar nucleotides could be transferred onto adsorbed γ-globulin and fibrinogen but not onto adsorbed albumin. *In vivo* experiments quantitated platelets adhering to Silastic Rubber tubing coated with proteins and subjected to treatment to remove different carbohydrate residues. These data, plus those of other workers, support the concept that the protein adsorption occurring when polymer contacts blood controls the platelet adhesion and further thrombus formation.

Our studies (19) indicated that proteins were readily adsorbed from aqueous solution onto hydrophobic polymer surfaces with Langmuir type adsorption and that the rate of adsorption toward a plateau surface concentration depends on the polymer nature. In the study of competitive adsorption from a protein mixture solution (20), fibrinogen and γ-globulin adsorb onto FEP very rapidly compared with PEUU and SR. Therefore, the FEP surface in contact with blood has more acceptor sites for platelet adhesion than does the PEUU or SR surface.

In this sense, a non-thrombogenic polymer could be obtained if we could control the blood protein adsorption so as to give predominantly albumin adsorption with a high adsorption rate and high surface plateau concentration or to desorb adsorbed proteins from the surface, *i.e.*, a surface with minimal or probably zero protein adsorption. For example, the anticoagulant heparin (33) and hydrogels (34) are known to reduce significantly the extent of adsorption of blood proteins onto foreign surfaces.

Acknowledgment

We thank J. D. Andrade for his helpful discussions and C. Adamson for technical assistance.

Literature Cited

1. Patat, F., Killman, E., Schliebener, C., *Rubber Chem. and Tech.* (1966) **39**, 36.
2. Landsteiner, K., Uhliz, R., *Centr. Baskt. Parasitenk. Abt. Orig.* (1905) **40**, 265.
3. Blitz, W., Steiner, H., *Biochem. Z.* (1910) **23**, 27.
4. Hitchcock, E. I., *J. Gen. Physiol.* (1925) **8**, 61.
5. Kemp, L., Rideal, E. K., *Proc. Roy. Soc. (London)* (1934) **A147**, 1.
6. Bull, H. B., *Biochim. Biophys. Acta* (1956) **19**, 464.
7. Chattoraj, D. K., Bull, H. B., *J. Amer. Chem. Soc.* (1959) **81**, 5128.
8. Ghosh, S., Breese, K., Bull, H. B., *J. Colloid Sci.* (1964) **19**, 457.
9. Baier, R. E., Dutton, R. C., *J. Biomed. Mater. Res.* (1969) **3**, 191.
10. Vroman, L., Adams, A. L., *Surface Sci.* (1969) **16**, 438.
11. Smith, L. E., Fenstermaker, C. A., Stromberg, R. R., *Amer. Chem. Soc., Div. Polym. Chem., Prepr.* **11**, 1376 (Chicago, September, 1970).
12. Gorman, R. R., Stoner, G. E., Catlin, A., *J. Phys. Chem.* (1971) **75**, 2103.
13. Horbett, T. A., Ling, T., Hoffman, A. S., *Amer. Soc. Artif. Intern. Organs* (1973) Abstracts, p. 29.
14. Andrade, J. D., *Medical Instrumentation* (1973) **7**, 110.
15. Dillman, W. J., Miller, I. F., *J. Colloid Sci.* (1973) **44**, 221.
16. Mattson, J. S., Smith, C. A., *Science* (1973) **181**, 1055.
17. Levine, S. N., *J. Biomed. Mater. Res.* (1969) **3**, 83.
18. Morrissey, B. W., Stromberg, R. R., *J. Colloid Interface Sci.* (1974) **46**, 152.
19. Lee, R. G., Kim, S. W., *J. Biomed. Mater. Res.* (1974) **8**, 251.
20. Lee, R. G., Adamson, C., Kim, S. W., *Thrombosis Res.* (1974) **4**, 485.

21. Dutton, R. C., Webber, T. J., Johnson, S. A., Baier, R. E., *J. Biomed. Mater. Res.* (1969) **3**, 13.
22. Scarborough, D. E., Mason, R. G., Dalldort, F. G., Brinkhous, K. M., *Lab. Invest.* (1969) **20**, 164.
23. Lee, R. G., Kim, S. W., *J. Biomed. Mater. Res.* (in press).
24. Kim, S. W., Lee, R. G., Oster, H., Coleman, D., Andrade, J. D., Lentz, D. J., Olsen, D., *Trans. Amer. Soc. Artif. Int. Organs* (1974) **20** (in press).
25. Andrade, J. D., U.S. AEC Report **COO-2147-2**, May 1972.
26. Vroman, L., *Thrombo. Diath. Haem.* (1964) **10**, 455.
27. Vroman, L., "Blood Clotting Enzymology," W. H. Seegers, Ed., Academic, New York, 1967.
28. Lee, R. G., Adamson, C., Kim, S. W., Lyman, D. J., *Thrombosis Research* (1973) **3**, 87.
29. Vroman, L., Adams, A. L., Klings, M., *Fed. Proc. Fed. Amer. Soc. Exp. Biol.* (1971) **30**, 1494.
30. Conference on "Mechanical Surface and Gas Layer Effects on Moving Blood," *Fed. Proc. Fed. Amer. Soc. Exp. Biol.* (1971) **30**, 1485.
31. Packham, M. A., Evans, G., Glynn, M. G., Mustard, J. F., *J. Lab. Clin. Med.* (1969) **73**, 686.
32. Zucker, M. B., Vroman, L., *Proc. Soc. Exptl. Biol. Med.* (1969) **131**, 318.
33. Leininger, R. I., Epstein, M. M., Falb, R. D., Grode, G. A., *Trans. Amer. Soc. Artif. Intern. Organs* (1966) **12**, 151.
34. Andrade, J. D., Lee, II. B., John, M. S., Kim, S. W., Hibbs, J., *Trans. Amer. Soc. Artif. Intern. Organs* (1973) **19**, 1.

RECEIVED January 17, 1974. Work supported by National Science Foundation grant GH-38996X and National Heart and Lung Institute, grant HL-13733-03.

Bovine Plasma Protein Adsorption onto Radiation-Grafted Hydrogels Based on Hydroxyethyl Methacrylate and N-Vinylpyrrolidone

T. A. HORBETT and A. S. HOFFMAN

Department of Chemical Engineering and Center for Bioengineering, University of Washington, Seattle, Wash. 98195

Fibrinogen, γ-globulin, and albumin adsorption onto water-swollen synthetic polymers (hydrogels) were studied. These hydrogels were based on hydroxyethyl methacrylate (HEMA) and N-vinylpyrrolidone (NVP) and were radiation grafted onto silicone rubber. Adsorption from water, physiologic buffered saline, and blood plasma were studied. The intrinsically low protein adsorption onto poly(HEMA) can be overshadowed by adsorption due to low levels of ionic impurities. Fibrinogen adsorption isotherms at concentrations up to physiologic levels show maximum adsorptions between ca. 0.2 and 0.8 μg/cm² for the surfaces examined with adsorption onto poly(HEMA) being the least. These levels are five- to tenfold higher than those observed for fibrinogen adsorption from plasma. Albumin–fibrinogen and γ-globulin–fibrinogen competition experiments only partially explain the depression of plasma fibrinogen adsorption.

Hydrogels are a class of synthetic polymers of diverse chemical nature distinguished from other polymers by the capacity to imbibe relatively large amounts of water in their structure. The water content of these materials varies from about 30 to 90 wt % depending on both the chemical nature and physical structure of the polymer. Many natural or biocompatible polymers are also highly hydrated, *e.g.* 30–50 wt % water is bound by globular proteins (1). Partly for this reason, hydrogels have potential usefulness as biomaterials capable of minimizing the unfavorable reactions often induced by foreign materials in contact with blood or tissue (2). However, protein adsorption onto materials in contact with blood quickly modifies their interfacial properties, and it is believed to be one of the primary events leading to thrombus formation at the blood/material interface (3, 4). Thus, protein adsorption onto the hydrogels under development in this laboratory has been one of the primary areas of study in the assessment of this new class of potential biomaterials.

Hydrogels in aqueous media are mechanically weak and, for biomedical applications, they frequently require bonding to stronger substrate materials. A variety of techniques have been developed for this purpose including electron beam radiation-initiated graft polymerization (5), dip coating on sutures (6), active vapor or plasma initiated grafting onto arterial prostheses (7, 8, 9), and gamma-ray radiation-initiated graft polymerization onto several types of surfaces (10, 11, 12, 13).

The hydrogels studied here were made by radiation grafting HEMA or NVP onto silicone rubber (Silastic). The silicone rubber backbone of these grafted hydrogels overcomes the intrinsic mechanical weakness of the hydrated poly(HEMA) and poly(NVP) hydrogels (11, 12, 13). The resultant poly(HEMA)/Silastic and poly(NVP)/Silastic grafts retain about 30–60 wt % water, respectively, in contrast to the underlying silicone rubber which adsorbs less than 1% water (13). These materials have shown considerable blood compatibility in tests with the *in vivo* vena cava ring test (14).

Protein adsorption onto potential biomaterials has been studied in the past using two basic approaches, each with its limitations. Adsorption from whole, flowing blood clearly represents the most relevant situation since good blood-compatibility is an important characteristic for biomaterials. Studies of this type have been reported (15). It is extremely difficult to assess such studies, however, because the effects of blood flow rate, blood cell adhesion, degree of activation of the coagulation system, and specific protein adsorbed are not readily evaluated. Thus the studies have largely been limited to showing that any material in contact with blood becomes coated with protein in a very short time. Vroman's group has developed some potentially useful techniques by using specific antisera to identify individual proteins adsorbed from plasma onto oxidized metal surfaces (16, 17, 18, 19, 20). Another basic approach involves measuring the adsorption of a single purified protein onto the material from water or buffer (21, 22).

In our studies, a partial synthesis of these approaches has been attempted, on the premise that only a clear understanding of protein adsorption onto hydrogels from simple buffered saline could lead to an interpretation of a particular protein's adsorption behavior from blood or plasma. Thus, protein adsorption onto hydrogels from water, from buffered saline, from protein mixtures, and from plasma have been studied. The large number of experiments necessary in such a multi-faceted approach cannot be presented here in their entirety. Instead, several key experiments representative of each fact of this research are presented to provide an overview of these studies and to point out some critical aspects of measuring protein adsorption onto hydrogels.

Monomer purity has a large effect on protein adsorption onto hydrogels. The intrinsically low protein adsorption onto poly(HEMA) hydrogels can be overshadowed by adsorption caused by these impurities. However, radiation grafting of poly(HEMA) onto a hydrophobic support does not affect the adsorption of fibrinogen in comparison with ungrafted poly(HEMA). Protein adsorption isotherms for fibrinogen onto poly-(HEMA)/Silastic, poly(NVP)/Silastic, and Silastic alone at concentrations up to physiological levels show a basic similarity in shape, but the maximum adsorption levels vary by a factor of four, with fibrinogen adsorption onto poly(HEMA)/Silastic being the least. These maximum levels differ from those observed for fibrinogen adsorption from plasma onto these same materials. The different adsorption behavior observed in buffer and plasma are not entirely explained by competition curves performed with albumin–fibrinogen or γ-globulin–fibrinogen mixtures. These results emphasize the importance of as yet unknown plasma factors in modifying fibrinogen adsorption onto some of these surfaces.

Materials and Methods

Materials. Medical grade non-reinforced silicone rubber sheeting, Silastic brand, was obtained from Dow Corning. HEMA monomer was obtained from Borden Chemical Co. commercial quality, and from Hydron Laboratories, Inc., highly purified. Methacrylic acid (MAAc), NVP, and ethyleneglycol dimethacrylate (EGDMA) were purchased from Borden Chemical Co. Fibrinogen (90% clottable, bovine), Pentex brand, was purchased from Miles Laboratories, Kankakee, Ill. Albumin (crystalline, bovine) and γ-globulin (fraction II, bovine) were obtained from Nutritional Biochemicals Corp. N-2-hydroxyethyl piperazine-N'-2-ethanesulfonic acid (HEPES), A grade, and lactoperoxidase, lyophilized B grade, were purchased from Calbiochem, San Diego. Plasma, citrated, platelet poor, was prepared from freshly drawn bovine blood. Reagent grade ninhydrin and methyl Cellosolve (peroxide free) were obtained from Pierce Chemicals Co. Reagent grade $SnCl_2 \cdot 2H_2O$, acetic acid, and H_2O_2 were products of Mallinckrodt Chemical Works, St. Louis. Reagent citric and boric acids were purchased from Merck & Co., Rahway, N.J. $Na^{125}I$ was obtained from ICN Chemical and Radioisotope Division, Irvine, Calif. and from New England Nuclear, Boston, Mass. G-25 Sephadex was purchased from Pharmacia Chemical Co., Piscataway, N.J.

ICl was a gift of D. Lagunoff. Sodium azide was a product of J. T. Baker Chemical Co. All other compounds were reagent grade or the purest available commercially.

Film Preparation. Hema and NVP were radiation-grafted onto Silastic to make poly(HEMA)/Silastic and poly(NVP)/Silastic hydrogels as previously described (*13*).

^{125}**I Fibrogen.** Two methods were used to prepare ^{125}I-labeled fibrinogen. In both methods, unincorporated iodide was removed immediately after the reaction by gel filtration on a G-25 Sephadex column pre-equilibrated in $0.01M$ HEPES, $0.147M$ NaCl, 0.02% azide, pH 7.4. The first method uses catalysis by lactoperoxidase and is based on studies by Marchalonis (*23*), and the second method uses iodine monochloride (ICl) and is based on work by Helmkamp *et al.* (*24*).

Protein Adsorption. Films were kept submerged in buffer in individual bottles. Concentrated protein solution in the same buffer (and at the same temperature and pH) was added with a pipette to avoid exposing the films to the air/water interface. The solutions were mixed by swirling the films. In studies at 37°C, the solutions were kept in a water bath regulated to ±1°C.

After equilibration, the films were first rinsed quickly by a dilution and displacement technique which insures that the films are not exposed to the protein solution/air interface. The dilution–displacement rinse was done by running solvent through the equilibration bottle at about 400 ml/min for approximately 1 min using a two-hole rubber stopper fitted with two glass tubes, one for entrance and one for exit of buffer.

Fibrinogen Adsorption from Plasma. Films were submerged in 2 ml of $0.01M$ HEPES, $0.147M$ NaCl, 0.02% azide, pH 7.4. Eight ml of citrated plasma (pH 7.6) containing ^{125}I-fibrinogen was added, and the solutions were mixed by swirling the films.

Fibrinogen–Albumin and Fibrinogen–γ-Globulin Competition Experiments. Fibrinogen solutions containing ^{125}I-fibrinogen were mixed with albumin or γ-globulin solutions and buffer to give a final fibrinogen concentration of 0.02 mg/ml and final albumin or γ-globulin concentrations of between 0 and 20 mg/ml. Films were submerged in 5 ml of buffer, and 5 ml of the desired protein solution was added.

Ninhydrin Assay for Adsorbed Proteins. Measurements were made by a colorimetric procedure based on the reaction of ninhydrin with amino acids (*25*). The films were hydrolyzed in 5 ml of $2.5N$ NaOH for 2 hrs in capped plastic tubes in a boiling water bath. Then 1.5 ml of glacial acetic acid was added and mixed; next 1 ml of ninhydrin reagent was added and mixed. [The reagent was three times more concentrated in ninhydrin, SnCl$_2$, and citrate than prescribed by Moore and Stein (*25*)]. The tubes were capped and boiled 20 mins more. The solution was clarified by centrifugation, and the absorbance read immediately at 570 nm on a Beckman DB spectrophotometer. If necessary, the sample was diluted with 50–50 2-propanol–water. Calibration curves (absorbance *vs.* μg of protein) were constructed in the 0–30 and 0–100 μg range with known amounts of each type of protein subjected to this same analysis procedure.

Film and reagent blank determinations were always made and used to correct the data appropriately. The assay was linear throughout the range encountered here.

Other Methods. Protein concentrations were calculated from the 280 nm absorbance measured with a Beckman DB spectrophotometer. The pH of the protein solution was adjusted as necessary with a Radiometer GK2303c pH electrode and an Orion model 401 Ionalyzer. ^{125}I was measured with a pulse-height, analyzer-counter modular system consisting of models 40-12B, 49-25, 33-10, 30-19, 29-1, and 10-8 manufactured by Radiation Instrument Development Lab., Des Plaines, Ill.

Results and Discussion

The Effect of Monomer Purity on Protein Adsorption onto Poly-(HEMA). The importance of relatively minor contamination of the monomers used in formulating hydrogels to be used in biomedical applications has not been recognized widely as yet, although Bruck has referred to this problem in connection with the soft contact lens (26). Protein adsorption studies performed with hydrogels made with monomers of typical commercial quality illustrate this potential problem.

Table I.   γ-Globulin Adsorption onto Poly(HEMA)/Silastic [a]

Material	Protein Solvent	Amount Adsorbed ($\mu g/cm^2$)
Experiment A (commercial grade HEMA)		
untreated Silastic	H_2O	0.8
poly(HEMA)/Silastic (6% graft)	H_2O	12.7
poly(HEMA)/Silastic (28% graft)	H_2O	22.9
Experiment B (commercial grade HEMA)		
untreated Silastic	H_2O	0.8
untreated Silastic	buffered saline [b]	0.8
poly(HEMA)/Silastic (18% graft)	H_2O	12.6
poly(HEMA)/Silastic (18% graft)	buffered [c] saline	1.9
Experiment C (purified HEMA) [d]		
poly(HEMA)/Silastic (33% graft)	H_2O	1.54
poly(HEMA)/Silastic (33% graft)	buffered saline [b]	0.11

[a] All adsorption experiments were done in 0.5 mg/ml γ-globulin solutions at room temperature for at least 40 hrs, followed by at least 6 hrs of rinsing in the equilibration solvent. Adsorbed protein was determined using the ninhydrin assay.
[b] Buffered saline: $0.01M$ HEPES, $0.147M$ NaCl, pH 7.4.
[c] Buffered saline: $0.05M$ Imidazole, $0.112M$ NaCl, pH 7.4.
[d] Hydron HEMA is the purified HEMA referred to.

Table I summarizes these early studies in the form of three key experiments.

The very first measurements of protein adsorption onto these hydrogels revealed markedly greater adsorption onto these materials than onto the untreated Silastic and they showed that protein adsorption increased as the amount of poly(HEMA) grafted onto the Silastic increased (experiment A, Table I). These results were surprising because the apparent strong interaction of poly(HEMA) with proteins evidenced by such greatly enhanced adsorption did not agree with the expected low free energy at the hydrogel–solution interface.

The protein had been dissolved in distilled water in these initial adsorption studies in order to compare the results with the extensive data of Brash and Lyman (21) on protein adsorption from water onto a variety of materials—including Silastic, the standard used in these studies. The relatively close agreement of our data with those of Brash and Lyman (21) on γ-globulin and fibrinogen adsorption from water onto Silastic supported the validity of the adsorption and assay procedures used here. The maximum adsorption levels observed in both studies for plasma protein adsorption onto Silastic (0.8–1.8 $\mu g/cm^2$) are considerably higher than the levels corresponding to monolayer formation on a flat, smooth surface (calculated to be about 0.2 $\mu g/cm^2$ for these proteins). This result suggested that multilayers of protein were being formed. Since proteins are often less stable in pure distilled water than in solutions of physiologic ionic strength and pH, multilayer adsorption could occur from denaturation of the proteins. Experiments were begun using a solvent more closely related to the physiological situation with respect to ionic strength and pH.

Experiment B, Table I shows that adsorption onto Silastic was little affected by buffered saline, but a dramatic decrease in adsorption onto poly(HEMA)/Silastic occurred. This large effect of ionic strength suggested the presence of ionic impurities in the HEMA monomer used to make the poly(HEMA) and stressed the potential biological importance of any type of contamination of the monomer.

Fortunately, highly purified HEMA became available (from Hydro Med Sciences) about this time, and other monomers were readily purified by vacuum distillation. The poly(HEMA) hydrogels made with the purified HEMA showed far lower protein adsorption from either water or buffered saline than hydrogels made with the commercially available HEMA as experiment C, Table I shows. These results emphasize the biological importance of hydrogel composition in particular and biomaterial composition and purity in general.

A common contaminant in commercial HEMA is MAAc, e.g., it was present at about the 2% level in the unpurified HEMA used in the initial

experiments discussed above. The availability of the highly purified
HEMA (0.02% MAAc level) allowed us to investigate the effect of
MAAc levels in the range usually encountered in the use of unpurified
HEMA by simply adding known amounts of MAAc to the pure HEMA.
The amount of protein adsorption onto the resultant MAAc–HEMA/
Silastic hydrogels is shown in Figures 1 and 2 and Table II.

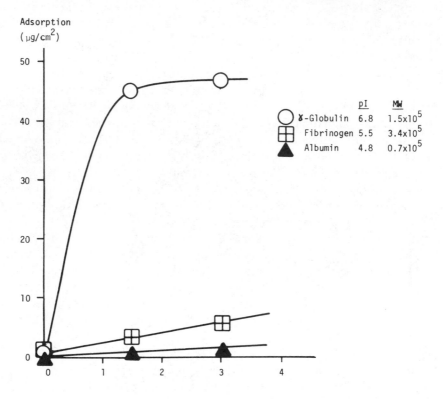

*Figure 1.   The effect of MAAc on protein adsorption onto poly(HEMA)/
Silastic at low ionic strength.   The solvent was 0.005M HEPES, pH 7.4.
Protein concentration was 0.5 mg/ml.   See Table II for other details.*

The expected effect of the MAAc is shown most clearly by the
measurements of protein adsorption at low ionic strength (0.005M
HEPES) listed in Table II and illustrated in Figure 1 since it is clear
from these that the protein with the most positive charge, *i.e.*, γ-globulin,
is adsorbed most strongly by the negatively charged MAAc groups. The
isoelectric pH's of γ-globulin, fibrinogen, and albumin are *ca.* 6.8, 5.5,

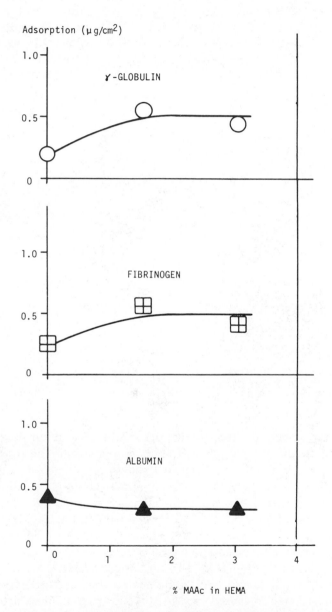

Figure 2.   The effect of MAAc on protein adsorption onto poly(HEMA)/Silastic at physiological ionic strength.   The solvent was 0.01M HEPES, 0.147M NaCl, pH 7.4.   Protein concentration was 0.5 mg/ml. See Table II for other details.

Table II.   Effect of MAAc on Protein Adsorption onto Poly(HEMA)/Silastic Hydrogels[a]

Protein	Solvent, pH 7.4	MAAc in Monomer Mixture, %	Amount Adsorbed, $\mu g/cm^2$
γ-Globulin	.005M HEPES	0.02	1.1
		1.5	45
		3.0	47
	.01M HEPES, 0.147M NaCl	0.02	0.18
		1.5	0.54
		3.0	0.43
Fibrinogen	.005M HEPES	0.02	0.08
		1.5	3.7
		3.0	5.6
	.01M HEPES, 0.147M NaCl	.02	0.25
		1.5	0.55
		3.0	0.40
Albumin	.005M HEPES	0.02	0.02
		1.5	0.68
		3.0	0.68
	.01M HEPES, 0.147M NaCl	0.02	0.39
		1.5	0.28
		3.0	0.28

[a] The listed amount of methacrylic acid was contained in the HEMA monomer used to form the poly(HEMA)/Silastic hydrogels, which had 20% grafted poly(HEMA). These hydrogels were equilibrated for 45 hrs at 37°C in 0.5mg/ml protein solutions in the listed solvents and then rinsed in the equilibration solvent, using decantation and dilution and 15 min of stirring. Ninhydrin assays were then used to determine absorbed protein.

and 4.8, respectively, so that at the pH of the adsorption experiments (7.4), the amount of adsorption of each protein would be expected to decrease in the same order, as Figure 1 shows.

The low ionic strength data also show that relatively small amounts of MAAc are necessary to cause great increases in γ-globulin adsorption, e.g., an increase of almost fiftyfold between 0.02 and 1.5% MAAc. This finding confirms calculations which indicate that MAAc content in this range could reasonably explain the amount of γ-globulin adsorption from water observed for hydrogels made with unpurified HEMA.

The data in Figure 1 for γ-globulin indicate that little further increase in adsorption is caused by MAAc levels past 1.5% which at first was puzzling However, the effective protein-binding capacities of ion exchangers is far below their theoretical capacities because of steric hindrance of the bulky protein molecules. For example, the number of titratable groups in CM-Sephadex C-50 indicates a capacity of 310 g hemoglobin/g gel whereas the actual capacity is 9 g/g (27). Thus, the leveling off of protein adsorption past 1.5% MAAc content is understandable if the 1.5% MAAc hydrogel is completely covered by protein so that additional MAAc groups would be inaccessible to protein.

In the presence of physiologic concentrations of salt, all protein adsorption onto these MAAc–HEMA hydrogels is small (Figure 2), so that it is clear that this type of hydrogel behaves as a typical ion exchange resin in its interactions with proteins. In this regard, although the three proteins studied here do not show much difference in adsorption from solutions at physiological ionic strength, many proteins do adsorb readily onto ion exchangers at this ionic strength. Lactoperoxidase, for example, binds to carboxymethylcellulose at $0.25M$ NaCl and pH 7.0 (28). Thus, preferential adsorption of certain proteins can be expected when MAAc–HEMA hydrogels are exposed to the *in vivo*-like environment. The adsorption of γ-globulin on to such hydrogels is large at low ionic strength, and it is probable that these results reflect a potential for similar behavior towards more positively charged proteins even at physiological ionic strength. Such potential for adsorption by hydrogels made with impure HEMA may well have undesirable effects when the materials are used as biomaterials in contact with blood or tissue.

Adsorption onto Ungrafted Poly(HEMA)

The low protein adsorption observed for poly(HEMA)/Silastic hydrogels made with purified HEMA called for a more sensitive assay for adsorbed proteins. The ninhydrin assay originally used is plagued by blank level variations when used to measure low levels of protein adsorption. The use of ^{125}I-radiolabeled proteins solved this problem and in addition opened up a large new area of study because the technique is non-destructive, specific, and flexible, especially with regard to protein mixtures. However, an additional more fundamental problem also became evident with the observation of low level protein adsorption for the grafted poly(HEMA)/Silastic hydrogels made with purified HEMA, *viz*: the possible presence of ungrafted patches of Silastic. Even slight exposure of the underlying Silastic might be responsible for the observed adsorption rather than the poly(HEMA) itself. The grafted hydrogels appear macroscopically and microscopically uniform in their coverage of the underlying Silastic, and the pore size of poly(HEMA) hydrogels is supposedly far too small for a protein to be able to enter the gel matrix (29). Thus it seemed unlikely that the adsorption onto poly(HEMA)/Silastic was perturbed by the underlying Silastic. Nonetheless, this was an important question central to all adsorption studies with grafted

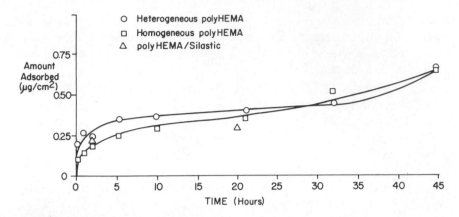

Figure 3. Fibrinogen adsorption onto ungrafted poly(HEMA). The poly-(HEMA) sheets [and grafted poly(HEMA)/Silastic] were equilibrated at 37°C in 1 mg/ml fibrinogen solution in 0.01M HEPES, 0.147M NaCl, 0.02% azide, pH 7.4 for the time depicted and then rinsed 60 sec with buffer at room temperature by the dilution displacement technique (see Methods).

poly(HEMA) hydrogels, so some fibrinogen adsorption studies with pure poly(HEMA) sheets were performed.

The poly(HEMA) sheets were prepared by B. Ratner using a special technique he developed. The HEMA solutions were poured between glass plates, and polymerization was chemically initiated. The chemical and physical properties of this material are very similar to those of radiation-grafted poly(HEMA) insofar as protein adsorption is concerned. Heterogeneous or homogeneous poly(HEMA) films were made by polymerization in solvents in which the poly(HEMA) is insoluble or soluble, respectively; the result is a white opaque material in the first case and a transparent material in the second case. The resulting films were washed free of excess monomer and then soaked in the buffer to be used in the fibrinogen adsorption experiment for 10 days at 37°C prior to the actual experiment.

The time course of fibrinogen adsorption onto the two types of poly-(HEMA) is depicted in Figure 3 which also includes representative points for poly(HEMA) grafted onto Silastic. The slow rise to the final adsorption level seen for both types of poly(HEMA) is very similar to the kinetics observed for grafted poly(HEMA), as is the actual amount of adsorption. The slight disparity between the poly(HEMA) types is probably related to the more open and thus rougher surface of the heterogeneous poly(HEMA).

The poly(HEMA) films used in this study were much thicker (0.1 cm) than those of grafted poly(HEMA) (about 0.001 cm). Assuming equivalent porosities with respect to proteins for both cast and grafted poly(HEMA), the thicker, cast poly(HEMA) should entrap more protein than the grafted poly(HEMA), if this porosity is sufficient to allow fibrinogen to enter the gel structure at all. The very close agreement in protein adsorption onto ungrafted and grafted poly(HEMA) strongly suggests that fibrinogen does not enter the gel structure and that adsorption onto poly(HEMA)/Silastic occurs on the poly(HEMA) surface and so cannot be substantially influenced by the underlying Silastic.

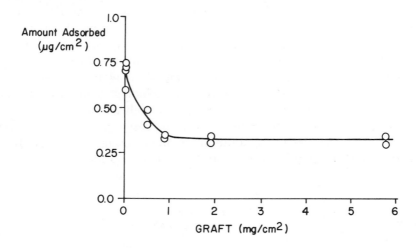

Figure 4.   Fibrinogen adsorption onto radiation-grafted poly(HEMA)/ Silastic: Effect of graft level. Films grafted to the degree depicted were equilibrated at 37°C for 20 hrs in 1 mg/ml fibrinogen solution in 0.01M HEPES, 0.147M NaCl, 0.02% azide, pH 7.4 and then rinsed by dilution displacement and extensive soaking with buffer at room temperature.

Another indication of full coverage of the substrate by poly(HEMA) on grafted poly(HEMA)/Silastic films is shown in Figure 4. Fibrinogen adsorption onto the most highly grafted films (5.8 mg/cm²) is the same as onto films having only about one-fifth the graft, but grafted film much below this point (1 mg/cm²) shows increased adsorption characteristic of the underlying Silastic. Thus, full coverage of the Silastic appears to occur at about 1 mg/cm² of graft so that the highly grafted films (∼ 5–6 mg/cm²) used in most protein adsorption studies can be assumed to possess a purely poly(HEMA) interface.

These results indicate how much graft is necessary to convert the interface from Silastic to poly(HEMA). This provides a guideline for other studies with surfaces which are more difficult to graft to the high levels obtainable with Silastic. Only about 1 mg/cm² of poly(HEMA) graft might be necessary to effect increased biocompatibility of a material.

Adsorption Isotherms at Physiologic Protein Concentrations

Most previous studies of plasma protein adsorption at the solid–solution interface have been restricted to protein concentration ranges far below those actually existing in plasma (21, 30). Protein adsorption data obtained at protein concentrations ranging from 1/100 to 1/10 the physiologic levels which are typical of such studies are not necessarily representative of the protein adsorption behavior at physiological concentrations. Several unique properties of protein solutions do not become quantitatively significant until the higher protein concentrations typical of the physiologic milieu are reached. For example, protein–protein interactions may influence multilayer formation at higher concentrations; osmotic pressure effects at high protein concentrations may be especially important with water-porous but protein-excluding materials such as hydrogels; and the approach to the natural solubility limit of the protein at physiologic concentrations may favor increased partitioning of the protein at the interface. In addition, in interpreting adsorption from plasma, experiments at high concentrations of protein are vital.

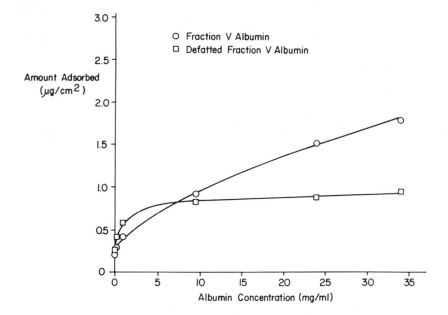

Figure 5. Albumin adsorption isotherms on Silastic before and after fat removal from the albumin. Films were equilibrated at 37°C for 20 hrs at the depicted albumin concentration in 0.01M HEPES, 0.147M NaCl, 0.02% azide, pH 7.4 and then rinsed with buffer at room temperature by the dilution displacement technique. Fraction V albumin (NBC #2106) was used untreated or defatted with charcoal (see Table III and text).

Table III.   Fatty Acid Content and Apparent Adsorption onto
Silastic of Various Albumin Preparations[a]

Albumin Preparation	Fatty Acid Content,[b] moles/mole albumin	Apparent Adsorption at	
		9.6 mg/ml	34mg/ml
Fraction V (NBC #2106)	4.08	0.93	1.76
Defatted fraction V (NBC #2106)[c]	—	0.83	0.93
Crystalline albumin (Pentex lot 24)	0.67	1.63	2.40
Defatted crystalline albumin (Pentex lot 24)[d]	0.05	0.96	1.3
Crystalline albumin (NBC lot X)	0.59	0.84	0.99
pH 3 treated crystalline albumin (NBC lot X)	0.65	0.81	1.07
Defatted crystalline albumin (NBC lot X)[c]	−0.01	0.84	1.17

[a] Adsorption (in $\mu g/cm^2$) was measured after 20 hrs at 37°C in $0.01M$ HEPES, $0.147M$ NaCl, 0.02% azide, pH 7.4.
[b] These determinations were made by the Chen method (31).
[c] Darco M charcoal was used to defat.
[d] Norit A charcoal was used to defat.

However, several problems may be encountered in using these higher protein concentration solutions. Most of the difficulties arise from the instability or impurity of the protein preparation. Even when the impurities are relatively small fractions of the total protein, they will begin to adsorb when the total protein concentration is high enough to raise the concentration of the impurity to a significant level.

This appears to be the case for albumin, probably because of the fatty acid content of this protein. Markedly different adsorption isotherms were observed for different albumin preparations, and some isotherms did not appear to reach a plateau adsorption level even at physiological albumin concentrations (*see* upper curve in Figure 5). Removal of the fatty acid by the acid charcoal method of Chen (31) appears to normalize the appearance of the isotherm (*see* the lower curve in Figure 5). Since the fatty acid content of various albumin preparations varies markedly, the variations in apparent adsorption behavior of different albumin preparations appears to be attributable to this cause (*see* Table III).

In the case of fibrinogen, the slow time course of adsorption onto poly(HEMA)/Silastic and Silastic make attempts at obtaining true equilibrium adsorption isotherms difficult because the fibrinogen solutions will undergo slow fiber formation unless protected from bacterial degradation. Also, storage of such solutions is a problem since fibrinogen

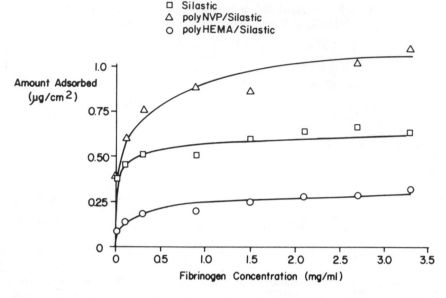

Figure 6. Fibrinogen adsorption isotherms on Silastic, poly(NVP)/Silastic, and poly(HEMA)/Silastic. Untreated Silastic and Silastic grafted with poly-(NVP) (2.7 mg/cm²) or poly(HEMA) (5.8 mg/cm²) were equilibrated at $37°C$ for 20 hrs in fibrinogen solution of the depicted concentration in $0.01M$ HEPES, $0.147M$ NaCl, 0.02% azide, pH 7.4 and then rinsed by dilution displacement and extensive soaking with buffer at room temperature.

solutions apparently have a minimum stability at $4°C$, the typical re-frigerator storage temperature. Fibrinogen solutions form a cryopre-cipitate at about $4°C$ and this material is very liable to proteolytic attack. A small amount of bacteriocide incorporated in the buffer and the use of freshly made (or stored frozen) fibrinogen solutions obviates these difficulties and enables collecting the data reported.

Adsorption isotherms for Silastic, poly(NVP)/Silastic, and poly-(HEMA)/Silastic obtained after 20 hrs of adsorption are shown in Figure 6. The films were rinsed quickly (*see* Methods) and then soaked for about 20 hrs to complete desorption. The shape of the isotherms does not appear to be significantly affected by the extensive rinse although initial adsorption levels are greater than the final levels in Figure 6, as Table IV shows. The standard error in the desorption is much smaller than the difference in average desorption among the three surfaces examined (*see* Table IV). Thus, despite the large range in fibrinogen concentrations used, the desorption is a relatively constant fraction of the amount adsorbed, but this average desorption differs for the three surfaces studied. These differences are probably related to differences in the ways in which fibrinogen interacts with each surface.

Comparison of the adsorption isotherm curves shows that the general shape of the isotherm is similar for all three materials. This indicates that the strength of fibrinogen interaction with these surfaces is similar in all cases since the similarity in the shape of the curves means that the protein concentration required to drive the adsorption to completion (as measured by the concentration required to achieve the half-maximal level of adsorption) is approximately the same for the three materials. However, the saturation level of adsorption varies by about a factor of four in the order poly(HEMA)/Silastic < Silastic < poly(NVP)/Silastic. This large a difference may be enough to affect the *in vivo* thrombogenicity of these materials. The apparently enhanced adsorption onto poly(NVP)/Silastic relative to Silastic occurs because the original area of the film is used to calculate the data in this report. The radiation grafting process causes the films to swell significantly. The maximum adsorption levels onto poly(NVP)/Silastic, Silastic, and poly(HEMA)/Silastic based on the swollen areas are 0.77, 0.62, and 0.23 $\mu g/cm^2$, respectively, in comparison with the values of 1.08, 0.62, and 0.32 in Figure 6.

The adsorption isotherms in Figure 6 also reveal that poly(NVP)/Silastic is much more similar to Silastic than poly(HEMA)/Silastic in its interactions with fibrinogen. This unexpected divergence in the properties of the two hydrogels is caused by certain unique properties of the poly(NVP)/Silastic hydrogels. Although the poly(NVP)/Silastic graft retains about 60 wt % water compared with about 30% for the poly-(HEMA) graft, the surface of the poly(NVP)/Silastic films does not appear wet as a result of the relatively high contact angle of water droplets adhering to the film. In contrast there is an absence of drop formation on the wettable poly(HEMA)/Silastic graft surface. Based on this and several other types of observations (*see* Refs. *11* and *13*), the poly-(NVP) graft is thought to be intermingled with the poly(dimethylsiloxane) chains of the silicone rubber matrix while the poly(HEMA) graft rests above and is separate from the underlying silicone rubber matrix.

Table IV. Desorption of Fibrinogen from Silastic, Poly(NVP)/Silastic, and Poly(HEMA)/Silastic[a]

Material	% Desorption[b]
Silastic	13 ± 4
Poly(NVP)/Silastic	22 ± 8
Poly(HEMA)/Silastic	38 ± 7

[a] Films were equilibrated with 0.01–3.3 mg/ml fibrinogen solution at 37°C for 20 hrs, rinsed by dilution displacement for 60 sec, then soaked for at least 20 hrs.

[b] The amount of fibrinogen remaining after the 20+ hrs soak divided by the amount present after the 60 sec rinse, times 100, gave the percent remaining, this was subtracted from 100 to give the percent desorption for each film. Desorption from a series of films exposed to various fibrinogen concentrations was measured, and the average and standard deviation of this series are listed.

Table V.   Fibrinogen Adsorption

	Adsorption from Buffer,[b] $\mu g/cm^2$	
Material	Initial[f]	Final[g]
Silastic [h]	0.78	0.65
Poly(HEMA)/Silastic [i]	0.50	0.30
Poly(NVP)/Silastic [j]	1.50	1.00

[a] Buffer solution contained $0.01M$ HEPES, $0.147M$ NaCl, 0.02% sodium azide, pH 7.4. Plasma (pH = 7.6) was obtained from citrated bovine blood by centrifugation at 5°C and was stored frozen. All adsorption experiments were done at 37°C while rinsing was done at room temperature with buffer.
[b] The saturation or plateau level adsorptions observed in 20-hr adsorption isotherm experiments are listed. The adsorption observed at the three highest fibrogen concentrations used in the adsorption isotherm (2.1, 2.7, and 3.3 mg/ml) were averaged to obtain the listed value.
[c] The final or equilibrium adsorption levels in time course experiments are listed. For Silastic, adsorption values obtained after 1 to 25 hrs were averaged to obtain the listed value. For poly(HEMA)/Silastic and poly(NVP)/Silastic, adsorption after 8 to 45 hrs were averaged to obtain the listed value.

Fibrinogen Adsorption from Plasma

The use of hydrogel-coated materials in contact with blood involves exposure to a complex mixture of proteins in the plasma. Thus, although it was shown that adsorption of proteins onto poly(HEMA) from buffer is intrinsically low, albeit not absent, the question remained as to the performance of these materials in contact with blood. For a variety of reasons, including interference from cell adhesion and clotting, it is extremely difficult to measure adsorption of plasma proteins from blood, so protein adsorption from citrated bovine plasma was performed instead. This approach seems suitable since all of the components most likely to affect the adsorption of one protein—namely, the other proteins—are present, and competition for adsorptive sites can occur. In this respect we follow Vroman's pioneering studies (16, 17, 18, 19, 20). Instead of using the antisera techniques of Vroman's group, however, high specific activity radiolabeled fibrinogen was added to the plasma in small amounts to label the plasma fibrinogen pool without significant perturbation of the total concentration of this or other plasma proteins. This approach provides a means for quantitative measurement of adsorption of this specific protein in the presence of competition from all other proteins. This technique requires measurement of the total fibrinogen content in the plasma in order to calculate the plasma fibrinogen specific activity, but this is readily done by established enzymatic technique (32).

Fibrinogen adsorption from citrated blood plasma onto Silastic, poly-(HEMA)/Silastic and poly(NVP)/Silastic have been measured twice with two separate plasma preparations made from the blood of separate

from Buffer and Plasma[a]

Adsorption from Plasma,[c] $\mu g/cm^2$

Experiment A[d]		Experiment B[e]	
Initial[f]	Final[g]	Initial[f]	Final[g]
0.063	0.058	0.11	0.098
0.23	0.052	0.24	0.16
—	—	0.43	0.20

[d] The final fibrinogen concentration in this experiment was 1.50 mg/ml. The citrated plasma had been made 1.5 years prior to its use in this experiment and had been stored frozen in polyethylene bottles.

[e] The final fibrinogen concentration in this experiment was 2.50 mg/ml. The citrated plasma had been made 1 week prior to its use in this experiment and had been stored frozen in polyethylene bottles.

[f] Initial adsorption was measured after a 1-min dilution displacement rinse.

[g] After the brief dilution displacement rinse, the films were soaked in buffer for at least 20 hrs prior to re-counting to determine the listed final adsorption.

[h] 10-mil thick medical grade silicone rubber sheets were obtained from Dow Corning.

[i] 10-mil Silastic + 20 HEMA, 40 MeOH, 40 H_2O $\xrightarrow{0.25 \text{ mrad}}$ 5.8 mg/cm², 31% H_2O.

[j] 10-mil Silastic + 20 NVP, 60 MeOH, 20 H_2O $\xrightarrow{0.25 \text{ mrad}}$ 3.0 mg/cm², 61% H_2O.

animals. Comparison of the results from the two plasma pools is complicated by the difference in fibrinogen concentrations because of routine variation in fibrinogen levels between animals. Thus, the first experiments were done at 1.50 fibrinogen/ml while the second experiments were done at 2.50 mg fibrinogen/ml. Both these concentrations are high

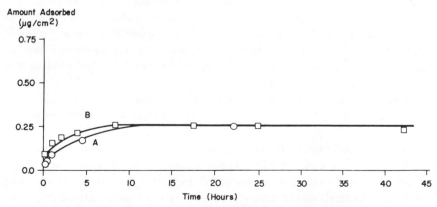

B □ I WEEK OLD PLASMA; 2.50 mg/ml FIBRINOGEN IN EQUILIBRATION
A ○ 1.5 YEAR OLD PLASMA; 1.50 mg/ml FIBRINOGEN IN EQUILIBRATION

Figure 7. Plasma fibrinogen adsorption onto poly(HEMA)/Silastic. Poly-(HEMA)-grafted Silastic (ca. 5 mg/cm²) films were equilibrated in a plasma solution (containing ^{125}I-fibrinogen) at 37°C for the time depicted and then rinsed briefly by the dilution displacement technique with 0.01M HEPES, 0.147M NaCl, pH 7.4, and then counted.

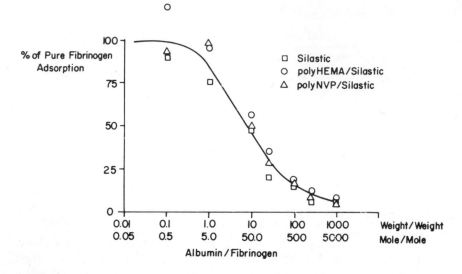

Figure 8.   Fibrinogen–albumin competition adsorption onto Silastic, poly(NVP)/ Silastic, and poly(HEMA)/Silastic. Untreated Silastic and films grafted with poly(NVP) (2.7 mg/cm²) or poly(HEMA) (5.8 mg/cm²) were equilibrated in solutions containing 0.01 mg fibrinogen/ml plus the concentration of albumin corresponding to the albumin/fibrinogen ratio depicted. The solvent was 0.01M HEPES, 0.147M NaCl, 0.02% azide, pH 7.4. Equilibration was carried out at 37°C for 20 hrs and was ended by dilution displacement rinse with buffer. The films were further rinsed by soaking for at least 20 hrs prior to counting to determine the fibrinogen on the film.

enough to cause maximal adsorption if the adsorption were done from buffer, so one might expect little difference in results between the two sets of experiments. Figure 7 shows this to be the case. The final adsorption onto poly(HEMA)/Silastic from the two plasma pools is virtually identical, but the final level is reached more quickly in experiment B. The increased rate of adsorption in this experiment is attributed to the higher fibrinogen concentration in this plasma.

Table V shows that the amount of adsorption onto Silastic from plasma is significantly depressed below its saturation value measured in buffer presumably because of competition from other components of the plasma. The difference in adsorption from the two plasma pools may result from the increased fibrinogen concentration in one pool which would allow more effective competition for adsorption onto Silastic and result in enhanced adsorption. Since the adsorption of fibrinogen onto poly(HEMA)/Silastic from plasma is not so greatly depressed relative to adsorption from buffer (*see* Table V), an increase in plasma fibrinogen concentration might not have so large an effect on adsorption onto poly-(HEMA)/Silastic as it apparently does on adsorption onto Silastic itself.

In any case, it is clear that the main findings in the two plasma experiments are the same: fibrinogen adsorption onto Silastic from plasma is less than onto poly(HEMA)/Silastic, which is the reverse of the situation for adsorption from buffer, as Table V shows. These results thus lead to the conclusion that other plasma constituents are very effective competitors for fibrinogen adsorption onto Silastic while adsorption of fibrinogen onto poly(HEMA)/Silastic from plasma and buffer is quantitatively and qualitatively much more similar.

Competition Adsorption of Albumin and γ-Globulin against Fibrinogen

The reduced adsorption of fibrinogen from plasma onto Silastic and poly(HEMA)/Silastic compared with that from pure buffered saline solutions could be caused by competition from other proteins for the adsorption sites. Albumin and γ-globulin are both present in plasma in relatively high concentrations (about 45 and 10 mg/ml, respectively, compared with *ca.* 3 mg/ml for fibrinogen), so either might compete effectively with fibrinogen for adsorption. To test this, mixtures of ^{125}I-fibrinogen

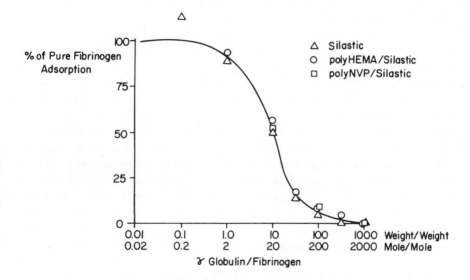

Figure 9.   Fibrinogen–γ-globulin competition adsorption onto Silastic, poly-(NVP)/Silastic, and poly(HEMA)/Silastic.   Untreated Silastic and films grafted with poly(NVP) (2.7 mg/cm²) or poly(HEMA) (5.8 mg/cm²) were equilibrated in solutions containing 0.01 mg fibrinogen/ml plus the concentration of γ-globulin corresponding to the γ-globulin/fibrinogen ratio depicted.   The solvent was 0.01M HEPES, 0.147M NaCl, 0.02% azide, pH 7.4.   Equilibration was carried out at 37°C for 20 hrs and was ended by a dilution displacement rinse with buffer.   The films were further rinsed by soaking for at least 20 hrs prior to counting to determine the fibrinogen on the film.

and albumin or γ-globulin were made in various ratios and then allowed to equilibrate with the film. In this way, fibrinogen adsorption (as ^{125}I) could be determined in the presence of the other proteins. The results are presented in Figures 8 and 9. The lower scales in the figures are moles of protein/mole of fibrinogen to allow comparison of the relative strength of competition of the proteins considered. The half-point for maximum fibrinogen adsorption is reached at about 20 moles γ-globulin/ mole fibrinogen, but it occurs at about 50 moles albumin/mole fibrinogen. Roughly speaking, albumin is several fold less effective a competitor than γ-globulin is against fibrinogen for adsorption sites onto Silastic, poly-(HEMA)/Silastic, or poly(NVP)/Silastic; both are adsorbed much more weakly than fibrinogen.

The albumin and γ-globulin vs. fibrinogen experiments suggest that competition from these proteins may be a quantitatively dominant factor in modifying surface adsorption of fibrinogen from plasma at normal concentrations of these proteins. The albumin/fibrinogen and γ-globulin/ fibrinogen molar concentration ratios in plasma are approximately 100 and 6, respectively. The competition experiments showed that about 60% and 30% reductions in plasma fibrinogen adsorption would be expected at these ratios of albumin and γ-globulin, respectively. The 90% reduction in fibrinogen adsorption expected from the combined competitive effect of these two proteins is apparently reflected in the roughly 90% reduction in fibrinogen adsorption onto Silastic from plasma in comparison with its adsorption from buffer (see Table V). However, neither protein competed so differently to Silastic and poly(HEMA)/ Silastic as to account for the very different behavior of these surfaces toward fibrinogen in plasma and in buffer. Thus, the plasma fibrinogen experiments as well as the albumin and γ-globulin competition experiments indicate the importance of as yet unknown factors in significantly modifying fibrinogen adsorption onto surfaces, a finding that agrees with the observations of modification of adsorbed fibrinogen by plasma factors in Croman's group (16, 17, 18, 19, 20). Lipid or lipo–protein interactions with surfaces seem a likely area for research on this problem especially since strong lipid interactions are expected with silicone rubber (33, 34, 35).

Competition studies by Lee et al. (30) complement those reported here. Adsorption rates and maximum adsorption levels were compared in single protein solutions and a mixture of albumin, γ-globulins and fibrinogen but only one concentration ratio was used in the mixture (0.25 mg albumin/ml, 0.15 mg γ-globulin/ml, and 0.075 mg fibrinogen/ml). Fibrinogen adsorption onto Silastic from the mixture was observed to be 67% of that for the fibrinogen adsorption from a pure fibrinogen solution. Using Figures 8 and 9 to predict the amount of pure fibrinogen adsorption onto Silastic expected in the mixture used by Lee *et al.* (*30*) gives a figure of 60% which is acceptable considering the different protein concentrations and buffers used in the two studies.

Lee *et al.* (*30*) observed effects attributable to the material on initial rates of adsorption in the mixture compared with single protein solutions. However, the final composition of the adsorbed protein from the mixture was virtually the same on all three surfaces examined (fluorinated ethylene/propylene copolymer; segmented copolyether–urethane–urea; and Silastic). In the studies reported here no significant difference in the competition curves of these same proteins has been observed for the three materials studied so far (poly(HEMA)/Silastic, poly(NVP)/Silastic, and Silastic). This is interpreted to mean that there are no significant compositional differences among the protein layers adsorbed onto these surfaces from a mixture. Thus, for five chemically rather diverse materials which represent a range of hydrophobic and hydrophilic interfacial properties, the composition of the protein layer adsorbed from a mixture of the three dominant plasma proteins appears to be very similar. Specific, preferential adsorption of certain of the plasma proteins is supported by both studies since in neither case was the adsorbed protein layer directly reflective of the composition of the protein mixture used in adsorption. Fibrinogen is greatly preferred by all surfaces studied. However, this preferential adsorption does not appear to depend significantly on the type of material since the competition curves reported here and the final composition of the adsorbed layer in the study of Lee *et al.* (*30*) were the same for each material tested. Properties specific to the proteins rather than to the surfaces seem to be the cause for the preferential adsorption of certain proteins by these materials. The similarity of these materials in regard to protein adsorption is an interesting parallel to the very similar degree of platelet adhesion observed for another equally diverse group of materials by Friedman *et al.* (*36*). On the other hand, the performance of the materials studied here and by Lee *et al.* (*30*) in the vena cava ring test (*14, 37*) is very different. Since the blood compatibility of these materials is believed to be determined by the composition of the adsorbed protein layer (*3, 38*), these results raise several important questions.

A simple explanation for the similar adsorption characteristics of materials with divergent chemical and thrombogenic properties is that the tests used to distinguish the interactions of these materials with proteins are too simple and limited in scope. Only three of the large number of plasma proteins were involved in the two studies discussed. In the work reported here, differences have been noted in the protein adsorption occurring from buffer, from simple protein mixtures, and from plasma. The plasma results particularly stress the large gap in our knowledge concerning other plasma factors which influence adsorption of proteins since fibrinogen adsorption from plasma onto Silastic was depressed (relative to adsorption from buffer) to a much greater degree than that onto poly(HEMA)/Silastic. The importance of other plasma factors in fibrinogen adsorption is also clearly revealed by the rapid modification of the antigenicity of adsorbed fibrinogen by exposure to plasma (16). Still further complications may occur in protein adsorption from flowing, whole blood, the actual conditions under which the blood compatibility differences among materials are clinically important. Differences in equilibrium protein adsorption onto different surfaces may be evident only under those complex but realistic circumstances.

The different blood compatibilities of the materials studied here and by Lee et al. (30) in the in vivo Gott ring test (14, 37), despite the similar composition of the adsorbed protein layer, has other explanations. Little is known about the configurational state of the proteins on the materials studied here and by Lee et al. (30), so it is possible that differences in the degree of denaturation of the proteins on the different surfaces is a more important determinant of thrombogenicity than simple composition. Also, only equilibrium adsorption results have been discussed, and it is quite possible that the composition of the adsorbed protein layer during the first few minutes is the crucial thrombogenic determinant (3). Finally, protein adsorption as such is not a thrombogenic event; it is the interaction of the protein coated solid/blood interface with other components of the blood which may generate thrombosis on the surface. Thus, although the amount of protein adsorbed, the composition of this layer as a function of time, and the configuration of the adsorbed proteins probably are all important parameters in determining the thrombogenicity of a material in contact with blood, these factors are only contributory to the much more complicated overall interaction of the interface with blood components which eventually may cause thrombosis. The relevance of protein adsorption measurements under the simple conditions so far studied to the overall interaction remains to be established.

Acknowledgment

We thank B. Ratner for preparation of the poly(HEMA) sheets.

Literature Cited

1. Kuntz, I. D., *J. Amer. Chem. Soc.* (1971) **93**, 514.
2. Wichterle, O., Lim, D., *Nature* (1960) **185**, 117.
3. Baier, R. E., Loeb, G. I., Wallace, G. T., *Fed. Proc. Fed. Amer. Soc. Exp. Biol.* (1971) **30**, 1523.
4. Salzman, E. W., *Bull. N.Y. Acad. Med.* (1972) **48**, 225.

5. Yasuda, H., Refojo, M. F., *J. Polym. Sci. Part A* (1964) **2**, 5093.
6. Tollar, M., Stol, G., Kliment, K., *J. Biomed. Mater. Res.* (1969) **3**, 305.
7. Bruck, S., *Biomat. Med. Dev. Art. Org.* (1973) **1**, 79.
8. Scott, H., Kronick, P. L., Hillman, E. E., Contract NIH-NHLI-71-2017, National Heart and Lung Institute, National Institutes of Health, Bethesda, Maryland, Ann. Rep. (Aug. 1971) **PB 206499**, (Sept. 1973) **PB 221846**, (Jan. 1974) **PB 236308.**
9. Halpern, R. B., McGonigal, P. J., Greenberg, H., Heldon, A., Contract PH-43-66-1124, National Heart and Lung Institute, National Institutes of Health, Bethesda, Maryland, Ann. Rep. (Jan. 1971) **PB 200987**, (Sept. 1972) **PB 212724**, (Feb. 1973) **PB 215886**, (Feb. 1974) **PB 230310.**
10. Lee, H. B., Shim, H. S., Andrade, J. D., *Amer. Chem. Soc., Div. Polym. Chem., Prepr.* **13** (2) 729 (New York, August 1972).
11. Hoffman, A. S., Harris, C., *Amer. Chem. Soc., Polym. Prepr.* **1**, 740 (1972).
12. Hoffman, A. S., Schmer, G., Harris, C., Kraft, W. G., *Trans. Amer. Soc. Artif. Intern. Organs* (1972) **18**, 10.
13. Ratner, B. D., Hoffman, A. S., *Amer. Chem. Soc., Div. Org. Coatings and Plastics Chemistry, Prepr.* **33**, 386 (1973).
14. Ratner, B. D., Hoffman, A. S., Whiffen, J. D., *Biomat. Med. Art. Org.* (1974) in press.
15. Baier, R. E., Gott, V. L., Feruse, A., *Trans. Amer. Soc. Artif. Intern. Organs* (1970) **16**, 50.
16. Vroman, L., Adams, A. L., *Thrombos. Diathes. Haemouh. (Stuttg.)* (1967) 510.
17. Vroman, L., Adams, A. L., *J. Biomed. Mater. Res.* (1969) **3**, 43.
18. *Ibid.* (1969) **3**, 669.
19. Vroman, L., Adams, A. L., *J. Polymer Sci. Part C* (1971) **34**, 159.
20. Vroman, L., Adams, A. L., Klings, M., *Fed. Proc. Fed. Amer. Soc. Exp. Biol.* (1971) **30**, 1494.
21. Brash, J. L., Lyman, D. J., *J. Biomed. Mater. Res.* (1969) **3**, 175.
22. Morrissey, G. W., Stromberg, R. B., *Amer. Chem. Soc., Div. Org. Coatings and Plastics Chemistry, Prepr.* **33**, 333 (1973).
23. Marchalonis, J. J., *Biochem. J.* (1969) **113**, 299.
24. Helmkamp, R. W., Goodland, R. L., Bale, W. F., Spar, I. L., Mutschler, L. E., *Cancer Res.* (1960) **20**, 1495.
25. Moore S., Stein, W. H., *J. Biol. Chem.* (1948) **176**, 367.
26. Bruck, S. D., Statement to Hearings before the Subcommittee on Government Regulation of the Select Committee on Small Business, U. S. Senate, on Safety, Efficacy, and Competition Problems of the Soft Contact Lens, July 6-7, 1972, U. S. Government Printing Office, **5270-01618.**
27. "Sephadex Ion Exchangers. A Guide to Ion Exchange Chromatography," p. 12, Pharmacia Fine Chemicals, Inc., Piscataway, N.J., 1970.
28. Morrison, M., Hultquist, D. W., *J. Biol. Chem.* (1963) **238**, 2847.
29. Bruck, S. D., *J. Biomed. Mater. Res.* (1973) **7**, 387.
30. Lee, R. G., Adamson, C., Kim, S. W., *Thrombosis Research* (1974) **4**, 485.
31. Chen, R. F., *J. Biol. Chem.* (1967) **242**, 173.
32. Clauss, A., *Acta Hemat.* (1957) **17**, 237.
33. Carmen, R., Kahn, P., *J. Assoc. Adv. Med. Instrumentation* (1969) **3**, 14.
34. Nyilas, E., Kupski, E. L., Burnett, P., Hagg, R. M., *J. Biomed. Mater. Res.* (1970) **4**, 369.
35. Brash, J. L., Lyman, D. J., "The Chemistry of Biosurface," M. L. Hair, Ed., vol. 1, p. 177, Marcel Dekker, New York, 1971.
36. Friedman, L. I., Liem, H., Grabowski, E. F., Leonard, E. F., McCord, C. W., *Trans. Amer. Soc. Artif. Intern. Organs* (1970) **16**, 63.

37. Gott, V. L., Feruse, A., *Fed. Proc. Fed. Amer. Soc. Exp. Biol.* (1971) **30**, 1679.
38. Vroman, L., *Bull. N.Y. Acad. Med.* (1972) **48**, 302.

RECEIVED June 7, 1974. This work was largely supported by the U.S. Atomic Energy Commission, Division of Biomedical and Environmental Research (Contract AT(45-1)-2225). The National Institutes of Health, Institute of General Medical Sciences (NIH) GMS Grant 16436-03 to 06 provided partial support.

Fibrinogen, Globulins, Albumin and Plasma at Interfaces

L. VROMAN, A. L. ADAMS, M. KLINGS, and G. FISCHER

Veterans Administration Hospital, Brooklyn, New York

Changes in plasma protein films preadsorbed and then exposed to plasma or deposited by plasma itself, onto various surfaces were studied using ellipsometry and other techniques to observe film thickness, antigenicity, and activity of adsorbates. Plasma deposited matter onto 7s gamma-globulins; if intact clotting factor XII was present in film or plasma, some removal followed from oxidized silicon substrate. Fibrinogen films were partially removed and, as did globulin to some extent, lost their antigenicity on exposure to intact plasma even if lacking factor XII. Antigenicity was maintained if the substrate had been non-wettable. Albumin was not adsorbed out of plasma though it competed well against purified proteins. Glass preexposed to proteins adsorbed factor XII out of plasma.

At an air interface, water arranges itself like a hydrophobic film (*1*); even in very dry air, a metallic surface will obtain such a low energy coat of water (*2, 3*). Coiled protein molecules also adhere to relatively simple, non-yielding surface (*4*) to display complex surface properties. Onto a hydrophobic solid they adhere with hydrophilic residues exposed to the aqueous phase while on a hydrophilic one their apolar residues will be exposed (*5*). However, the forces involved in adhesion to a simple surface such as Lucite may be far from simple (*6*). On more complex surfaces such as gels and cell membranes (*7*) and in multilayer adsorption (*8*), reaction rates will be incomputable. If only 14 residue segments of any protein molecule are involved in adsorption (*9*), the distortion needed to accomplish such adhesion must have varied and indirect effects (*10 11, 12, 13, 14*) rendering the adhering shape sometimes more, rather than less, antigenic (*15, 16*) or changing its enzymatic activity which is especially sensitive to orientation (*17*). Neither a hard surface such as germanium (18) nor a hydrophobic one (19) need destroy the globular shape of the adsorbed protein molecule. Thus the antigenicity of adsorbed antigen can be retained sufficiently for use in immunoassays (20, 21) on a variety of hard, high energy, materials and with a variety of techniques such as allowing water to condense on glass (22), ellipsometry on silicon crystals (23), or interference color pattern formation on anodized tantalum (24).

Effects of Adsorption on Plasmatic Clotting Factors

In blood clotting, where activity rather than antigenicity is important, both the adsorbent and the adsorbate influence events. Various proteolytic enzymes become active in clotting if allowed to form complexes with the proper phospholipid micelles (25). It is the arrangement of phospholipid molecules that induces the right pairs of proteinaceous clotting factors, IX with VIII and then X with V, to combine forces at the interface. Here the enzymes (factors IX and X) are held at polar groups while their cofactors (VIII and V) are held at apolar ones in the micelle (26). Perhaps the initiating factors XII and XI of this chain can interact at any interface that initiates clotting (27), although purified factor XII, once activated, activates purified factor XI without forming a functional complex with it (28). However, under more physiological conditions things may be different. Collagen failed to cover itself with clot-promoting material when exposed to plasma that lacked either factor XII or factor XI; in addition, the collagen molecule needed its negatively charged groups as well as its helical structure to activate the factors XII and XI (29).

Ramifications of Factor XII Activation

Activation of factor XII leads to complex events (30) connecting several areas of physiology. Interfaces accidentally introduced are sometimes ignored. For example, the surface of bacteria may activate factor XII (31) and thus be responsible for the finding that immune complex activates factor XII directly. There is the lack of strict specificity in the actions of several enzymes. For example, factor XII is needed for the conversion of prekallikrein (perhaps identical with Fletcher factor) to kallikrein by glass (32). Then conversion of kininogen to kinin by kallikrein follows; but several precursors may activate themselves as well as others (33).

The pain-producing factor formed by diluting plasma may be identical with activated factor XII (XIIa) and splits into prekallikrein activator with factor XII activity and some fragments (34). Collagen gains kinin-generating (kallikrein-like) activity upon exposure to plasma. Most significant however is that maximal activity is gained on shortest exposure (15 sec) to plasma (35). Yet, in most laboratories the removal of kininogenase from plasma by quartz (36) or by Celite (37) is carried out without considering that short rather than long exposure can be most effective. On some surfaces normal plasma deposits fibrinogen in two seconds and converts it (renders it unable any longer to attract matter out of anti-human fibrinogen sera) within twenty seconds if the plasma is intact (had not been exposed to surface that activates factor XII). Factor XII seems to have no role in this converting activity, but only in the presence of intact factor XII did plasma remove some of the converted matter (39).

Antibody/antigen complex itself may not adsorb and activate factor XII (31), but its formation leads to activation of complement factor C_6 which in turn may activate clotting (40, 41). Here too, some products appear poorly soluble and may offer physical interfaces rather than chemical activity.

Formation of the proteolytic plasma enzyme, plasmin, out of its precursor, plasminogen, by adding kaolin or chloroform to plasma requires factor XII (42) as well as at least one cofactor. Chloroform may well create an insoluble protein film at the plasma/chloroform interface, again interposing more physical events. At least one Hageman factor cofactor has been purified (43). It is removed from plasma by activating powders such as glass and is perhaps responsible for the lysine esterase activity present in unactivated plasma (44). Thus, particular amounts of certain interfaces may simultaneously add and remove enzyme activities.

Relationships between Adsorption of Plasma Proteins and Adhesion of Platelets

Activation of Clotting and Adhesion of Platelets. It has been reported (45) that all uncharged hydrophobic surfaces adsorb the same proteins from plasma in the same undestructive way, but the adsorbed proteins are yet unidentified.

An experiment in which a test tube first receives hemoglobin and then blood has led to the proper conclusion that hemoglobin inhibits surface activation of clotting (46), but a wide range of other partially positively charged substances will compete just as successfully for the negatively charged activating sites on a solid (47)—especially if given a chance to get there before factor XII does. Blood plasma contains some components able to attach themselves to positive sites on collagen, which then becomes unable to aggregate platelets (48). Not only the polarity of surface groups (49) but also their ability to act specifically as a substrate for a platelet enzyme must be essential in binding platelet surface to non-platelet surface (50). Therefore the process would be affected by specific rather than by general properties of activators and inhibitors. For example, only a specific group of lipoproteins, perhaps by masking certain sites on platelets, enhances their aggregation (51); certain basic polymers promote aggregation and the release of platelet material but do require adenosine diphosphate (ADP) specifically (52). It seems likely that all surface-related properties of platelets: adhesion, aggregation, release (53), and clot retraction (54) have their own requirements. Where adhesion or aggregation of platelets is followed by release of clotting factor material, new pathways may lead to initiation of clotting while bypassing factors XII and XI (55, 56, 57). With the view to link clotting and platelets with thrombosis, efforts have been made to show that activation of factor XII leads to platelet adhesion *in vivo*, but evidence is still insufficient (58). Also, other factors may precede factor XII in such a chain; Fletcher factor deficiency is corrected *in vitro* by activated factor XII (59), and perhaps this factor precedes factor XII at the interface. Therefore, diagrams of clotting may appear complete (60), but there is always room at the top. Elsewhere, even a common enzyme may penetrate the system; for example, trypsin can activate factor XI (61).

Adsorption of Other Proteins, and Adhesion *vs.* Aggregation of Platelets. We have been interested in proteins that plasma may deposit along with factor XII (and perhaps Fletcher factor and factor XI) on various surfaces. Fibrinogen seems to play a peculiar role (38, 39, 62–83). Onto a glass-like suface, normal intact plasma deposits factor XII and fibrinogen within a few seconds at a rate that seems independent of the partner protein's presence in solution (62). Does either of these cause platelet adhesion *in vitro* or thrombosis *in vivo?* The properties of a surface that determine its thrombogenicity *in vivo* may not be related to its ability to make platelets adhere *in vitro*. To form a thrombus, platelets must adhere to platelets (63). Much work has been done to find a specific protein that would act as a glue on either the platelet or the solid surface. Patients lacking fibrinogen congenitally (64) or perhaps as a result of streptokinase injections (65) have platelets with poor adhesion to glass, but these experiments failed to determine whether the glue protein must be located on platelet or solid. It may become active or adhesive under the influence of distorting forces at the interface (19, 66). The distortions that proteins other than fibrinogen may undergo at the site of a wound appear insufficient to cause platelet adhesion and aggregation: the bleeding time of afibrinogenemia patients is prolonged and corrected by infusion of fibrinogen (67).

It has been suggested that cationic proteins—*i.e.*, fibrinogen, factor XII, and gamma-globulins—can all restore the aggregating ability of washed platelets and do so by reducing the net negative charge on the platelet membrane (*68*). That platelets will also adhere to negatively charged surfaces, such as those coated with heparin, is explained by assuming that the heparinized surface first adsorbs a protein from the plasma (*69*). Others have suggested it is fibrinogen especially which is needed for platelet aggregation (*70, 71*) but found that at least in a solution containing gelatin glass coated with globulins rather than with fibrinogen caused platelets not only to adhere but then to release serotonin (*72*). Polystyrene in two configurations appeared to adsorb more globulin and hence cause more platelet aggregation in one configuration than in the other (*73*). Similar efforts to see if conformation changes in more natural substrates, such as collagen, fibrin, and fibrin-coated collagen (*74*) have significant consequences and have been unsuccessful.

Adsorption of Proteins out of Plasma; Relationship of Their Fate to Adhesion of Platelets. Platelets suspended in serum, the liquid in blood left after clotting and therefore lacking fibrinogen, or in the plasma of a patient suffering from afibrinogenemia adhered only to glass that had adsorbed fibrinogen on it. The fibrinogen film could be deposited either by exposure of the glass to fibrinogen solution, or to normal plasma during exposure of less than about 5 sec (*75*). Longer exposure, causing the plasma to convert the adsorbed fibrinogen, left a film to which platelets would no longer adhere. In a range of 0.5 to 10 mg % (a normal value being about 300 mg %), increased concentration of fibrinogen in plasma was reflected by increased numbers of platelets per unit surface area of glass found to adhere (*76*). Platelets adhered more on collagen in presence of low concentrations of fibrinogen than of other proteins tested (*77*), though albumin at physiological concentrations (about 10 times those of fibrinogen) was also effective. While ADP was found necessary for fibrinogen to cause platelet aggregation, it was not needed for adhesion (*78*). This is evidence that a difference exists between platelet-to-platelet and platelet-to-solid adhesion.

Could platelet surface fibrinogen form dimers with fibrinogen adsorbed elsewhere? Platelets do adsorb fibrinogen; both *in vivo* and *in vitro*. The less dense ones adsorb most (*79*), but their ability to adsorb injected fibrinogen *in vivo* at all would suggest a rapid turnover on the platelet membrane. Also, the amounts of fibrinogen that will aid aggregation provoked by other means are very small (*77, 80*). Further studies are required to assemble these facts into a coherent theory. Certain platelets under certain conditions will adhere to fibrinogen-coated surfaces without releasing their aggregating material, while on gamma-globulin-coated surfaces release, and then aggregation will follow (81). Platelet fibrinogen being somewhat different from plasma fibrinogen (82) and occurring mostly in platelet granules (83) may behave unlike the fibrinogen we work with or may not be available to react at all. Gamma-globulins adsorbed onto latex (84) or glass (72) caused platelet aggregation, but on glass this was true only if the platelets were suspended in an unphysiological medium. In their own plasma, they adhered, though without aggregating, to fibrinogen-coated surfaces.

Since all of these proteins are available to the platelets in their native medium (the plasma) itself, such a protein (as well as any specific platelet enzyme substrate) would have to change considerably to become a glue. Perhaps the normally dissolved molecules simply become abnormally concentrated, oriented, or packed at the solid interface, or they undergo a change in conformation forcing them to expose something attractive. For example, as clotting follows surface activation of factor XII, so a chain of reactions can follow the formation of a complex between antigen (such as a foreign protein) and antibody. Platelets adhere to a surface that has been exposed to such immune reactions (85) once complement component C_3 had joined. When complement component C_6 also joined, the platelets lysed (86). On the other hand, no complement was needed for gamma-globulins on latex to make platelets release their nucleotides (87); the globulins had merely to be in the adsorbed state. Whether or not the complement components themselves are sticky, it has seemed that only those surfaces able to activate the chain of complement reactions can cause platelets to aggregate and release matter (88, 89).

The platelet enzymes involved (90), the complexity of their interactions and of others (91), and their relationships, both physiologically and perhaps philogenetically, with the reactions of certain white blood cells (90, 92) help us to see platelets as living particles. Their aggregation is used as a quantitative aid in immunoassay (93) in a technique that may well be applicable to quantitation of heat aggregatable gamma-globulins (94) and of specific immunoglobulin G (95).

The Search for Surfaces to Which Platelets Will Not Adhere

It is desirable to find a simple surface property that will induce an equally simple and hence predictable behavior of plasma and its proteins and then of blood and its platelets. Are simple guidelines for building non-thrombogenic materials available or even possible? Wettability (96), flow (97, 98, 99, 100), and the effects of air/liquid interfaces (101) all seem to be relatively simple, physical factors with a clear effect on platelet adhesion. Physical, hydrophobic bonding, e.g., a force imposed by the complexities of water, may be involved in platelet aggregation (102) as well as in adhesion. Charge distributions on the platelet surface are probably also required (103) and may well be so specific that they act rather like chemically acute and discriminating sense organs with the ability to receive a sharp image of their natural world and of its abilities to change. For platelets not to adhere, aggregate, and cause thrombosis, a non-thrombogenic surface may have to react as well as look natural (104). Heparin has been thought a natural antithrombotic surface constituent, but on a surface this anticoagulant may cause more rather than less adsorption of proteins and activation of factor XII (105), depending on the substrate carrying the heparin (106).

Albumin. This protein may form a more non-thrombogenic surface than does heparin. It is the most abundant plasma protein. Albumins have been studied extensively in surface chemistry laboratories. Bovine serum was found to be adsorbed most strongly onto hydrophobic surfaces (107) and in turn binds aromatic compounds mostly by hydrophobic bonding (108). Under a variety of conditions, platelets do not adhere to albumin-coated glass (72, 75), but the hydrophobic nature *per se* of the albumin film facing the platelets may not be responsible. Albumin forming a protective coating on platelets in certain experiments (109) may do so by using hydrophobic bonding precisely to adhere to the platelet membrane. Considering the high concentration of albumin in plasma, and the high availability of its hydrophobic, though perhaps often occupied, sites, our observation that plasma never deposited albumin even onto hydrophobic surfaces (23, 38, 110, 111, 112, 113, 114), is surprising.

Present Efforts to Unravel Interactions Among Plasma and Its Proteins at an Interface

It is clear that: a) plasma deposits several proteins onto one material; b) it then affects each deposited protein differently; c) neither competition among these proteins for a surface nor their subsequent fate in the plasma/solid interface is simply physical. To separate these overlapping events, each purified protein type was pre-adsorbed singly onto a very simple surface and then exposed to plasma. The results were recorded in terms of thickness and antigenicity. The sequences were put together to model the behavior of whole plasma meeting this simple surface. One such surface was that of an acid-treated silicon crystal slice. It also served to answer questions such as: does albumin compete as poorly against separate proteins as it does in whole plasma? Do highly charged molecules such as heparin affect the fate of pre-adsorbed protein films? As in the past (22, 23, 24, 38, 75, 106, 111, 112, 113, 114), our methods include ellipsometry, anodized tantalum interference color observation, water vapor condensation pattern and Coomassie Blue studies, adhesion of platelets, and the ability of plasma deposits to correct the clotting of factor XII deficient plasma. The latter two methods may give a measure of the significance of a deposit to platelets and clotting under our experimental conditions.

Although the data show a relationship between the ability of a surface to adsorb fibrinogen out of plasma, the plasma's inability to convert this fibrinogen, and the tendency of platelets to adhere, it does not confirm that adsorption of fibrinogen must precede adhesion of platelets to all kinds of surface, or that surfaces which adsorb fibrinogen under these conditions will be bad biomaterials *in vivo* as heart valves, blood vessels, or canulae, or *ex vivo* as artificial kidney membranes—even though the latter are most likely to be impeded by any adsorbate.

Data presented here represent mostly single observations further supported in subsequent modifications of each experiment. Because ellipsometry recordings are time consuming, a number of variables, such as pH and composition of buffer solution, have been kept constant.

Experimental

Materials. The materials used in these experiments are listed with their source and preparatory details.

Silicon crystal slices, n-type, oxidized (SiO), are used as described (23). Anodized tantalum sputtered glass (TaO) was obtained from Millis Research, Millis, Mass. Buffer (VS), isotonic Veronal solution was adjusted to pH 7.4 and diluted 1:4 with 0.85% NaCl. 15 ml were used per ellipsometer cuvet. Albumin, human, crystalline, 100% (Mann Research Labs., New York, N. Y.), was prepared in solution of 2 mg/ml VS. 0.2 ml were used per experiment. Fibrinogen, human, lyophilized, plasminogen-free, protein 96% clottable containing 0.4% sodium citrate and 0.9% NaCl (from Alan J. Johnson: New York University) was prepared as 3 mg/ml VS. 0.4 ml was used per experiment unless otherwise indicated. Fraction I (fibrinogen), human (Mann Research Labs.) was prepared in solution of 3 mg/ml VS and 4 ml was added per experiment. 7s gamma-globulins, human, chromatographically isolated 100% (Mann Research Labs.) in concentration of 8 mg/ml VS; 0.2 ml was added per experiment. Fraction III-I, human (Hyland-div. of Travenol, Calif.) in concentration of 3 mg/ml VS; 0.2 ml was added per experiment.,

Epsilon-Amino-n-Caproic Acid (EACA), C.P., homogeneous (Mann Research Labs.) in concentration of 0.1M; 0.4 ml per experiment was added. Plasminogen, human, purified (Mann Research Labs.) in concentration of 1 mg/ml VS (15 RPMI units/mg); 0.4 ml was added per experiment. Approximately 44–45 unit amounts (dissolved in VS) of Streptokinase (SK), 100,000 Christensen units/vial. 0.1 mg = 400 units (Lederle Laboratories) was used per experiment. Trypsin Inhibitor, Soybean (SBTI), chromatographic (component VI, salt free, lyoph.), 10,000 BAEE units of inhibition per mg (Mann Research Labs.), was prepared as 2 mg/ml VS. Platelin (General Diagnostics Div. of Warner-Chilcott, Morris Plains, N. J.) was used as directed by the manufacturer.

Stock solution of Coomassie Brilliant Blue R (Sigma Chemical Co., St. Louis) was prepared by dissolving 2.5 mg in 50 vol % methanol and 10 vol % acetic acid q.s. to 100 cc. For staining, the stock was diluted 20-fold with a solution of 40 vol % methanol, 5 vol % acetic acid, and 2.5 vol % glycerol. Heparin, Na, 100,000 international units/gm (K&K Labs., New York, N. Y.) in concentration of 14 mg/ml VS; 0.1 ml was added per experiment.

Protamine (Salmine sulfate) (PS) (General Biochemicals, Chagrin Falls, Ohio) in concentration of 2 mg/ml VS; 0.4 ml was added per experiment.

Normal intact plasma, collected in 0.1 vol. ACD, rendered platelet-poor by centrifugation, and stored in polystyrene tubes (Falcon Plastics Div., Becton, Dickinson, & Co., Oxnard, Calif.) at −40°C was used within 4 hrs after thawing.

Activated plasma was prepared by exposing normal intact plasma to 60 mg Speedex (Great Lakes Carbon Corp., Los Angeles) per ml plasma for 10 min, centrifuging, and collecting the supernatant. Factor XII-deficient plasma was collected in a 3% sodium citrate from a patient with severe factor XII deficiency.

Normal serum was pooled from normal donors, stored for several hours at 37°C, and frozen in glass tubes. Rabbit antisera to the following human proteins: From Hyland, antisera to albumin (list #071-107), fibrinogen (list #071-108), human fraction III-I (list #071-103), β-lipoproteins (list #071-113), human serum (list #071-121), Fab (071-258), Fc (071-259). From Mann Research Labs., antisera to 7s gamma-globulins (cat. #231) and total gamma-globulins (cat. #8090). From Hoechst, Woodbury, N. Y.: antiserum to pre-albumin (code #8506).

Techniques. Ellipsometry was carried out as follows: an SiO crystal slice covered with test material is placed in 15 ml VS which is stirred by vertical motion and kept at 37°C. Readings are taken, and the recording was started. Often, 0.1 ml normal intact plasma is added to the VS, and the mixture is replaced by fresh VS twice, after 2 min, and in other series after 45 min. Readings are repeated, and the differences between first (blank) and second readings of minimum light transmitting positions of the analyzer (for the ellipsometer with quarter wave plate placed after reflecting surface) or polarizer (for the ellipsometer with quarter wave plate placed before reflecting surface) is reported. 0.1 ml of anti-human fibrinogen, anti-human 7s gamma-globulins or anti-human albumin serum is added, followed 45 min later by replacing cuvet contents with fresh VS, and readings are taken. Various sequences of the antisera additions were used since we found antigenicity of a protein film may be destroyed by exposing it to a non-matching antiserum. The final antiserum was one against total human serum.

TaO interference colors are observed on a 1 × 3- or 2 × 3-inch TaO slide covered with test material. Some VS is dropped on the slide and some normal intact plasma is placed into this drop. After 10 min, another drop of VS is placed elsewhere on the slide and permitted to run into the VS–plasma mixture. As soon as the drops flow together, the entire slide

Table I. Behavior of Normal Intact Plasma at 7s

Normal Intact Plasma

Exp. No.	Globulin	Top time min	Top meas.	Bottom time min	Bottom meas.	EACA	SK
1	2.30	10	0.97	—	—	—	—
2	2.85	15	1.18	—	—	—	—
3	2.91	20	1.10	—	—	—	—
4	2.93	10	1.60	180	−1.86	—	—
5	2.81	10	2.28	210	−1.93	—	—
6	2.25	20	0.87[a]	40	−1.02	—	—
7	2.04	—	1.21[b]	150	−0.73	—	—
8	2.48	25	1.44[c]	—	—	—	—
9	2.23	60	1.67[c]	120	−0.39	—	—
10	3.09	10	1.13	220	−1.24	—	—
11	2.33	45	0.38[d]	—	—	—	0.15
12	2.39	10	1.24	20	−0.09[e]	−0.54	—
13	2.94	20	1.36	140	−1.08	—	—
14	2.46	20	1.43[f]	180	−0.89	—	—
15	1.83	2	0.26	—	—	—	—

[a] Plasma premixed 1:1 with SK.
[b] Plasma premixed 1:2 with EACA.
[c] Plasma premixed 1:1 with PS (1 mg/ml VS).

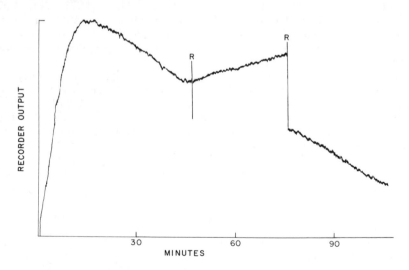

Figure 1. Tracing of ellipsometer recording, starting when 0.1 ml normal intact plasma had been added to the 15 ml buffer containing a slice of oxidized silicon on which 7s gamma-globulins had been adsorbed.

Curve shows adsorption of about 60 A, followed by desorption. Between R and R, curve was reversed by reversing analyzer deviation from minimal light transmitting position.

Gamma-Globulin-Coated Oxidized Silicon Surfaces

Normal Intact Plasma					Antiserum To			
Top		Bottom						
time min	meas.	time min	meas.	SK	Glob-ulin	Fr. 111_1	Fibrino-gen	Plas-minogen
—	—	—	—	—	—	—	—	—
—	—	—	—	—	—	—	—	—
—	—	—	—	—	—	3.92	—	—
—	—	—	—	—	—	2.81	—	—
—	—	—	—	—	—	—	—	—
—	—	—	—	—	—	—	—	—
—	—	—	—	—	—	—	−0.29	—
—	—	—	—	—	—	—	0.04	—
10	0.50	840	−0.66	—	1.31	—	—	—
10	1.01	35	−0.09	−1.66	—	—	—	—
10	0.62	840	−0.84	—	—	—	—	—
12	1.04	100	−0.07	−0.07	—	—	—	−0.61
10	0.25	60	−0.41	—	—	—	—	—
—	—	—	—	—	—	—	0.21	—

[d] Plasma premixed with antiplasminogen.
[e] Cuvet contents changed 10 min after the start of removal.
[f] Plasma premixed 1:1 with soybean trypsin inhibitor (2 mg/ml VS).

is rinsed with more VS and then with distilled water after which it is allowed to drip dry. The slide thus consists of 2 areas, one area where plasma only resided a few seconds or less and another where it resided for 10 min. On both areas, drops of various antisera are placed, rinsed off with VS, and then with water after 1 or 2 min. On bare TaO surfaces, the interference colors of natural, vertically (normal) reflected light are: bronze for the untreated surface, reddish purple for surface exposed to plasma or protein, and deep violet for sites where matching antisera had resided. On TaO surfaces pretreated with a biomaterial which itself causes a shift in color, a shift toward blue will be noted in subsequent deposits.

Glass surfaces were used in the same manner as were TaO surfaces. These surfaces were observed for antigen/antibody reaction sites either by vapor pattern technique or by staining. Water vapor pattern allows detection of wettable reaction sites, as reported (22).

For staining, slides are covered with Coomassie Brilliant Blue R solution while wet and rinsed with water 5 min later. Dried surfaces were placed face down on yellow paper for observation.

Results and Discussion

Ellipsometry Data. Most experiments were carried out in 15 ml of VS at 37°C on SiO surfaces. Values in the tables usually represent single experiments; exceptions are noted. In all tables values are given in chronological sequence of additives from left to right. The readings represent

Table II.   Behavior of Factor XII Deficient and of Activated

| | | Normal Intact Plasma | | XII-Deficient Plasma | | | |
| | | Top | | Top | | Bottom | |
Exp. No.	Glob-ulin	time min	meas.	time min	meas.	time min	meas.
1	2.83	10	0.84	10	0.06	220	−0.81
2	2.90	5	1.12	—	—	—	—
3	2.60	—	—	10	1.38	120	−0.17
4	2.36	—	—	10	1.42[a]	110	−2.80
	—	—	—	10	2.39[b]	110	−0.15
5	2.34	—	—	—	—	—	—
6	2.93	—	—	—	—	—	—

[a] Streptokinase (44–45 u) added to the plasma without rinsing after the curve leveled off.

a peak or equilibrium point. When recordings showed a down-slope, suggesting desorption, the polarizer or analyzer (depending on the ellipsometer used) was turned beyond its minimum light transmitting position to a point where the recorder pen had been immediately before. The angle of the element was noted allowing additional checks of optical thickness change as well as reversal of the curve if true desorption was taking place. Curves are not reversed by this manipulation but continue to drop if an air bubble or an increase in turbidity of the solution were the cause.

Types of antisera were added in different sequences following otherwise identical experiments, *e.g.* in Table VI, Exp. No. 1, anti-fibrinogen preceded anti-globulin serum, and in Exp. No. 2 the sequence was reversed. On the basis of previous calibrations with step-coated surfaces (*110*), the entries expressed in degrees can be multiplied by about 30 to obtain thickness in angstroms for a refractive index of about 1.60.

BEHAVIOR OF PLASMA AT 7s GAMMA-GLOBULIN-COATED OXIDIZED SILICON. Films of 7s gamma-globulins were prepared in the ellipsometer cuvet while recording, by addition of 0.4 ml of 8 mg globulin per ml VS to the 15 ml VS in the cuvet containing the SiO slice. After 90 min the solution was replaced three times by fresh VS, and readings were done when the temperature (37°C) had come to equilibrium. Films were about 2° to 3° thick (*see* Tables I–IV). Onto these, 0.1 ml normal intact plasma deposited about 1° to 2° of matter within about 10 min after it had been added to the VS (Table I, Exp. No. 1–7), and then removed nearly all and sometimes more than it had deposited (Exp. No. 4–7). A tracing of an actual recording is shown (Figure 1). Another addition of plasma recreated to some degree this adsorption and desorption sequence (Exp.

Plasma at 7s Gamma-Globulin-Coated Oxidized Silicon Surfaces

Normal Activated Plasma				Normal Intact Plasma				Anti-7s Gamma-Globulin
Top		Bottom		Top		Bottom		
time min	*meas.*	*time min*	*meas.*	*time min*	*meas.*	*time min*	*meas.*	*ulin*
—	—	—	—	—	—	—	—	—
10	0.61	180	−0.94	—	—	—	—	—
—	—	—	—	10	0.44	840	−1.93	—
—	—	—	—	—	—	—	—	0.27
—	—	—	—	—	0.00	—	—	—
120	1.42	—	0.00	—	—	—	—	—
10	1.57	120	−0.25	15	0.44	120	−0.77	1.97

b. Plasma premixed 1:1 with SK.

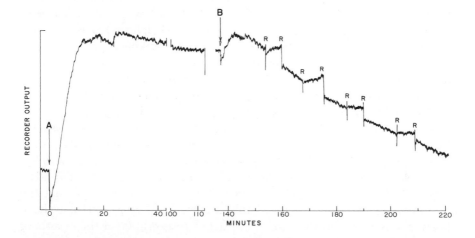

Figure 2. Tracing of ellipsometer recording; uneventful sections omitted. To buffer containing oxidized silicon slice on which 7s gamma-globulins had been adsorbed, 0.1 ml of normal activated plasma was added at A, causing adsorption. After 135 min, removal still being minimal, cuvet contents were replaced twice by fresh buffer, and at B, 0.1 ml intact plasma was added, resulting in removal. R: reversal (see Figure 1).

No. 10–14). Neither streptokinase nor 2ACA appeared to affect this event (exp. no. 6 and 7); therefore, it seems neither activation nor inhibition of plasmin influenced this desorption. Plasmin *per se* may not be involved in it at all. Protamine sulfate did inhibit it (Exp. No. 9; also, *see* below and Table IV).

Factor XII-deficient plasma, like normal plasma, did deposit matter onto the gamma-globulin films but was unable to remove it (Table II,

Table III.   Behavior of Variously Anticoagulated Plasmas at

Normal Intact Plasma

		Citrated			
		Top		Bottom	
Exp. No.	Globulin	time min	meas.	time min	meas.
1	2.75	—	1.63	—	−0.98
2	2.94	20	1.36	240	−1.08
3	3.09	10	1.13	220	−1.24
4	2.81	—	—	—	—
5	3.12	—	—	—	—
6	26.3	—	—	—	—

Exp. No. 3). However, it did remove matter left by normal plasma on brief contact with a globulin film (Table II, Exp. No. 1). In this aspect, normal activated plasma behaved like factor XII deficient plasma (Exp. No. 5 and 6 vs. Exp. No. 2) (see Figure 2). Streptokinase premixed with factor XII-deficient plasma caused it to remove a very large amount of matter after depositing some (Exp. No. 4), probably by activating plasminogen in the plasma. The antigenicity of the underlying globulin was destroyed by this activity (compare Exp. No. 4 and 6).

Thus it appears that something intact is needed either while plasma is forming its deposit or later while the deposit is exposed to plasma, to allow removal other than by plasminogen activation. The inability of activated plasma to remove its own deposit on the globulin films suggests that proteolytic activity, created by the activation, was not the desorbing agent. It also suggests that a cofactor of factor XII may have been removed by activation and is at least co-responsible for the desorbing activity of intact plasma (42).

Table IV.   Effects of Protamine Sulfate and of Heparin

Exp. No.	Globulin	PS	Heparin	Globulin
1	2.77	0.07	—	—
2	2.34	0.14	—	—
3	2.65	0.04	—	0.76
4	2.41	0.14	—	0.66
5	2.43	0.21	—	0.91
6	2.85	0.10	−0.25	0.54
7	2.68	0.06	0.02	0.25

7s Gamma-Globulin-Coated Oxidized Silicon Surfaces

Normal Intact Plasma

Heparinized				EDTA			
Top		Bottom		Top		Bottom	
time min	meas.	time min	meas.	time min	meas.	time min	meas.
—	—	—	—	—	—	—	—
—	—	—	—	—	—	—	—
—	—	—	—	—	—	—	—
10	1.72	240	−1.18	—	—	—	—
—	—	—	—	2–10	0.32	120	−0.17
—	—	—	—	10	0.44	60	−0.77

Neither the presence of heparin nor that of EDTA in solution inhibited desorption by intact plasma (Table III). On the other hand, when a preformed globulin film was first exposed to protamine sulfate, though this exposure affected film thickness very little (Table IV), normal intact plasma could no longer remove matter but only deposit it (Table IV, Exp. No. 2).

The matter deposited by normal plasma onto globulin films that had (Table IV, Exp. No. 1) or had not been pre-exposed to protamine sulfate (not listed) did not adsorb matter out of anti-human fibrinogen serum and therefore was probably not fibrinogen.

When exposure of a globulin film to protamine sulfate was followed by renewed exposure to globulin, more of the latter was deposited after which normal intact plasma would deposit more matter but remove little (Table IV, Exp. No. 4 and 5). If, however, exposure of the globulin film to protamine sulfate was followed by exposure to heparin (Exp. No. 6 and 7) again causing minimal changes in film thickness, the resulting surface

on 7s Gamma Globulin and its Interaction with Plasma

Normal Intact Plasma

Top		Bottom		Antiserum to	
time min	meas.	time min	meas.	Fibrinogen	Globulin
2	0.15	—	—	0.20	2.30
10	1.84	180	−0.09	—	—
—	—	—	—	—	—
10	1.64	90	−0.23	—	—
10	1.83	180	−0.29	—	—
10	1.28	130	−1.49	—	—
10	1.56	840	−1.34	—	—

would adsorb only little additional globulin. In contrast, normal plasma was now able to deposit and then to remove large amounts of matter, suggesting that the inhibitory action of the protamine sulfate on this activity had been neutralized by heparin at the interface. In most experiments successive additives do not meet in solution; after each added antiserum cuvet contents were replaced by fresh VS once, and after each of the other additives, at least twice.

Table V. Effects of Substrate and of Drying on Conversion of Pre-adsorbed Fibrinogen and on Desorption by Plasma

| Exp. No. | Fibrinogen | | | | Normal Intact Plasma | AF |
| | Wettable Slide | | Nonwettable Slide | | | |
	Dry	Wet	Dry	Wet		
Silicon slide						
1	1.64	—	—	—	—	2.70
2	1.60	—	—	—	0.29	2.29
3	—	1.99	—	—	—	3.36
4	—	1.83	—	—	0.04	0.17
5	—	—	2.98	—	—	3.56
6	—	—	2.18	—	−0.43	2.09
7	—	—	—	1.80	—	3.47
8	—	—	—	2.85	−0.04	3.89
Oxidized tantalum slide						
9	1.62	—	—	—	—	1.40
10	1.77	—	—	—	−0.45	1.34
11	—	1.46	—	—	—	1.58
12	—	1.47	—	—	−0.53	0.56
13	—	—	1.67	—	—	1.40
14	—	—	1.60	—	−0.64	1.76
15	—	—	—	1.74	—	1.48
16	—	—	—	1.76	−0.90	1.19

BEHAVIOR OF PLASMA AT FIBRINOGEN-COATED OXIDIZED AND UNOXIDIZED SILICON. In a large number of experiments (22, 23, 24, 38, 39) fibrinogen preadsorbed onto SiO, as well as fibrinogen deposited by plasma itself is affected by intact and factor-XII-deficient plasma in such a way that its ability to combine with anti-fibrinogen is lost. This conversion is slower on non-wettable surfaces and also on either wettable or non-wettable anodized tantalum surfaces. In presence of intact factor XII, it is followed by some desorption. To obtain some impression of the effects that conformation changes imposed on the adsorbed fibrinogen could have on the ability of plasma to convert and desorb it, the following experiments were done. A solution of 3 mg fibrinogen per ml VS was applied directly to dry surfaces of the following: unoxidized, nonwettable silicon (*23, 112*); oxidized, wettable silicon; anodized tantalum-sputtered glass (TaO); and the same rubbed with ferric stearate (TaN) to render it nonwettable. A study to be published elsewhere suggests that a fibrinogen solution, or even plasma itself, forms a fibrinogen film at the liquid/air interface which, when carried by a drop rolling over a hydrophobic surface, is transferred at the air/solid/liquid interface in a form that cannot be converted by intact plasma. In the present series, after 1 min. of exposure to the fibrinogen solution, the slides and slices were rinsed with VS and either placed in to the 15 ml VS in the ellipsometer or first rinsed with water, air dried, and then placed in VS in the ellipsometer. After readings were performed on the silicon set, 0.1 ml normal intact plasma was added; 45 min later the cuvet contents were replaced by fresh VS and readings were repeated. The anodized tantalum-sputtered slides, allowing slower or less conversion of preadsorbed fibrinogen, were exposed to 0.2 ml rather than to 0.1 ml plasma for 90 min rather than for 45 min. In all of these tests, 0.1 ml of anti-human fibrinogen serum was added next, and readings were performed 45 min later. The results on silicon substrates as listed (Table V) are averages of 2 tests for each experiment, while those on anodized tantalum substrates represent single experiments. The data indicate that dried fibrinogen was not converted well on any surface (Exp. No. 2, 6, 10, and 14), and less conversion took place on nonwettable than on wettable surfaces (compare Exp. No. 4 and 8 with Exp. No. 12 and 16), perhaps because of the air interfaces allowed during deposition. Conversion on wettable anodized tantalum was less than on oxidized silicon, even though removal was greater (Column 5) in this prolonged exposure to plasma, than it was on SiO. These data fit the qualitative observations of interference colors resulting from exposure of fibrinogen adsorbed onto TaO slides and exposed while wet with VS to undiluted citrated or heparinized plasma. Fibrinogen-coated areas that had been exposed to plasma are paler (thinner) but turn purple when exposed to anti-fibrinogen sera as do the areas not exposed to plasma. Thus, the data obtained on TaO indicate that removal of the adsorbed film by plasma need not be preceded by conversion. The possibility remained that fibrinogen itself could be variably antigenic depending on the wettability of its substrate and on its own state of drying. However, in a series of six tests (not listed), no significant differences were caused by these variables, fibrinogen films all causing a deposit of about 2.7°–3.5° by anti-fibrinogen serum.

THE BEHAVIOR OF PLASMA TOWARDS ITS OWN DEPOSITS OF FIBRINOGEN AND 7S GAMMA-GLOBULINS. The various fates of preadsorbed fibrinogen on exposure to plasma and of fibrinogen and 7s gamma-globulins deposited by plasma itself depended on the state of activation in the plasma. Both anti-fibrinogen and anti-7s gamma-globulin sera deposited much

Table VI.   Effects of Activation of the Ability of Plasma to Convert

Exp. No.	Fibrinogen	Intact Plasma		Activated Plasma	
		90 sec	45 min	90 sec	45 min
1	—	1.61	—	—	—
2	—	1.53	—	—	—
3	—	—	1.61	—	—
4	—	1.54	—	—	—
5	—	1.32	—	—	—
6	—	1.34	—	—	—
7	—	—	—	1.37	—
8	—	—	—	—	2.61
9	—	—	—	—	2.40
10	2.19	—	−0.56	—	—
11	2.28	—	—	—	−0.14
12	1.83	—	—	—	0.01 [b]
13	1.77	—	—	—	0.26
14	—	1.38	—	—	0.49
15 [c]	—	—	0.38	1.72	—

[a] Antiserum to fibrinogen added after antiserum to globulin.
[b] 14 min exposure, instead of 45 min.

matter onto films left by normal intact plasma in 1½ min of contact with oxidized silicon (Table VI, Exp. No. 1 and 2) but not onto films left by the plasma after 45 min of contact (Exp. No. 3). This conversion was obtained with 0.1 ml intact plasma being added to the 15 ml VS. An earlier study (39) showed that plasma converts its own fibrinogen deposit within 30 sec if undiluted, even at room temperature.

In a preliminary set of single experiments, we found no clear evidence of orientation in the globulin films left by plasma on 1½ min of contact with SiO. Reactivity with antisera to globulin fragments was slight and uniform. Conversion of globulins did not appear as complete as that of fibrinogen. Again we could show that activated plasma (Exp. No. 8 and 9) was less able to convert fibrinogen, but the fibrinogen deposited by intact plasma could be converted by activated plasma (Exp. No. 14) and *vice versa* (Exp. No. 15). Conversion of preadsorbed fibrinogen on SiO by intact plasma (Exp. No. 10) was always less complete than that of the fibrinogen deposited by plasma itself (Exp. No. 3), though in both situations the remaining film adsorbs much less than the 2.7°–3° adsorbed out of anti-fibrinogen serum by fibrinogen films that had not been exposed to intact plasma.

THE BEHAVIOR OF ALBUMIN IN PLASMA AT THE OXIDIZED SILICON SURFACE. A chance for albumin to be adsorbed out of plasma, not yet encountered on any of the materials tested, could be provided by removing fibrinogen and some globulins from competition. Some normal intact plasma was

Fibrinogen and 7s Gamma-Globulins at the Oxidized Silicon Surface

Antiserum To

Fab Frag.	Fc Frag.	IgG H&Lch	Fibrinogen	Globulin	Human Serum
—	—	—	1.25	1.50	—
—	—	—	0.80[a]	1.28	—
—	—	—	−0.12	0.32	—
0.41	—	—	—	—	—
—	0.93	—	—	—	—
—	—	0.80	—	—	—
—	—	—	0.95[a]	1.40	—
—	—	—	0.40	0.64	—
—	—	—	0.42[a]	0.42	—
—	—	—	0.73	0.76	1.13
—	—	—	2.81	—	—
—	—	—	2.01	1.96	1.92
—	—	—	1.90	2.59	2.06
—	—	—	0.27	0.54	—
—	—	—	0.04	—	—

[a] Activated plasma was added before intact plasma.

heated to 56°C, allowed to cool to room temperature (about 25°C) after 1 hr, and centrifuged to remove the precipitated fibrinogen. Plasma thus treated did not deposit detectable fibrinogen onto oxidized silicon (Table VII, Exp. No. 1) but did deposit some globulins which on prolonged exposure perhaps did undergo some conversion (compare Exp. No. 1 and 3). Some other, single experiments suggest that fibrinogen preadsorbed out of its solution (Exp. No. 4 and 5) or out of normal plasma (Exp. No. 6) was at least partially converted by heated plasma. The ability of heated plasma to deposit albumin (Table VIII) onto SiO was enhanced and

Table VII.   Effects of Heating on Intact Plasma

		Intact Plasma					
		Unheated		*Heated*[a]		*Antiserum to*	
Exp. No.	*Fibrino- gen*	*90 sec*	*45 min*	*90 sec*	*45 min*	*Fibrino- gen*	*Glob- ulin*
1	—	—	—	1.26	—	0.32	1.05
2	—	—	—	1.31	—	−0.02	1.19[b]
3	—	—	—	—	2.40	0.00	0.58
4	2.19	—	−0.56	—	—	0.73	0.76
5	2.38	—	—	—	−0.04	0.94	0.73
6	—	0.86	—	—	1.21	0.02	0.48

[a] Plasma maintained at 56°C for 1 hr and centrifuged.
[b] Antiserum to globulin added before antiserum to fibrinogen.

Table VIII.   Effects of Heating on Subsequent Behavior of Plasma Albumin on Oxidized Silicon

Exp. No.	Intact		Activated		Antiserum to	
	90 sec	45 min	90 sec	45 min	Albumin	Globulin
Heated Plasma						
1	1.73	—	—	—	2.65	1.12
2	1.61	—	—	—	2.82[a]	1.34
3	—	2.56	—	—	2.54	1.07
4	—	2.77	—	—	2.03[a]	0.77
5	—	—	1.41	—	3.20	—
6	—	—	1.40	—	—	1.32
7	—	—	—	1.99	2.74	—
8	—	—	—	1.99	—	1.18
Unheated plasma						
9	1.46	—	—	—	0.54	—
10	—	1.60	—	—	0.11	—

[a] Antiserum to albumin added after antiserum to globulin.

remained so even with addition of fibrinogen to the heated plasma. In 9 tests, one volume of normal activated or normal intact plasma was diluted with an equal volume of VS, kept at 65°C for 2 hrs, and centrifuged. To one volume of the supernatant, an equal volume of 3 mg/ml VS fibrinogen

Table IX.   Competition between

Exp. No.	% wt/wt in Soln		Film Meas.	Antiserum To:	
	Albumin	Fibrinogen		Fibrinogen	Albumin
1	100[a]	—	1.09	0.16	3.41
2	100	—	0.69	0.66	1.39
3	100	—	0.94	0.77[b]	3.85
4	98.2	1.8	1.15	0.52	3.32
5	93.0	7.0	1.13	1.41	2.65
6	93.0	7.0	1.49	0.93[b]	2.10
7	92.8	7.2	0.83	1.69	2.26
8	88.9	11.1	0.89	1.73	2.55
9	80.0	20.0	1.00	1.59	1.74
10	80.0	20.0	1.08	0.94[b]	0.78
11	57.1	42.9	1.07	2.15	1.86
12	57.1	42.9	1.43	1.26[b]	0.51
13	40.0	60.0	1.53	2.23[b]	0.26
14	25.0	75.0	1.94	3.16	1.35
15	10.0	90.0	2.05	3.28	0.77
16	—	100	1.71	3.18	0.29
17	—	100	1.67	1.51[b]	−0.17

[a] In this case only, 40 mg albumin per ml VS was used.

was added and then 0.1 ml of this mixture to the 15 ml VS at 37°C in the ellipsometer. Measurements, followed by exposure to antisera, were carried out as described. They showed that the films deposited by the mixtures of fibrinogen and heated plasma were 1.30°–2.27° thick, adsorbed 0.14°–1.35° (averaging 0.61°) out of anti-fibrinogen serum, and then 1.27°–2.75° (averaging 2.12°) out of anti-albumin serum.

Human serum, another mixture that lacks the competing fibrinogen, deposited films under the same experimental conditions. The films were less able to adsorb matter out of anti-albumin serum than was heated plasma; neither heated nor unheated serum showed clear ability to convert fibrinogen (data not listed).

COMPETITION BETWEEN PROTEINS. Competition between 7s gamma-globulins and fibrinogen, albumin and fibrinogen, and albumin and 7s gamma-globulins for the SiO surface was observed by placing 0.4 ml of a mixture into the 15 ml VS of the ellipsometer at 37°C allowing adsorption for 90 min, replacing the solution 3 times with fresh VS, and exposing to successive antisera as usual, each for 45 min. 60 tests were made; in many, the sequence of addition of antisera was reversed. One set of data is given here as an example (Table IX). Graphs are presented (Figures 3 and 4) in which the percent of matter adsorbed out of the first antiserum added per total amount adsorbed out of both antisera, is plotted against relative concentration, wt/wt, of the two constituents in solution.

Albumin and Fibrinogen

Total Antisera	1 Antiserum, % of Total Sera	
	Fibrinogen	Albumin
3.57	4.5	95.5
2.05	30.7	69.3
4.62	16.7[b]	83.3
3.84	13.5	86.5
4.06	34.7	65.3
3.03	30.7[b]	69.3
3.95	42.8	57.2
4.28	40.4	59.6
3.33	47.7	52.3
1.72	54.6[b]	45.4
4.01	53.6	46.4
1.77	71.9[b]	28.1
2.49	89.2[b]	10.8
4.51	70.0	30.0
4.05	89.0	11.0
3.47	91.6	8.4
1.34	100[b]	0.0

[b] Antiserum to fibrinogen added after antiserum to albumin.

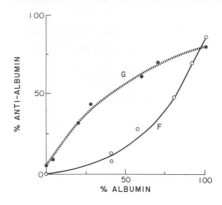

Figure 3. Amount (degrees, ellipsometer readings) adsorbed out of anti-albumin serum, in % of total adsorbed out of it and out of anti-7s gamma-globulin serum (curve G) or anti-fibrinogen serum (curve F), plotted against % of albumin (wt/wt) in 7s gamma-globulin (curve G) or fibrinogen solution (curve F). Intercept of curve G with Y axis shows some matter was adsorbed out of anti-albumin serum onto albumin-free globulin film.

Exposure of a protein film to one antiserum may inhibit adsorption out of a second one. The first, non-matching antiserum may merely fill up holes in the protein film, but events are probably more complex. For example, onto a thick albumin film adsorbed out of a concentrated solution (Table IX, Exp. No. 1), anti-fibrinogen serum deposited less matter than onto a thinner film (Line 2). However, it also deposited matter when following rather than preceding anti-albumin serum (Exp. No. 3). When fibrinogen was actually present in solution, much more anti-fibrinogen is deposited on the first rather than the second, addition of antiserum (compare Exp. No. 9 and 10, 11 and 12, 16 and 17), as if the anti-albumin serum converted the human fibrinogen that had been deposited. Competition may best be expressed by the wt/wt proportion between the two contestants that results in a film which adsorbs a film out of one antiserum that is half as thick as is the total film formed by both antisera. Thus (*see* Table IX, Exp. No. 9 and 10), under our experimental conditions, albumin would have to be four times as concentrated in a solution as fibrinogen to have an equal chance of adsorption. Similarly, 60% of

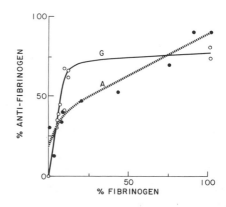

Figure 4. Amount of anti-fibrinogen serum deposit (% of total adsorbed out of two antisera) plotted against wt % fibrinogen in solution with second protein. The latter was either 7s gamma-globulin (curve G) or albumin (curve A).

The intercept of curve A with the Y axis at about 20, indicates that after a certain amount of anti-fibrinogen serum matter had been deposited onto a film consisting entirely of albumin, anti-albumin serum deposited about 4 times more than had anti-fibrinogen serum.

an albumin/7s gamma-globulin mixture must be albumin to obtain a film that adsorbed equal amounts of matter from anti-albumin and anti-7s gamma-globulin serum. Since films deposited by solutions containing 100% of a constituent (films formed without competition) all adsorbed about 3° to 4° of the matching antiserum, the potency of antisera or the molecular dimensions of their antibody varied too little to disturb these crude calculations. Albumin, occurring in normal plasma in a concentration that should allow good competition against the combined concentrations of all globulins and fibrinogen, did not seem able to attach itself at various surfaces when facing them in its natural milieu. That it did compete well in heated plasma to which fibrinogen had been returned suggests that heating not only removes more than fibrinogen from the plasma but alters interrelationships among the plasma proteins of which we are not yet aware.

Observation of Anodized Tantalum Interference Colors. Though most of our observations of optical changes caused by protein adsorption and interaction on these surfaces were qualitative, ellipsometric measurements done in parallel support our conclusions. From two tantalum sputtered glass slides of $1 \times 5\frac{1}{2}$ in., we have prepared a gauge of thickness by anodizing them in steps of 0.5 V over a range of 15–55 5 V in 0.01% HNO_3 and then adsorbing a protein (fibrinogen) over a longitundinal half (at right angles to the steps). This runner caused on each step a color shift corresponding to further anodization of 0.5 V the color of each step when protein-coated became like the color of uncoated area of the next step. In the following descriptive report, we shall simply call the color shift from a background bronze to pale beet red to deep purple to violet to deep and then pale sky blue an increase in thickness. A few experiments are selected which illustrate their use, add to, or deviate markedly from our ellipsometry data. Slides will be referred to as TaW (wettable) and TaN [TaW slides polished with Kimwipes (Kimberly-Clark Corp., Neenah, Wisc.)] after exposure to ferric stearate in petroleum ether) (non-wettable). Air/liquid/solid interfaces were avoided. A section of a TaW slide was wetted with VS, and a drop of plasma was placed on the wet area while the slide was kept slightly tilted to prevent spreading. Twelve min later (the slide now being kept horizontally under a petri dish with moist cotton) VS was placed on the bare dry area of the slide and allowed to conflue with the plasma–VS mixture. Immediately afterward the entire slide was rinsed with VS, then with water, and then air dried. Now, drops of anti-7s gamma-globulin serum and of anti-fibrinogen serum were placed and spread carefully over parts of the areas where the plasma had resided for 12 min and where it had merely passed while being washed away. Two min later the entire slide was rinsed again

with VS and water and air dried. The results (*see* Figure 5) showed
that only where the plasma had resided quite briefly did the antisera
deposit a considerable amount of matter. A trace of film had also been
deposited by the anti-globulin serum where the plasma had resided for
12 min. This single observation implies that plasma, if intact, deposits
fibrinogen and globulins onto the TaW surface within seconds and converts fibrinogen completely, and the globulin deposits partially on longer
exposure.

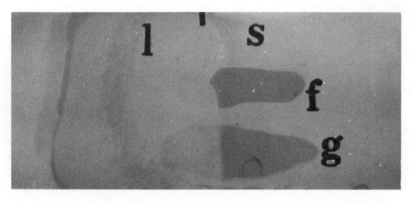

*Figure 5.   Anodized tantalum-sputtered glass slide on which normal
intact plasma had resided for 12 min (more or less vertical zone from
1 to ') and where it had resided for a few seconds (area s, from ' to right
end of field).*

*After rinse and air drying, slide was covered over one horizontal area with
anti-fibrinogen serum (f), over another with anti-7s gamma-globulin serum.
Both were rinsed off 2 min later (see text). Note area of g on one is clearly
visible, indicating the plasma had not fully converted its own globulin deposit
in 12 min.*

Fibrinogen Films. Films of fibrinogen preadsorbed onto TaW and
either kept wet with VS or rinsed and dried as described were partially
removed by drops of normal intact plasma even if the fibrinogen in solution had been pre-mixed with 8 parts of 7s gamma-globulins to 3 parts
fibrinogen by weight. When antisera were applied next, as described,
anti-7s gamma-globulin serum still deposited matter while the anti-
fibrinogen serum was unable to do so. Therefore, either the plasma had
removed fibrinogen selectively out of the mixed fibrinogen/globulin
film, it had converted the fibrinogen selectively, or perhaps it even had
removed some of both protein types and replaced some of the fibrinogen
with globulins. These results were obtained by exposing the preadsorbed
films to nearly undiluted intact plasma for as long as ½–1 hr to obtain
a significant degree of conversion. On a TaN slide, conversion was slower
than on TaW slides.

One anti-human fibrinogen serum (Schwarz-Mann) out of several tested deposited large amounts of matter onto fibrinogen films (not on other surfaces), but on exposure to a vigorous flow of VS or a light brushing with detergent, this antiserum deposit could be removed along with some of the underlying fibrinogen while elsewhere the thickness of film remained unaffected. Therefore, the following experiment was carried out. A TaW slide was allowed to adsorb fibrinogen over one area which was next rinsed with VS, then with water, and tilted so that the upper-fibrinogen-coated area could drip dry, while the bottom one still wet with water was rinsed with VS again. The bottom layer and the uncoated VS-wetted area adjoining it were then covered with a suspension of platelets in serum. Ten min later the slide was rinsed with VS, with formalin, and after 2 min again with water, then air dried. A drop of the Schwarz-Mann antiserum was placed where a bare TaW area, one area covered with fibrinogen, and one covered with fibrinogen and platelets as well (no platelets had adhered to bare TaW) all met. Two min later the slide was rinsed with VS and with water and air dried. The antiserum had deposited a heavy film only where fibrinogen had been adsorbed on the TaW, whether or not the fibrinogen had been exposed to platelets; but where the antiserum had resided on adhering platelets, the latter had been removed, leaving spots that appeared thinner than did the fibrinogen film elsewhere if observed under a microscope with a vertical illuminator. This experiment suggests that the fibrinogen was still present under the adhering platelets and available to the antiserum which could detach it along with its platelets under the proper flow of VS.

Globulins Films. Films of 7s gamma-globulins on TaW could be shown to adsorb matter out of normal intact plasma, but the subsequent desorption observed in the ellipsometer on silicon oxide, could not be shown to occur on TaW. In another experiment, a TaW slide was first exposed to 8 mg 7s gamma-globulin per ml VS, rinsed with VS, and then exposed to normal intact plasma. After 20 min, when in other experiments the plasma had been shown to deposit no additional matter, the slide was rinsed with VS and water and air dried. Drops of antisera were placed on the dry surface as described. No reactions were seen with antisera against human albumin, antitrypsin, chymotrypsin, alpha-2N-glycoprotein, C'5, transferrin, hemopexin, alpha-2HS-glycoprotein, and alpha-1-lipoprotein. There was a slight reaction with anti-alpha 1 beta-glycoprotein serum. The antiserum against human serum premixed with an equal volume of 7s gamma-globulin solution also deposited some matter. Our failure thus far to identify the protein left after prolonged exposure of either an oxidized silicon crystal or a TaW surface to plasma is given added significance by the following experiment. Antisera (Hyland) were produced by injecting rabbits with eluates from Speedex (Dicalite from Great Lakes

Carbon Corp.) that had been exposed to normal intact plasma for various times. On a TaW slide that had been exposed to normal intact plasma for seconds or for many minutes, these antisera would still deposit a significant amount of matter, confirming parallel data obtained by ellipsometry on oxidized silicon. In other words, the unknown proteins adsorbed onto Speedex were to some degree identical to those left after conversion on TaW surfaces.

Coomassie Brilliant Blue R Staining. Other than by its simplicity, and its applicability to glass as well as to plastic surfaces, results of Coomassie Blue staining were not surprising. Deposition and conversion of fibrinogen and globulin by intact plasma on glass, conversion of pre-adsorbed fibrinogen by plasma in 30 min, and conversion of 7s gamma-globulins by intact plasma in 1 hr, all seemed to parallel the behavior of plasma on oxidized silicon. The technique could also be applied to certain polyurethanes and, probably, to various artificial kidney membrane materials.

Water Vapor Condensation Patterns

The wettability of sites where presumably antibody had been deposited on an antigenic film allowed rapid identification on proteins adsorbed on surfaces such as unoxidized metal or on others that were unfit for interference color or Coomassie Blue observation. Since all data confirmed those obtained by other means they will not be listed. Some details are of interest. Wherever water drops condensed and were allowed to evaporate, a dot of matter presumably transported by the moving air/water boundary was deposited in the center of each drop during evaporation. With reexposure to air saturated with water, condensation would start on each dot and result in a pattern identical to the first one. Coomassie Blue staining, or exposure to metal oxide suspensions (*110*), would show a reticulum of protein concentrated between the water drop sites.

Factor XII Activity and Plasma Proteins at One Interface

Thirteen glass test tubes were wetted with VS; protein solutions as described were poured into 12 of them, three each receiving the same solution. All were emptied after 10 min and, along with a blank glass tube, were rinsed eight times with VS. The blank tube and one tube of each set received no further surface treatment, while two tubes of each set received 0.5 ml normal intact plasma. Contact in all of these surface treatments was promoted by gentle rotation. After 10 min the tubes containing plasma were drained and rinsed eight times with VS. One-tenth ml Platelin suspension and 0.2 ml factor XII deficient plasma were placed into each tube, and 0.2 ml 0.032M calcium chloride was added; clotting times were measured at 25°C. The following proteins were used: fibrinogen, albumin, and 7s gamma-globulin (lot R2606 and lot Y3167). Results can be summarized as follows. Clotting times among tubes coated with proteins only: 570 sec (globulin lot R2606, and fibrinogen coatings), 450 sec (globulin lot Y3167), and 630 sec (albumin coating). Clotting times in all tubes that had subsequently been exposed to normal plasma were approximately 90 sec. The clotting time in the blank tube was 790 sec. We must conclude that factor XII competed successfully with albumin, fibrinogen, and globulins. In many other experiments, a film deposited by the normal plasma on glass would also bring the clotting time of factor XII deficient plasma down to about 90 sec. Successful competition, however, need not imply replacement. There may be sites on glass less easily occupied by other plasma proteins than factor XII. We had found earlier (*62*) that the presence of fibrinogen on glass neither helped nor hindered the very rapid adsorption of factor XII activity. Whether this activity is actually factor XII or a product of subsequent interaction with and activation of factor XI was not yet determined.

The Significance to Blood Platelets

Certain plasma proteins preadsorbed on glass may interact with other constituents in the plasma and thus become adhesive to platelets. In several series of experiments (*113*), normal platelets washed twice in saline and suspended in serum or in plasma that had been freed of fibrinogen by heating and centrifugation adhered to a film of fibrinogen on glass if the film had been exposed briefly to intact plasma but not if the exposure was 10 min or longer. Only few platelets would adhere to films of albumin or of globulin lot R2606 whether or not these films had first been exposed for either 5 sec or 10 min to normal intact plasma. Quite recently another lot of 7s gamma-globulins from the same manufacturer as the first (lot Y3167) formed a film on glass that caused platelets to adhere and to expand. Both this lot and another recent one, instead of forming a clear solution in VS, yielded an opalescent one and may be compared with the heat-aggregated preparations that affect platelets profoundly.

In recent, unpublished experiments, the protein that platelets were suspended in appeared to determine to which film of protein they would adhere. For example, platelets suspended in albumin may adhere less to fibrinogen than to gamma-globulin films on glass or anodized tantalum, but when suspended in afibrinogenemic or other fibrinogen-low plasmas or serum they adhere most to fibrinogen.

Platelets at air interfaces may adhere to a film of protein formed by the plasma. This film appears to contain fibrinogen in a state that will not allow the plasma to convert it (114); thus, more platelets may adhere to this unconvertable fibrinogen than elsewhere. On a hydrophobic solid, an advancing drop of blood would transfer both the protein film and the platelets on it; hence the great loss of platelets found at hydrophobic surfaces in presence of an air interface (101).

Conclusions

The term conversion indicates some way in which some unknown, intact factors in citrated or heparinized plasma (or even whole intact native blood) can destroy the antigenicity of a protein film if adsorbed on certain substrates. Though conversion occurs under quite artificial, extracorporeal conditions, it is yet another response of blood when it meets a large enough interface. A large change within a short time on a sufficiently large amount of blood can synchronize reactions among enough molecules to end up with a visible and therefore measurable clotting time. The same kind of artifact is needed to observe immune reactions, activation of fibrinolysis, and other events that are possible reactions which occur in the body as the result of minute causes as the rupture of a single blood cell or theslight bruising of a capillary. It is reasonable for molecular reactions to occur normally on a molecular (or 2-molecule) scale even though their combined effects on the body is trivial.

With the above in mind, the present data can be interpreted as follows.

When a blood–surface interface forms that resembles oxidized silicon, the passing plasma will deposit fibrinogen on it and within 30 sec transform it so a platelet passing by will not stick. If intact factor XII is present, some will be adsorbed and it or its dissolved mates or cofactor molecule will cause some removal of the film. Conversion need not preceed this removal. Some 7s gamma-globulin molecules will have adhered along with fibrinogen ones and will also be somewhat converted by the plasma. In addition, the plasma will deposit some matter on them and then remove it again if either the deposit or the plasma itself contains intact factor XII. No platelet will adhere to this globulin film or to the matter plasma deposits on it.

Where a surface forms in the blood that resembles anodized tantalum more than oxidized silicon, fibrinogen will also be deposited, and some of it, if the plasma remains intact, will be removed. However, conversion will not take place and a platelet may adhere. On the molecular scale, the surface of the adhering platelet itself will suddenly change its mosaic with patches of molecules facing the blood plasma that had never done so before.

No surface was found on which plasma would deposit albumin preferentially, even though albumin is its most abundant component. The present study of heated plasma to which fibrinogen had then been added and of the behavior of albumin in presence of another protein suggests that in normal plasma there is something more than the sum of competing proteins that keeps albumin from being adsorbed.

The methods used are applicable to some problems. In air, a silicon crystal slice covered with 400 to 800 A of oxide gave quite sensitive changes in polarizer and analyzer readings upon adsorption of protein and antiserum which, upon computation, yielded (for example) an antibody thickness of 57 A and a refractive index of 1.4771. Computation of imaginary data for protein films in this range on silicon that had been oxidized in varying degrees (but as if observed in buffer) yielded values that when plotted for two refractive indexes of protein, demonstrated that these two could not be distinguished well. Yet, overall sensitivity to optical thickness change, a product of both thickness and refractive index, remained great. As a consequence, no distinction can be made between an adsorbed film developing bare spots and one becoming thinner overall. Other methods are needed to determine if conversion involves distortion or displacement of antigenic molecules. The problem may require a different viewpoint rather than a new technique.

Acknowledgment

The authors wish to thank B. Vromen, I.B.M., Poughkeepsie, N. Y.

Literature Cited

1. Hauxwell, F., Ottewill, R. H., "A Study of the Surface of Water by Hydrocarbon Adsorption," *J. Colloid Interface Sci.* (1970) **34**, 473.
2. Bernett, M. K., Zisman, W. A., "Effect of Adsorbed Water on the Critical Surface Tension of Wetting on Metal Surfaces," *J. Colloid Interface Sci.* (1968) **28**, 243.
3. Shafrin, E. G., Zisman, W. A., "Effect of Adsorbed Water on the Spreading of Organic Liquids on Soda–Lime Glass," *J. Amer. Ceramic Soc.* (1967) **50**, 478.
4. Hagler, A. T., Scheraga, H. A., Nemethy, G., "Current Status of the Water–Structure Problem; Application to Proteins," *Ann. N.Y. Acad. Sci.* (1973) **204**, 51.
5. Vroman, L., "Effect of Adsorbed Proteins on the Wettability of Hydrophilic and Hydrophobic Solids," *Nature* (1962) **196**, 476.
6. Ghosh, S., Breese, K., Bull, H. B., "Hydrophobic Properties of Adsorbed Protein," *J. Colloid Sci.* (1964) **19**, 457.

7. Allan, B. D., Norman, R. L., "The Characterization of Liquids in Contact with High Surface Area Materials," *Ann. N.Y. Acad. Sci.* (1973) **204**, 150.
8. Silberberg, A., "Multilayer Adsorption of Macromolecules," *J. Colloid Interface Sci.* (1972) **38**, 217.
9. Gonzalez, G., MacRitchie, F., "Equilibrium Adsorption of Proteins," *J. Colloid Interface Sci.* (1970) **32**, 55.
10. Evans, M. T. A., Mitchell, J., Mussellwhite, P. R., Irons, L., "The Effect of the Modification of Protein Structure on the Properties of Proteins Spread and Adsorbed at the Air–Water Interface," *Surface Chem. Biol. Systems*, Advan. Exp. Med. Biol.," M. Blank, Ed. (1970) **7**, 1-22.
11. Baier, R. E., Loeb, G. I., Wallace, G. T., "Role of an Artificial Boundary in Modifying Blood Proteins," *Fed. Proc.* (1971) **30**, 1523.
12. Lumry, R., Rajender, S., "Enthalpy–Entropy Compensation Phenomena in Water Solutions of Proteins and Small Molecules: a Ubiquitious Property of Water," *Biopolymers* (1970) **9**, 1125.
13. Rosenberg, B., Postow, E., "Semi-conduction in Proteins and Lipids; its Possible Biological Import," *Ann. N.Y. Acad. Sci.* (1969) **158**, 161.
14. Frölich, H., "Long Range Coherence and the Action of Enzymes," *Nature* (1970) **228**, 1093.
15. Stern, I. J., Kapsalis, A. A., Neil, B. L., "Immunogenic Effects of Materials on Plasma Proteins," *Conf. Proc. Artif. Heart Program, Natl. Heart Inst.* (1969) 259–267.
16. Peters, J. H., Goetzl, E., "Recovery of Greater Antigenic Reactivity in Conformationally Altered Albumin," *J. Biol. Chem.* (1969) **244**, 2068.
17. Koshland, D. E., Jr., Carraway, K. W., Dafforn, G. A., Gass, J. D., Storm, D. R., "The Importance of Orientation Factors in Enzymatic Reactions," *Symp. on Quantit. Biol.*, Cold Spring Harbor (1971) **36**, 13.
18. Baier, R. E., Dutton, R. C., "Initial Events in Interactions of Blood with a Foreign Surface," *J. Biomed. Mater. Res.* (1969) **3**, 191.
19. Brash, J. L., Lyman, D. J., "Adsorption of Plasma Proteins in Solution to Uncharged Hydrophobic Polymer Surfaces," *J. Biomed. Mater. Res.* (1969) **3**, 175.
20. Hutchinson, H. D., Ziegler, D. W., "Simplified Radioimmunoassay for Diagnostic Serology," *Appl. Microbiol.* (1972) **24**, 742.
21. Engvall, E., Perlmann, P., "Enzyme-Linked Immunosorbent Assay, Elisa. III. Quantitation of Specific Antibodies by Enzyme-Labeled Anti-immunoglobulin in Antigen-Coated Tubes," *J. Immunol.* (1972) **109**, 129.
22. Vroman, L., Adams, A. L., "Identification of Adsorbed Protein Films by Exposure to Antisera and Water Vapor," *J. Biomed. Mater. Res.* (1969) **3**, 669.
23. Vroman, L., Adams, A. L., Klings, M., "Interactions among Blood Proteins at Interfaces," *Fed. Proc.* (1971) **30**, 1494.
24. Adams, A. L., Klings, M., Fischer, G. C., Vroman, L., "Three Simple Ways to Detect Antibody/Antigen Complex on Flat Surfaces," *J. Immunol. Methods* (1973) **3**, 227.
25. Kazal, L. A., "Interactions of Phospholipids with Lipoproteins, with Serum and its Proteins, and with proteolytic and Nonproteolytic Enzymes in Blood Clotting," *Trans. N.Y. Acad. Sci.* (1965) **27**, 613.
26. Hemker, H. C., Kahn, M. J. P., Devilee, P. P., "The Adsorption of Coagulation Factors onto Phospholipids," *Thrombos. Diathes. Haemorrh.* (1970) **24**, 214.
27. Schoenmakers, J. G. G., "De Hageman Factor" (1965) Thesis, Univ. Nymegen, Netherlands, 155 pp.

28. Ratnoff, O. D., "Studies on the Product of the Reaction between Activated Hageman Factor (Factor XII) and Plasma Thromboplastin Antecedent (Factor XI)," *J. Lab. Clin. Med.* (1972) **80**, 704.
29. Wilner, G. D., Nossel, H. L., LeRoy, E. C., "Activation of Hageman Factor by Collagen," *J. Clin. Invest.* (1968) **47**, 2608.
30. Ratnoff, O. D., "A Tangled Web. The Interdependence of Mechanisms of Blood Clotting, Fibrinolysis, Immunity and Inflammation," *Thrombos. Diathes. Haemorrh. Suppl.* (1971) **45**, 109.
31. Cochrane, C. G., Wuepper, K. D., Aiken, B. S., Revak, S. D., Spiegelberg, H. L., "The Interaction of Hageman Factor and Immune Complexes," *J. Clin. Invest.* (1972) **51**, 2736.
32. Davies, G. E., Holman, G., Lowe, J. S., "Role of Hageman Factor in the Activation of Guinea-Pig Pre-kallikrein," *Brit. J. Pharmacol. Chemother.* (1967) **29**, 55.
33. Eisen, V., "Enzymic Aspects of Plasma Kinin Formation," *Proc. Royal Soc. Ser. B* (1969) **173**, 351.
34. Kaplan, A. P., Austen, K. F., "A Pre-albumin Activator of Prekallikrein," *J. Immunol.* (1970) **105**, 802.
35. Harpel, P. C., "Studies of the Interaction between Collagen and a Plasma Kallikrein-Like Activity," *J. Clin. Invest.* (1972) **51**, 1813.
36. Wendel, U., Vogt, W., Seidel, G., "Purification and Some Properties of a Kininogenase from Human Plasma Activated by Surface Contact," *Hoppe-Seylers Zschr. Physiol. Chem.* (1972) **353**, 1591.
37. Özge-Anwar, A. H., Movat, H. Z., Scott, J. G., "The Kinin System of Human Plasma. IV. The Interrelationship between the Contact Phase of Blood Coagulation and the Plasma Kinin System in Man," *Thrombos. Diathes. Haemorrh.* (1972) **27**, 141.
38. Vroman, L., Adams, A. L., "Identification of Rapid Changes at Plasma–Solid Interfaces," *J. Biomed. Mater. Res.* (1969) **3**, 43.
39. Vroman, L., Adams, A. L., "Findings with the Recording Ellipsometer Suggesting Rapid Exchange of Specific Plasma Proteins at Liquid/Solid Interfaces," *Surface Sci.* (1969) **16**, 438.
40. Zimmerman, T. S., Muller-Eberhard, H. J., "Blood Coagulation Initiation by a Complement-Mediated Pathway," *J. Exper. Med.* (1971) **134**, 1601.
41. Götze, O., Muller-Eberhard, H. J., "The C3-Activator System: an Alternate Pathway of Complement Activation," *J. Exper. Med.* (1971) **134**, 90s.
42. Ogston, D., Ogston, C. M., Ratnoff, O. D., Forbes, C. D., "Studies on a Complex Mechanism for the Activation of Plasminogen by Kaolin and by Chloroform: the Participation of Hageman Factor and Additional Cofactors," *J. Clin. Invest.* (1969) **48**, 1786.
43. Herbert, R. J., Ogston, D., Douglas, S., "Further Purification and Properties of Hageman Factor Cofactor," *Biochim. Biophys. Acta* (1972) **271**, 371.
44. Sherry, S., Alkjaersig, N. K., Fletcher, A. P., "Observations on the Spontaneous Arginine and Lysine Esterase Activity of Human Plasma, and their Relation to Hageman Factor," *Thrombos. Diathes. Haemorrh. Suppl.* (1966) **20**, 243.
45. Lyman, D. J., Brash, J. L., Chaikin, S. W., Klein, K. G., Carini, M., "The Effect of Chemical Structure and Surface Properties of Synthetic Polymers on the Coagulation of Blood. II. Protein and Platelet Interaction with Polymer Surfaces," *Trans. Amer. Soc. Artif. Int. Organs* (1968) **14**, 250.
46. Triantaphyllopoulos, E., "Hemoglobulin, Inhibitor of the Contact Phase of Blood Coagulation," *Life Sci.* (1971) **10**, 813.
47. Nossel, H. L., Rubin, H., Drillings, M., Hsieh, R., "Inhibition of Hageman Factor Activation," *J. Clin. Invest.* (1968) **47**, 1172.

48. Nossel, H. L., Wilner, G. D., Drillings, M., "Inhibition of Collagen-Induced Platelet Aggregation by Normal Plasma," *J. Clin. Invest.* (1971) **50**, 2168.

49. Nossel, H. L., Wilner, G. D., LeRoy, E. C., "Importance of Polar Groups for Initiating Blood Coagulation and Aggregating Platelets," *Nature* (1969) **221**, 75.

50. Hugues, J., "Mode of Action of Inhibitors of Platelet Adhesion to Collagen," *Pathol. Biol.* (1972) **20**, 65.

51. Farbiszewski, R., Skrzydlewski, Z., Worowski, K., "The Effect of Lipoprotein Fractions on Adhesiveness and Aggregation of Blood Platelets," *Thrombos. Diathes. Haemorrh.* (1969) **21**, 89.

52. Massini, P., Luscher, E. F., "On the Mechanisms by which Cell Contact Induces the Release Reaction of Blood Platelets; the Effect of Cationic Polymers," *Thrombos. Diathes. Haemorrh.* (1972) **27**, 121.

53. Spaet, T. H., Stemerman, M. B., "Platelet Adhesion," *Ann. N.Y. Acad. Sci.* (1972) **201**, 13.

54. Mason, R. G., Read, M. S., Saba, S. R., Shermer, R. W., "Apparent Similarity of Mechanisms of Platelet Adhesion and Aggregation. Differentiation of these Functions from Clot Retraction," *Thrombos. Diathes. Haemorrh.* (1972) **27**, 134.

55. Walsh, P. N., "The Effect of Dilution of Plasma on Coagulation: the Significance of the Dilution–Activation Phenomenon for the Study of Platelet Coagulant Activities," *Brit. J. Haematol.* (1972) **22**, 219.

56. Walsh, P. N., "The Effects of Collagen and Kaolin on the Intrinsic Coagulant Activity of Platelets: Evidence for an Alternative Pathway in Intrinsic Coagulation Not Requiring Factor XII," *Brit. J. Haematol.* (1972) **22**, 393.

57. Schiffman, S., Rapaport, S. I., Chong, M. M. Y., "Platelets and Initiation of Intrinsic Clotting," *Brit. J. Haematol.* (1973) **24**, 633.

58. Nossel, H. L., "Activation of Factors XII and XI in Thrombogenesis," *Bull. N.Y. Acad. Med.* (1972) **48**, 281.

59. Hathaway, W. E., Alsever, J., "The Relation of 'Fletcher Factor' to Factors XI and XII," *Brit. J. Haematol.* (1970) **18**, 161.

60. Vroman, L., "A Resemblance between the Clotting of Blood Plasma and the Breakdown of Cytoplasm," *Nature* (1965) **205**, 496.

61. Saito, H., Ratnoff, O. D., Marshall, J. S., Pensky, J., "Partial Purification of Plasma Thromboplastin Antecedent (Factor XI) and Its Activation by Trypsin," *J. Clin. Invest.* (1973) **52**, 850.

62. Vroman, L., Adams, A. L., "Peculiar Behavior of Blood at Solid Interfaces," *J. Polym. Sci.* (1971) **34**, 159.

63. Vroman, L. (moderator), "Panel Discussion: Section I," *Bull. N.Y. Acad. Med.* (1972) **48**, 313.

64. Inceman, S., Caen, J., Bernard, J., "Aggregation, Adhesion and Viscous Metamorphosis of Platelets in Congenital Fibrinogen Deficiencies," *J. Lab. Clin. Med.* (1966) **68**, 21.

65. Cronberg, S., "Effect of Fibrinolysis on Adhesion and Aggregation of Human Platelets," *Thrombos. Diathes. Haemorrh.* (1968) **19**, 474.

66. Baier, R. E., "The Role of Surface Energy in Thrombogenesis," *Bull. N.Y. Acad. Med.* (1972) **48**, 257.

67. Caen, J., Castaldi, P., Inceman, S., "Le rôle du fibrinogène dans l'hémostase primaire," *Nouvelle Rev. Franc. d'Hématol.* (1965) **5**, 327.

68. Bang, N. U., Heidenreich, R. O., Trygstad, C. W., "Plasma Protein Requirements for Human Platelet Aggregation," *Ann. N.Y. Acad. Sci.* (1972) **201**, 280.

69. Salzman, E. W., Merrill, E. W., Binder, A., Wolf, C. F. W., Ashford, T. P., Austen, W. G., "Protein–Platelet Interaction on Heparinized Surfaces," *J. Biomed. Mater. Res.* (1969) **3**, 69.

70. Breddin, K., Krywanek, H. J., "Plasma Factors for Platelet Aggregation," *Acta Med. Scandin. Suppl.* (1970) **525**, 15.
71. Vermylen, J., Donati, M. B., de Gaetano, G., "Protein Requirement for Platelet Aggregation," *Acta Med. Scandin. Suppl.* (1970) **525**, 19.
72. Packham, M. A., Evans, G., Glynn, M. F., Mustard, J. F., "The Effect of Plasma Proteins on the Interaction of Platelets with Glass Surfaces," *J. Lab. Clin. Med.* (1969) **73**, 686.
73. Mustard, J. F., Glynn, M. F., Nishizawa, E. E., Packham, M. A., "Platelet–Surface Interactions: Relationship to Thrombosis and Hemostasis," *Fed. Proc.* (1967) **26**, 106.
74. Hovig, T., Jørgensen, L., Packham, M. A., Mustard, J. F., "Platlet Adherence to Fibrin and Collagen," *J. Lab. Clin. Med.* (1968) **71**, 29.
75. Zucker, M. B., Vroman, L., "Platelet Adhesion Induced by Fibrinogen Adsorbed onto Glass," *Proc. Soc. Exper. Biol. Med.* (1969) **131**, 320.
76. Mason, R. G., Read, M. S., Brinkhous, K. M., "Effect of Fibrinogen Concentration on Platelet Adhesion to Glass," *Proc. Soc. Exper. Biol. Med.* (1971) **137**, 680.
77. Lyman, B., Rosenberg, L., Karpatkin, S., "Biochemical and Biophysical Aspects of Human Platelet Adhesion to Collagen Fibers," *J. Clin. Invest.* (1971) **50**, 1854.
78. Niewiarowski, S., Regoeczi, E., Mustard, J. F., "Platelet Interaction with Fibrinogen and Fibrin: Comparison of the Interaction of Platelets with that of Fibrinoblasts, Leukocytes, and Erythrocytes," *Ann. N.Y. Acad. Sci.* (1972) **201**, 72.
79. Vainer, H., Ardaillou, N., "Membrane Receptors and Platelet Function. IV. Interaction of Platelet Populations with ^{125}I-Fibrinogen," *Thrombosis Res.* (1972) **1**, 283.
80. Harbury, C. B., Hershgold, J. E., Schrier, S. L., "Requirements for Aggregation of Washed Human Platelets Suspended in Buffercd Salt Solutions," *Thrombos. Diathes. Haemorrh.* (1972) **28**, 2.
81. Jenkins, S. P., Packham, M. A., Guccione, M. A., Mustard, J. F., "Modification of Platelet Adherence to Protein-Coated Surfaces," *J. Lab. Clin. Med.* (1973) **81**, 280.
82. Ganguly, P., "Isolation and Some Properties of Fibrinogen from Human Blood Platelets," *J. Biol. Chem.* (1972) **247**, 1809.
83. Day, H. J., Solun, N. O., "Fibrinogen Associated with Subcellular Platelet Particles," *Scandin. J. Haematol.* (1973) **10**, 136.
84. Lüscher, E. F., "Immune Complexes and Platelet Aggregation," *Acta Med. Scandin. Suppl.* (1970) **525**, 151.
85. Henson, P. M., "The Adherence of Leucocytes and Platelets Induced by Fixed IgG Antibody or Complement," *Immunology* (1969) **16**, 107.
86. Henson, P. M., "Mechanisms of Release of Constituents from Rabbit Platelets by Antigen–Antibody Complexes and Complement. I. Lytic and Nonlytic Reactions," *J. Immunol.* (1970) **105**, 476.
87. Mueller-Eckhardt, C., Lüscher, E. F., "Immune Reactions of Human Blood Platelets. II. The Effect of Latex Particles Coated with Gamma-globulin in Relation to Complement Activation," *Thrombos. Diathes. Haemorrh.* (1968) **20**, 168.
88. Jobin, F., Tremblay, F., "Platelet Reactions and Immune Processes. I. Activation of Complement by Several Platelet-Activating Surfaces when Coated with Gammaglobulin; Mechanism of Complement Fixation," *Thrombos. Diathes. Haemorrh.* (1969) **22**, 450.
89. Jobin, F., Tremblay, F., "Platelet Reactions and Immune Processes. II. The Inhibition of Platelet Aggregation by Complement Inhibitors," *Thrombos. Diathes. Haemorrh.* (1969) **22**, 460.

90. Jobin, F., Tremblay, F., Morissette, M., "Platelet Reactions and Immune Processes. III. The Inhibition of Platelet Aggregation by Chymotrypsin Substrates and Inhibitors," *Thrombos. Diathes. Haemorrh.* (1970) **23**, 110.

91. Pfueller, S. L., Lüscher, E. F., "The Effects of Immune Complexes on Blood Platelets and Their Relationship to Complement Activation," *Immunochemistry* (1972) **9**, 1151.

92. Henson, P. M., "Mechanisms of Release of Constituents from Rabbit Platelets by Antigen–Antibody Complexes and Complement. II. Interaction of Platelets with Neutrophils," *J. Immunol.* (1970) **105**, 490.

93. Dolbeare, F. A., "Platelet Aggregation as a Quantitative Immunologic Technique," *Immunol. Communs.* (1973) **2**, 65.

94. Davis, R. B., Holtz, G. C., "Clumping of Blood Platelets by Heat Aggregated Gammaglobulin—Turbidimetric Characteristics and Synergistic Effects of Vasoactive Amines," *Thrombos. Diathes. Haemorrh.* (1969) **21**, 65.

95. De Gaetano, G., Vermylen, J., Verstraete, M., "A Specific Human Immunoglobulin G Preparation Causing Platelet Clumping through a Release Reaction," *Acta Med. Scandin. Suppl.* (1971) **525**, 157.

96. Baier, R. E., Shafrin, E. G., Zisman, W. A., "Adhesion: Mechanisms that Assist or Impede It," *Science* (1968) **162**, 1360.

97. Richardson, P. D., "The Performance of Membranes in Vascular Prosthetic Devices," *Bull. N.Y. Acad. Med.* (1972) **48**, 379.

98. Friedman, L. I., Leonard, E. F., "Platelet Adhesion to Artificial Surfaces: Consequences of Flow, Exposure Time, Blood Condition, and Surface Nature," *Fed. Proc.* (1971) **30**, 1641.

99. Leonard, E. F., Grabowski, E. F., Turitto, V. T., "The Role of Convection and Diffusion on Platelet Adhesion and Aggregation," *Ann. N.Y. Acad. Sci.* (1972) **201**, 329.

100. Dintenfass, L., Rozenberg, M. C., "Effect of Temperature, Velocity Gradient and I.V. Heparin on *In Vitro* Blood Coagulation and Platelet Aggregation," *Thrombos. Diathes. Haemorrh.* (1967) **17**, 112.

101. Lyman, D. J., Klein, K. G., Brash, J. L., Fritzinger, B. K., "The Interaction of Platelets with Polymer Surfaces. I. Uncharged Hydrophobic Polymer Surfaces," *Thrombos. Diathes. Haemorrh.* (1970) **23**, 120.

102. Lycette, R. M., Danforth, W. F., Koppel, J. L., Olwin, J. H., "Aggregation of Human Blood Platelets and Its Correlation with Their Uptake of Phospholipid-Specific Dyes," *Amer. J. Clin. Pathol.* (1970) **54**, 692.

103. Mehrishi, J. N., "The Human Blood Platelets as a Molecular Mosaic—Its Role in Aggregation and Thrombus Formation," *Thrombos. Diathes. Haemorrh.* (1971) **26**, 377.

104. Salzman, E. W., "The Events that Lead to Thrombosis," *Bull. N.Y. Acad. Med.* (1972) **43**, 225.

105. Falb, R. D., Takahashi, M. T., Grode, G. A., Leininger, R. I., "Studies on the Stability and Protein Adsorption Characteristics of Heparinized Polymer Surfaces by Radioisotope Labeling Techniques," *J. Biomed. Mater. Res.* (1967) **1**, 239.

106. Klings, M., Adams, A. L., Vroman, L., "Effects of Protamine Sulfate, Polybrene and Heparin on the Behavior of Plasma, Plasma Proteins, Platelets and Factor XII Activity at Interfaces," *Thrombosis Res.* (1972) **1**, 507.

107. MacRitchie, F., "The Adsorption of Proteins at the Solid/Liquid Interface," *J. Colloid Interface Sci.* (1972) **38**, 484.

108. Sahyun, M. R. V., "Binding of Aromatic Compounds to Bovine Serum Albumin," *Nature* (1966) **209**, 613.

109. Rossi, E. C., "The Effect of Albumin Upon the Loss of Enzymes from Washed Platelets," *J. Lab. Clin. Med.* (1972) **79**, 240.

110. Vroman, L., "Effects of Hydrophobic Surfaces Upon Blood Coagulation," *Thrombos. Diathes. Haemorrh.* (1964) **10**, 455.
111. Vroman, L., "What Factors Determine Thrombogenicity?," *Bull. N.Y. Acad. Med.* (1972) **48**, 302.
112. Vroman, L., "Surface Activity in Blood Coagulation," in "Blood Clotting Enzymology," W. H. Seegers, Ed., p. 279, Academic, New York, 1967.
113. Vroman, L., "Correlation between Behavior of Plasma, Proteins and Platelets on Biomaterials," *6th Annual Contractors' Conf. Artif. Kidney-Chronic Uremia Progr., Nation. Inst. Arthr., Metab. & Digest. Dis.,* 1973, 130.
114. Vroman, L., Adams, A. L., Klings, M., Fischer, G., "Blood and Air," *Proc. Symp. Blood Bubble Interaction in Decompression Sickness,* K. N. Ackles, Ed., Defense & Civil Inst. Environmental Med. (DCIEM 73-CP-960) (1973) 49–70.

RECEIVED December 4, 1973. Work supported in part by agreement NIH-IA #02-88038 with the Artificial Kidney-Chronic Uremia Program, National Institute of Arthritic, Metabolic, and Digestive Diseases, NIH, Bethesda, Md.

13

Surface Chemistry of Dental Integuments

RONALD P. QUINTANA

Department of Medicinal Chemistry, College of Pharmacy, University of
Tennessee Center for the Health Sciences, Memphis, Tenn. 38163

*The acquired pellicle, a proteinaceous film adsorbed onto
tooth surfaces, plays a significant role in the development of
dental plaque which is a prime etiologic factor in dental
caries and in periodontal disease. The chemical nature and
function of pellicle in plaque genesis are reviewed along
with the interactivities contributing to the integrity of
plaque. Among antiplaque agents under current investigation is chlorhexidine, 1,1'-hexamethylenebis[5-(p-chloro-
phenyl)biguanide]. The relevance of surface chemistry to
the compound's efficacy is summarized.*

A recently published monograph (*1*) dealt with the relevance of surface chemistry and physics in the control of dental-deposit-mediated
diseases, with the characteristics and nature of dental deposits, with
their predilection for pathogenesis, with demographic and prevalence
patterns of disorders associated with dental integuments, and with con-

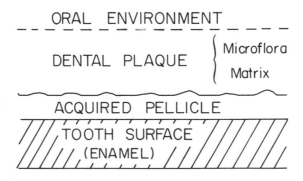

*Figure 1. Schematic representation of dental
integuments*

ventional prophylactic practices. This paper summarizes the chemistry of dental integuments, including pertinent new findings.

While there has been considerable confusion in regard to nomenclature of dental integuments, Figure 1 delineates the terms acquired pellicle and dental plaque as they are used in this chapter. The acquired pellicle is a structureless, colorless, translucent, almost invisible film, 0.05–1μ thick, which forms rapidly and spontaneously on clean tooth surfaces and adheres there rather tenaciously; it has a salivary origin and a protein nature. Its formation is independent of bacteria. While it may be removed by abrasives or by scaling, it reforms quickly (*2, 3, 4, 5, 6*). Dental plaque, on the other hand, constitutes "a soft amorphous granular deposit which accumulates on the surfaces of teeth" (*3*) (*see* Figure 2). When mature, it comprises "a myriad of microorganisms embedded in a relatively insoluble matrix that is largely of microbial origin, but is at least partially of salivary origin. It contains little food debris and only a few epithelial cells" (*2*).

A   B

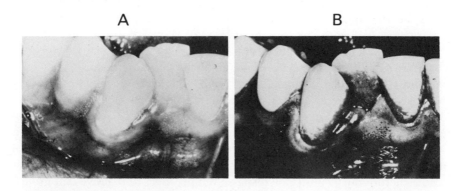

Figure 2.   Dental plaque on human teeth: A, unstained dentition; B, the same teeth stained with solution disclosing the presence of plaque (dark areas on tooth surfaces)

Many factors affect the development of plaque on a tooth, including (a) the location of the tooth in the mouth, (b) the anatomy of the tooth and that of surrounding tissues, (c) the nature of the tooth surface, (d) the presence of nutrients from the diet, saliva, and gingival fluid, (e) the time of exposure of the tooth in the oral environment, etc. (*7*). While plaque forms on almost any smooth enamel surface where abrasion is minimal, it seems to accumulate more readily in interproximal areas adjacent to gingival margins, in enamel defects, and at other sites which are not self-cleansing. Its adherence to the underlying surface is relatively strong. Rinsing or spraying with water will not remove it completely, but mechanical cleansing such as toothbrushing with dentifrices (2, 3) can.

Formation of Acquired Pellicle

Pellicle is believed to result from selective adsorption of salivary proteins and glycoproteins onto the hydroxyapatite crystallites comprising the enamel surface. Pure hydroxyapatite has the composition $Ca_{10}(PO_4)_6(OH)_2$ (see Ref. 8 for details of its crystal structure and properties). While many details of the composition of pellicle require further investigation, a sialic-acid-containing glycoprotein in human saliva appears to have a high affinity for hydroxyapatite, even in competition with other glycoproteins (9) (cf. Ref. 5). Moreover, Rölla and Sönju reported sulfated glycoproteins to be present in pellicle formed on teeth in vivo; evidence was already available for the occurrence of such compounds in saliva and for their affinity for hydroxyapatite (10). Considering the prominent adsorption characteristics of acidic proteins or glycoproteins, an important factor in the adsorption process may be the affinity between calcium of hydroxyapatite and negatively charged carboxylate (or sulfate) groups of saliva glycoproteins (9, 11).

Analyses for the carbohydrate components of pellicle glycoprotein have revealed the presence of sialic acid, fucose, glucose, galactose, mannose, N-acetylglucosamine, and N-acetylgalactosamine (2, 4). Sialic acid and fucose typically occur in terminal positions of the oligosaccharide chain, whereas N-acetylglucosamine or N-acetylgalactosamine moieties link the saccharide segment to the protein core (2). Schrager and Oates' studies on the principal glycoprotein from human mixed saliva have indicated the latter to possess "a basic homogeneous composition and structure but [to be] polydisperse with respect to reactive end groups [sulfate; sialic acid] and charge" (12). It was also noted that qualitative and quantitative differences exist among carbohydrate components of the principal salivary glycoprotein and those derived from plasma. Galactosamine was found in the salivary glycoprotein but not in plasma glycoproteins; conversely, mannose was absent in salivary but present in plasma glycoproteins. Thus pellicle glycoproteins comprise both salivary and plasma types.

The amino acid composition of the protein portion of a 2-hour pellicle acquired in vivo on human teeth was the subject of a communication by Sönju and Rölla (13). Their data (Table I) are significant because, unlike previous studies in which pellicle films were obtained by acid demineralization of tooth enamel, their pellicle preparation was removed by careful scaling only. No significant differences were observed in findings for various teeth, an observation that supports selective adsorption of the salivary proteins. Small amounts of glucosamine and galactosamine, indicative of glycoproteins, were also detected; no diaminopimelic acid or muramic acid was found, however, pointing to the absence of bacterial contamination. Sönju and Rölla further observed that the amount of pellicle deposited on human teeth continued to increase for 1.5 hrs and then leveled off. This is consistent with prior findings of others that the film adjacent to the enamel surface had a maximum thickness of about 1μ even after several days.

Table I.   Amino Acid Composition of 2-Hour Pellicle[a]

Amino Acid	*Content,[b] moles/100 moles*
Asp	7.3
Thr	3.7
Ser	9.6
Glu	12.8
Pro	2.2
Gly	17.0
Ala	7.3
Val	4.1
Cys	0.9
Ile	2.9
Leu	6.1
Tyr	1.6
Phe	2.9
Orn	6.4
Lys	7.0
His	4.1
Arg	4.1

[a] From Ref. *13*.
[b] Average values.

Using contact angle measurements, Baier determined the critical surface tension of the spontaneously acquired films deposited on a clean inorganic solid placed in human mouths for periods of 30 sec, 2 min, and 15 min (*14*) (*see* Table II). Concurrent determination of multiple attenuated internal reflection IR spectra of the adsorbed films provided evidence for the presence of protein material. While the spectra revealed

Table II.   Data From Contact Angle Experiments on Surface Film Adsorbed onto Clean Solid Placed in the Mouth[a]

Exposure in Mouth, min	*Critical Surface Tension,[b] dynes/cm*	*Slope,[b] cm/dyne*
0.5	33.3	−0.014
2	34.3	−0.012
15	36.1	−0.008

[a] From Ref. *14*.
[b] Determined from Zisman plots (*14*).

no major changes in composition, the progressive surface-chemical changes—*i.e.*, increase in critical surface tension, decrease in slope of the Zisman plots—were interpreted to mean that the configuration rather than the composition of the film components was changing. Since the observed changes reflected a trend towards generally greater surface free energies, it was suggested "that dental integuments become more and more a 'comfortable' substrate for attachment, adhesion, and colonization by oral bacteria as time lapses."

Baier also conducted preliminary contact-angle studies on human teeth *in situ* (*14*). He determined values of critical surface tension to be approximately 32 dynes/cm for teeth which had been recently cleaned with toothpaste. A similar value was obtained for teeth not as recently cleaned. Taking other factors into consideration, a surface energy of 30 to 40 ergs/cm^2 represents the natural situation. The exact values are contingent upon individual eating and oral hygiene practices.

Development of Dental Plaque

Subsequent to the deposition of pellicle, predominantly coccoidal bacteria appear and proliferate. They originate, apparently, from enamel defects, from adjacent oral tissues, and from the saliva (*2*, *15*). In this connection, scanning electron microscope studies (*16*) suggest that gingival fluid is important for tooth-surface colonization. Bacterial accumulations occurred on a cleaned tooth within five minutes when fluid was readily available (from severely inflamed gingivae) whereas, in other instances, microorganisms were evident after a few hours, "small colonies of organisms [being] readily observed after 24 hours of continuous exposure to the oral environment."

More specifically, streptococci are among the first organisms to colonize on the tooth surface. In the oral cavity of man, *Streptococcus mutans* is particularly important since some correlations between caries activity and the presence of this organism in plaque have been made (*17*).

As noted by Gibbons and Spinell, two modes of adhesion occur in the development of dental plaque: (a) adhesion of the microorganisms to the tooth surface (or to the acquired pellicle) and (b) adhesion of the microbial cells to one another (*18*). One of the most distinguishing characteristics of S. *mutans* is the ability to produce extracellular polysaccharides from sucrose, and available data suggest that these compounds play a significant role in plaque genesis. The mechanism requires elucidation, but one explanation proposes that the ability of high-molecular-weight dextrans to trigger agglutination of S. *mutans* is involved (*19*).

Since low-molecular-weight dextrans do not promote aggregation, the process requires molecules of dimensions so as to effect binding of more than one streptococcal cell per molecule of dextran. The nature of the receptor sites on the surface of the microorganisms is not clear, but they may be associated with dextransucrase, the enzyme responsible for dextran synthesis. It is present as a cell-associated form and is known to have a high affinity for dextrans (19). Liljemark and Schauer's finding that treatment of oral streptococci with proteolytic enzymes reduced the microorganisms' adherence to saliva-coated or dextran-coated hydroxyapatite supports protein involvement on the bacterial cell surface in the reactive site binding the bacteria to hydroxyapatite (20). Their experiments also demonstrated that microorganisms not subjected to any pretreatment adhered considerably better to saliva-coated and to dextran-coated hydroxyapatite than to uncoated hydroxyapatite.

According to Rölla, ionic bonds are important in the associations between bacterial polysaccharides and protein-coated tooth surfaces (21). This was based on *in vitro* experiments on the affinity of dextran for hydroxyapatite powder coated with salivary glycoprotein; specifically, adsorption of dextran was inhibited by $0.5M$. Prior treatment of the coated hydroxyapatite with neuraminidase also reduced adsorption of dextran. Neuraminidase would be expected to reduce the negative charge of the protein coat by removing ionized sialic acid moieties. Of course, reduced adsorption of dextran could result from conformational changes induced in the pellicle protein by the neuraminidase treatment, as was apparently effected by $4M$ or $8M$ urea, in other experiments.

While important, dextran-mediated adhesion is not the only factor involved in dental plaque formation. In fact, most types of plaque bacteria neither elaborate dextran nor are agglutinated by it. Thus, in exploring other modes of interbacterial adhesion in the evolving plaque, Gibbons and Spinell have obtained evidence suggesting a role for salivary proteins (18). The salivary polymers they studied interacted with the microorganisms' surfaces while effecting aggregation. The composition of the compounds and the mechanism of their interactions were thought to be "vital for an understanding of plaque formation."

Thus, in plaque, both carbohydrate and protein material contribute to the matrix. While the origins of these components have been indicated, other hypotheses were advanced to explain incorporation of salivary proteins in plaque (18). For example, Leach has proposed that plaque proteins arise as a result of the action of glycosidases (*e.g.*, neuraminidase) on salivary glycoproteins (5). Data of Briscoe *et al.* suggest, however, that neuraminidase does not modify adsorption behavior of salivary proteins (22).

Surface-Chemical Aspects of Chlorhexidine in Plaque Control

Dental plaque constitutes a primary etiologic factor in both dental caries and in periodontal diseases; and since these conditions are among the most prevalent affecting mankind (23), one could not overestimate the value of truly effective plaque-control agents (24, 25, 26). Among compounds under current investigation for clinical use is the bisbiguanide chlorhexidine (Structure I), 1,1'-hexamethylenebis[5-*p*-(chlorophenyl)-biguanide] (*e.g.*, 24, 27–37). It is normally used as a digluconate, diacetate, or dihydrochloride salt.

$$Cl-\langle\bigcirc\rangle-N-\overset{\overset{NH}{\|}}{C}-N-\overset{\overset{NH}{\|}}{C}-N-CH_2CH_2CH_2CH_2CH_2CH_2-N-\overset{\overset{NH}{\|}}{C}-N-\overset{\overset{NH}{\|}}{C}-N-\langle\bigcirc\rangle-Cl$$

I

Chlorhexidine has a broad spectrum of antimicrobial action which, undoubtedly, is associated with its antiplaque effects. The mechanism has been studied extensively by Hugo and Longworth who report that chlorhexidine behaves similarly to cationic antibacterial agents such as surface-active quaternary ammonium compounds (38, 39, 40; *cf.* 41). The primary action involves adsorption of the compound onto the cell surface, the reactive sites being anionic. This is followed by a disruptive effect on the cytoplasmic membrane, changes in permeability of the latter, and leakage of intracellular components. High concentrations of chlorhexidine, however, may inhibit leakage by inhibiting autolytic enzymes, by forming a sealing layer (or layers) on the cell surface, by congealing the cytoplasmic membrane, or by precipitating intracellular components (proteins, nucleic acid).

Other studies have shown appropriate concentrations of chlorhexidine to inhibit adenosine triphosphatase of *Streptococcus faecalis* membrane. Electron micrographs have shown a so-called blistering of bacterial cell walls (42) by chlorhexidine. This was attributed to "cellular extrusion or to the accumulation of drug aggregates on the cell surface" (43).

In addition to antimicrobial activity, continuing investigation reveals other important modes of antiplaque efficacy. For example:

(a) Chlorhexidine adsorbs onto tooth surfaces, pellicle and plaque, and provides an antimicrobial milieu for a period of time through gradual release of the compound (28).

(b) The agent may also affect adherence of oral streptococci to teeth, as Kornman *et al.* have found that chlorhexidine significantly reduced affinity of dextran-encapsulated S. *mutans* for protein-coated hydroxyapatite (44). The compound may react with anionic functions of the protein, blocking sites of attachment normally available to the dextran-encapsulated bacteria. Chlorhexidine, adsorbed onto tooth surfaces, may act as a surface-modifying agent (24).

(c) In view of the importance of salivary glycoproteins in plaque-development, the coincidence (45) of chlorhexidine concentrations precipitating salivary glycoproteins and those effecting clinical efficacy is significant. The pH-dependence of chlorhexidine–protein interaction suggests the importance of the net negative charge of the proteins.

(d) In addition to chlorhexidine's efficacy in preventing accumulations of plaque and calculus, the compound has capabilities of affecting already formed plaque on teeth (24). In this connection, Tanzer *et al.* noted partial disruption of *in vitro* plaque preparations when these were treated with chlorhexidine (46).

A detailed study of chlorhexidine's surface-chemical characteristics was undertaken in our laboratories. This included work with a series of carefully selected analogs comprising segments of the parent molecule (Structure II; R = n-hexyl, n-propyl, or H) or extensions of these (Structure II; R = n-octyl or n-dodecyl).

II

The gradual changes in the chemical structure and physical properties of the congeners facilitated identification of molecular functions and/or physicochemical characteristics associated with chlorhexidine's surface-active behavior. While the study explored principally interactions between the evaluant entities and monomolecular films providing carboxyl, hydroxyl, and amide groups at the monolayer/water interface, the compounds' comparative effects at air/water, n-hexane/water, and hydroxyapatite/water interfaces were also investigated (47, 48, 49).

The work with monolayer systems (stearic acid, stearyl alcohol, N-octadecylacetamide) revealed the importance of interactions between dicationic chlorhexidine molecules and anionic carboxylate groups in effecting substantial increases in the surface pressure of the films, and, in the instance of the chlorhexidine analogs, also pointed to the significance of a sufficiently lengthy [5]N-alkyl substituent (Structure II, R = n-hexyl) in effecting film penetration. Lacking the structure of a classical surfactant, chlorhexidine, with its alternating hydrophobic and hydrophilic moieties, has been designated as a specific surface-active agent, some specific group(s) being responsible for its accumulation at interfaces (50). The comparative evaluation of the effect of chlorhexidine and the delineated congeners on surface and interfacial tension and on adsorption onto hydroxyapatite suggested that the hexamethylene chain is a major contributor to the surface activity of chlorhexidine. Details of studies on the compounds' antiplaque efficacy *in vitro* (51) will be reported elsewhere.

Acknowledgment

The author thanks James W. Clark for the photographs constituting Figure 2.

Literature Cited

1. Lasslo, A., Quintana, R. P., Eds., "Surface Chemistry and Dental Integuments," Charles C. Thomas, Springfield, Illinois, 1973.
2. Burnett, G. W., Pennel, B. M., "Surface Chemistry and Dental Integuments," A. Lasslo and R. P. Quintana, Eds., pp. 3–73, Charles C. Thomas, Springfield, Illinois, 1973.
3. Glickman, I., "Clinical Periodontology," 4th ed., pp. 291–314, W. B. Saunders, Philadelphia, 1972.
4. Mayhall, C. W., *Aeromed. Rev.*, Review 5–71, United States Air Force School of Aerospace Medicine, Aerospace Medical Division (AFSC), Brooks Air Force Base, Texas, 1971.
5. Leach, S. A., *Ala. J. Med. Sci.* (1968) 5, 247–255.
6. Meckel, A. H., *Arch. Oral Biol.* (1965) 10, 585–597.
7. Egelberg, J., "Dental Plaque," W. D. McHugh, Ed., pp. 9–16, E. and S. Livingstone, Edinburgh, 1970.
8. Zimmerman, S., "Dental Biochemistry," E. P. Lazzari, Ed., pp. 70–91, Lea and Febiger, Philadelphia, 1968.
9. Rölla, G., Mathiesen, P., "Dental Plaque," W. D. McHugh, Ed., pp. 129–140, E. and S. Livingstone, Edinburgh, 1970.
10. Rölla, G., Sönju, T., *J. Dent. Res.* (Special Issue) (1973) 52, 133, 193.
11. Bernardi, G., Kawasaki, T., *Biochim. Biophys. Acta* (1968) 160, 301–310.
12. Schrager, J., Oates, M. D. G., *Arch. Oral Biol.* (1971) 16, 287–303.
13. Sönju, T., Rölla, G., *Caries Res.* (1973) 7, 30–38.
14. Baier, R. E., "Surface Chemistry and Dental Integuments," A. Lasslo and R. P. Quintana, Eds., pp. 337–391, Charles C. Thomas, Springfield, Illinois, 1973.
15. Kleinberg, I., "Advances in Oral Biology," Vol. 4, P. H. Staple, Ed., pp. 43–90, Academic, New York, 1970.
16. Saxton, C. A., *Caries Res.* (1973) 7, 102–119.
17. Krasse, B., *Ala. J. Med. Sci.* (1968) 5, 267–268.
18. Gibbons, R. J., Spinell, D. M., "Dental Plaque," W. D. McHugh, Ed., pp. 207–215, E. and S. Livingstone, Edinburgh, 1970.
19. Gibbons, R. J., Fitzgerald, R. J., *J. Bacteriol.* (1969) 98, 341–346.
20. Liljemark, W. F., Schauer, S. V., *J. Dent. Res.* (Special Issue) (1973) 52, 130.
21. Rölla, G., *Arch. Oral Biol.* (1971) 16, 527–533.
22. Briscoe, Jr., J. M., Pruitt, K. M., Caldwell, R. C., *J. Dent. Res.* (1972) 51, 819–824.

23. Donnelly, C. J., "Surface Chemistry and Dental Integuments," A. Lasslo and R. P. Quintana, Eds., pp. 74–192, Charles C. Thomas, Springfield, Illinois, 1973.
24. Quintana, R. P., Lasslo, A., Clark, J. W., Baier, R. E., "Surface Chemistry and Dental Integuments," A. Lasslo and R. P. Quintana, Eds., pp. 392–413, Charles C. Thomas, Springfield, Illinois, 1973.
25. Clark, J. W., Jurand, J. G., Miller, C. D., "Surface Chemistry and Dental Integuments," A. Lasslo and R. P. Quintana, Eds., pp. 193–275, Charles C. Thomas, Springfield, Illinois, 1973.
26. McHugh, W. D., *J. Periodont. Res.* (1972) Suppl. 10, 40–41.
27. Ochsenbein, H., *Schweiz. Mschr. Zahnheilk.* (1973) **83**, 819–827.
28. Hamp, S., Lindhe, J., Löe, H., *J. Periodont. Res.* (1973) **8**, 63–70.
29. Warner, V. D., Mirth, D. B., Turesky, S. S., Glickman, I., *J. Pharm. Sci.* (1973) **62**, 1189–1191.
30. Turesky, S., Glickman, I., Sandberg, R., *J. Periodontol.* (1972) **43**, 263–269.
31. Cumming, B. R., Löe, H., *J. Periodont. Res.* (1973) **8**, 57–62.
32. Löe, H., Mandell, M., Derry, A., Schiøtt, C. R., *J. Periodont. Res.* (1971) **6**, 312–314.
33. Löe, H., *Int. Dent. J.* (1971) **21**, 41–45.
34. Löe, H., Schiøtt, C. R., *J. Periodont. Res.* (1970) **5**, 79–83.
35. Schiøtt, C. R., Löe, H., Jensen, S. B., Kilian, M., Davies, R. M., Glavind, K., *J. Periodont. Res.* (1970) **5**, 84–89.
36. Rölla, G., Löe, H., Schiøtt, C. R., *J. Periodont. Res.* (1970) **5**, 90–95.
37. Davies, R. M., Jensen, S. B., Schiøtt, C. R., Löe, H., *J. Periodont. Res.* (1970) **5**, 96–101.
38. Hugo, W. B., Longworth, A. R., *J. Pharm. Pharmacol.* (1966) **18**, 569–578.
39. *Ibid.* (1964) **16**, 751–758.
40. *Ibid.* (1964) **16**, 655–662.
41. Emilson, C. G., Ericson, T., Heyden, G., Lilja, J., *J. Periodont. Res.* (1972) 189–191.
42. Franklin, T. J., Snow, G. A., "Biochemistry of Antimicrobial Action," pp. 51–53, Academic, New York, 1971.
43. Hugo, W. B., Longworth, A. R., *J. Pharm. Pharmacol.* (1965) **17**, 28–32.
44. Kornman, K. S., Clark, W. B., Kreitzman, S. N., *J. Periodont. Res.* (1972) Suppl. 10, 33–34.
45. Hjeljord, L. G., Rölla, G., Bonesvoll, P., *J. Dent Res.* (Special Issue) (1973) **52**, 191.
46. Tanzer, J. M., Reid, Y., Reid, W., *Antimicrob. Ag. Chemother.* (1972) **1**, 376–380.
47. Quintana, R. P., Fisher, R. G., Lasslo, A., *J. Dent. Res.* (1972) **51**, 1687.
48. Fisher, R. G., Quintana, R. P., Boulware, M. A., *J. Dent. Res.*, in press.
49. Fisher, R. G., Quintana, R P., *J. Dent. Res.*, in press.
50. Heard, D. D., Ashworth, R. W., *J. Pharm. Pharmacol.* (1968) **20**, 505–512.
51. Dr. J. M. Tanzer, Department of General Dentistry, School of Dental Medicine, University of Connecticut Health Center, private communication.

RECEIVED December 4, 1973. This work was supported by the National Institute of Dental Research, National Institutes of Health, through Grant DE-03139.

14

Demonstration of the Involvement of Adsorbed Proteins in Cell Adhesion and Cell Growth on Solid Surfaces

R. E. BAIER and L. WEISS

Department of Biophysics, Roswell Park Memorial Institute, Buffalo, N. Y.

Experiments with solids immersed in cell culture media, with and without living cells present, convincingly demonstrate the absolute condition that protein-dominated films accumulate at the solid–liquid boundaries prior to cell adhesion. The reality and speed of this spontaneous, adsorptive event are documented by using infrared-transmitting, multiple internal reflection plates as the immersed solid with care to prevent film transfer from gas–liquid interfaces.

This brief report highlights the involvement of adsorbed (conditioning) films of proteins in the behavior of cells at solid surfaces, particularly during routine experiments designed to investigate the cell adhesion, cell migration, cell growth, or cell aggregation process. Adsorbable macromolecules, used in supplements to otherwise well-defined culture media, affect such experiments greatly.

All our methods have been reported (*1, 2, 3*). The cells were derived from Ehrlich–Lettre' hyperdiploid ascites tumors (EAT) kept in culture at the Roswell Park Memorial Institute. For details of the culture method, cultured cells, and culture media, *see* Refs. *4* and *5*.

Typical Results

An IR spectrum of freshly prepared RPMI 1630 cell culture medium supplemented with 5% calf serum, the medium used routinely to culture viable cells, is presented in Figure 1. Figure 1 is an IR spectrum of centrifugally concentrated and washed EAT cells which were cultured in this medium. The IR spectrum of the salt-free residue from the supernatant liquid obtained after centrifuging the cells constitutes Figure

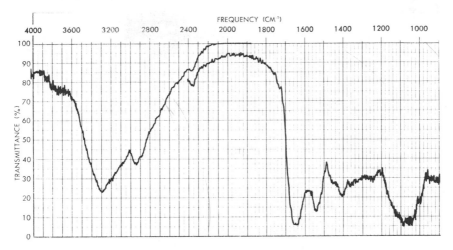

Figure 1.   IR spectrum of freshly prepared RPMI 1630 cell culture medium (+ 5% calf serum supplement) with no cells present

3. Cell lines usually cannot be successfully propagated in various synthetic media without supplemental (and, unfortunately, poorly characterized) serum components. This requirement for supplemental protein materials probably reflects—at least initially—the need for adsorbable proteinaceous constituents which will spontaneously accumulate at and favorably modify (for cell adhesion and propagation) the surfaces of the culture containers. There was essentially no spontaneous film adsorption from freshly prepared RPMI 1630 culture medium without

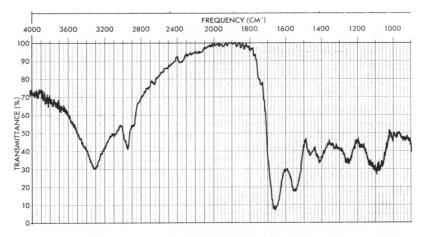

Figure 2.   IR spectrum of centrifugally concentrated and washed cells which had been cultured in RPMI 1630 medium (+ 5% calf serum supplement)

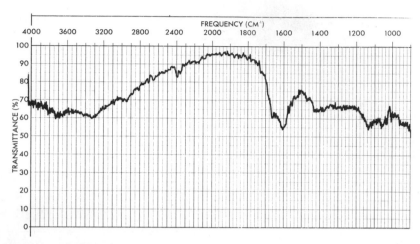

Figure 3.   IR spectrum of saltfree residue from cellfree supernatant of centrifuged RPMI 1630 medium (+ 5% calf serum supplement) in which cells had been cultured

added calf serum during a 10-min immersion of an internal reflection prism with which even monolayer amounts of organic contamination can be detected, and there were no significant absorption bands in the IR spectrum (Figure 4). By contrast, an organic film was spontaneously adsorbed onto another prism during a 10-min immersion in a freshly pre-pared cell culture medium with a 5% calf serum supplement (Figure 5). This spontaneously adsorbed film was dominated by a glycoprotein com-

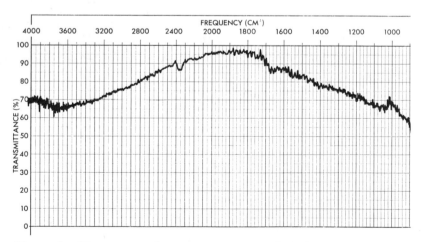

Figure 4.   IR spectrum demonstrating essential absence of spontaneous film adsorption from freshly prepared RPMI 1630 cell culture medium (without added calf serum) during 10-min immersion with no cells present

Technique: MAIR, sensitive to even a monolayer of organic contamination

ponent. Once adsorbed, it was completely insoluble in water and in saline media; thus adsorption had denatured the film sufficiently so that its original solubility characteristics had been lost. Water or saline extraction usually removes glycoprotein or proteoglycan films which are not strongly bound to the solid–liquid boundary (3). Similarly a film was spontaneously adsorbed during a 10-min prism immersion in used (often considered exhausted) culture medium in which cells had been propagated for some time (Figure 6); glycoprotein adsorption, even from the exhausted medium, was rapid and irreversible.

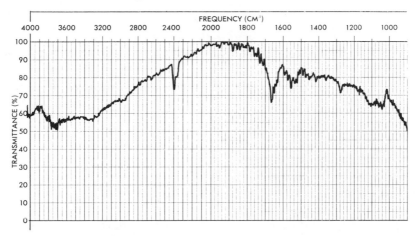

Figure 5. IR spectrum of film spontaneously adsorbed during 10-min immersion in freshly prepared cell culture medium (RPMI 1630 + 5% calf serum supplement) with no cells present

The presence of these spontaneously adsorbed (*i.e.* adsorbed during minimum exposure times), thin films on polymeric substrates such as polystyrene culture dishes and glass plates usually cannot be demonstrated by other spectroscopic methods. For example, the modified internal reflection spectroscopic technique, in which an auxiliary salt prism (usually the malleable salt KRS-5) is pressed against the demonstrably (by other techniques) protein-coated substrates, always fails; this technique is not sensitive enough for the near monolayer ranges required for this demonstration. The required sensitivity is achieved only when adsorption occurs directly on the face of a clean, or thin film-coated, internal reflection element.

A current goal is to establish long term biocompatibility of engineering and structural materials, especially for prosthetic devices, by promoting the rapid attachment and proliferation of living cells compatible with the site of ultimate implantation. Fine fabric meshes are of great utility in this application. The great sensitivity (and simplicity) of

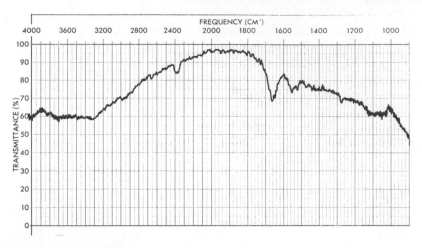

Figure 7.   IR spectrum of cell culture on 5-μ thick, polypropylene-based, microfabric substrate

monitoring cell adhesive events within a microfabric layer by internal reflection spectroscopic methods (*1, 2, 3*) is illustrated by Figures 7 and 8. Figure 7 is the IR spectrum after 2 hrs of cell culture within a 5-μ thick, polypropylene fiber, microfabric substrate (applied to a germanium internal reflection prism at Union Carbide Corp. through the courtesy of Joseph Byck). Light microscopy revealed that the large protein spectral bands resulted from extensive cellular coverage; thus, even during this time, cell attachment and growth (as indicated by the abundance of cellular material) were excellent. The cells were extremely resistant to

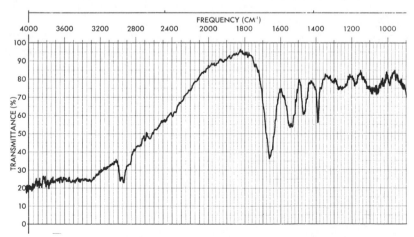

Figure 6.   IR spectrum of film spontaneously adsorbed during 10-min immersion in used culture medium (RPMI 1630) in which cell growth had occurred after removal of cells by centrifugation

detachment even by lengthy and vigorous rinsing in saline media and water. By contrast, in experiments to culture the same cell line on bare germanium prisms, there was little permanent cell attachment in two hours. Despite apparent success in using many microfabrics to promote cell attachment and cell growth, our experiments also demonstrated some worrisome features of this material. For example, simple soaking in distilled water seriously leached residual microfabric contaminants from the polypropylene mesh (Figure 8); depending on their composition, such components leached into a delicate cell culture medium might seriously interfere with attachment or propagation of less viable cell lines.

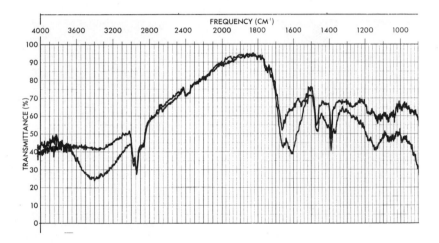

Figure 8.   IR spectra illustrating severe leaching of microfabric contaminants

Lower trace, 5-μ *thick polypropylene-based microfabric as prepared; and* upper trace, *same fabric after soaking in distilled water*

Discussion

Only the methods used in this study (*1, 2, 3*) can detect and follow the initial events at the boundary between a solid substrate and the biological milieu. This extraordinary sensitivity derives from the ability to monitor events from within the substrate itself by making the substrate capable of supporting multiple attenuated internal reflection (MAIR) at a variety of useful spectroscopic wavelengths. In other studies, it was concluded that adsorbed protein on solid substrates is of essentially native (*i.e.* solution or volume phase) configuration and present in significant thickness (*6*). Those studies used a different version of the internal reflection spectroscopic method wherein a prism element was forcefully pressed against a plastic or foil surface which had accumulated, after very long immersion periods, equilibrium amounts of adsorbed protein. Further, those studies were generally conducted in the absence of the normally attaching cells which would have modified the process of film buildup before it reached equilibrium. Such studies detected only the nature of the protein adsorbed onto other layers of protein already at the solid–solution interface. The primary, interfacially localized layers cannot be detected by methods other than those used here. There is little relevance to speculation about the effect of these secondary protein layers which are probably not even present when cells first arrive and attach to solid surfaces.

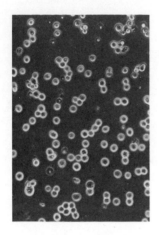 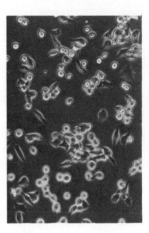

Figure 9.   Phase contrast micrographs of live cells adherent to glass
Culture medium, RPMI 1640; left, *with no added protein; and* right, *with 20% fetal calf serum supplement*

Work is just now beginning on the effect of substrate chemistry on protein adsorption from supplemented culture media and on the secondary effect of these layers on cell adhesion processes (7). In preliminary work, we determined that a marked differentiation in cell morphology reflects the presence of adsorbed proteinaceous components at a substrate–solution interface. For example, when living cells settle to clean glass surfaces in the absence of supplemental protein, the cells remain rounded and poorly adhesive, and they are unable to develop good cell monolayer coverage of the surface (*see* Figure 9). On the other hand, the presence of supplemental protein in the medium induces greater cell adhesion; increased adhesion is apparently mediated by extensive morphological changes in the cell (many elongated forms), and eventually the cells completely cover the solid surface.

Literature Cited

1. Baier, R. E., Loeb, G. I., "Multiple Parameters Characterizing Interfacial Films of a Protein Analogue," in "Polymer Characterization: Interdisciplinary Approaches," C. D. Craver, Ed., pp. 79–96, Plenum, New York, 1971.
2. Baier, R. E., Dutton, R. C., "Initial Events in Interactions of Blood with a Foreign Surface," *J. Biomed. Mater. Res.* (1969) **3**, 191–206.
3. Baier, R. E., "Applied Chemistry at Protein Interfaces," ADVAN. CHEM. SER. (1975) **145**, 1.
4. Moore, G. E., Sandburg, A. A., Ulrich, K., "Suspension Cell Culture and *In Vivo* and *In Vitro* Chromosome Constitution of Mouse Leukemia L-1210," *J. Nat. Cancer Inst.* (1966) **36**, 405–413.
5. Mayhew, E., "Electrophoretic Mobility of Ehrlich Ascites Carcinoma Cells Grown *In Vitro* and *In Vivo*," *Cancer Res.* (1968) **28**, 1590–1595.
6. Lyman, D. J., Brash, J. L., Chaikin, S. W., Klein, K. G., Carini, M., "Protein and Platelet Interaction with Polymer Surfaces," *Trans. Amer. Soc. Artif. Intern. Organs* (1968) **14**, 250–255.
7. Wilkins, J., "Adhesion of Ehrlich Ascites Tumor Cells to Surface-Modified Glass," Ph.D. Dissertation, State University of New York at Buffalo, 1974.

RECEIVED June 7, 1974.

15

Glycoprotein Adsorption in Intrauterine Foreign Bodies

R. E. BAIER

Department of Biophysics, State University of New York at Buffalo, Buffalo, New York

J. LIPPES

Department of Gynecology-Obstetrics, State University of New York at Buffalo, Buffalo, New York

Surface chemical evaluation of plastic and metal intrauterine devices removed after years of trouble-free use revealed the presence of a similar glycoproteinaceous coating in all cases. Spontaneous adsorption of such coatings probably precedes the mobilization of macrophages to the devices location and triggers the antifertility effect noted. Further research required on the isolation of the spontaneously accumulated glycoprotein and the relationship of this process to the original surface properties of the devices is outlined.

Intrauterine foreign bodies are effective in regulating fertility, and it is expected that they will become increasingly important in worldwide population control (*1, 2*). This investigation inquired into the potential modifications of the true surfaces of such intrauterine foreign bodies. These surface modifications may be responsible for initiation of secondary biological effects, and they may provide a general explanation of foreign body efficacy in fertility regulation. Intrauterine devices (IUDs) in humans mobilize macrophages to the endometrium. These cells may secrete substances which prevent fertilization of nidation, or they may prevent pregnancy by a combination of physiologic effects.

It has been suggested that a nonspecific foreign body reaction is the unifying principle in the mechanism of intrauterine foreign body action. In primates, the mechanism of action remains unknown; a number of subsequent biological changes which have been described are secondary phenomena resulting over the longer term from the initial foreign body reaction (*3*).

A recent review (*4*) considers a foreign body reaction induced in soft tissues by the presence of artificial implants to be a chronic inflammatory response; it was noted specifically that denaturation of proteins at the surface of the implanted material may evoke an immunologic response. This possibility has already been examined experimentally with respect to materials in contact with blood. Stern and co-workers (*5*) reasoned as follows:

If autologous protein were adsorbed and denatured by materials implanted in a major route of blood flow, they might present a constant antigenic stimulus with development of antibodies to the adsorbed protein coat or to small amounts of desorbed, denatured protein released into the circulation. Certain materials might be considered to have undesirably intense immunogenic properties to the extent that alteration of the surface stereochemistry during adsorption might expose greater numbers of antigenic sites in autologous proteins . . .

Our experiments, begun in 1968, have demonstrated the reality of such adsorption on the surfaces of intrauterine contraceptives; qualitative analyses were made of the spontaneously deposited material. Although evidence for protein denaturation and antibody response was not sought in this study, we do speculate on the importance of such a mechanism in fertility regulation.

Methods and Materials

The intrauterine foreign bodies examined in this study were mostly low-density polyethylene in the double-loop configuration known as the Lippes loop D. They were selected after voluntary removal from the wearers for the expressed purpose of allowing a planned pregnancy, or for personal convenience, after having served for as long as five years as effective regulators of fertility. All the devices were obtained from the normal clinical supply, and they were both inserted and removed according to standard gynecological office practice with no special handling at either insertion or withdrawal. Removed rings and loops were allowed to air dry and were then stored in small paper or gauze wrappings until selection for investigation. At that time, the individual devices were used directly as substrates for contact angle measurements according to the Zisman method (*6*), or for film and particle extraction for spectroscopic analysis. It was quite easy to scrape or flake off some of the dried, clear coating which was usually completely free of red blood cells in the regions sampled (or over the whole device). The scrapings were analyzed directly in standard x-ray diffraction equipment or by IR spectroscopy (*7*) after they were spread in a drop of distilled water over the surface of an internal reflection prism. All IR spectra reproduced in this report

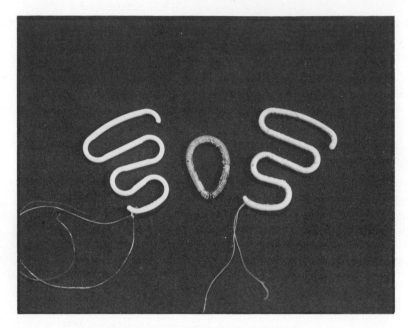

Figure 1.   Intrauterine foreign bodies

*Left: Lippes loop D prior to insertion; center: Hall Stone ring device
removed from uterus after 5 yrs of trouble-free use; and right: Lippes
loop D removed from uterus after 5 yrs of trouble-free use*

were originally recorded on Perkin-Elmer models 21 and 700 dual-beam
spectrophotometers.

Results

Figure 1 is a photograph of a Lippes loop prior to insertion, a
stainless-steel Hall Stone ring as it appeared after removal from a wearer
who had used it from 1964 through 1969 with no apparent complaints,
and a Lippes loop voluntarily removed from another patient for the purpose of a planned pregnancy after five years of use.

In Table I are listed the slowly advancing contact angles of a variety
of highly purified, diagnostic wetting liquids applied directly to the surfaces of the unimplanted loop, the loop removed after five trouble-free
years in the intrauterine cavity, and the spontaneously spread films of
distilled-water extracts of the surface coatings from the stainless-steel
spring device and the removed loop. In plots of the contact angle data
in the standard Zisman format (6) (Figure 2), these data are extrapolated
to critical surface tension values at the cosine-contact-angle-equals-one
axis. It is clear (cf. Table I and Figure 2) that, when judged in the dry
state as necessitated by contact angle techniques, the wetting properties
of the unused and the worn Lippes loops were considerably dissimilar,

which emphasizes the difference in true interfacial chemistry. Furthermore, the distilled-water-spread coatings from the biologically exposed, plastic and stainless-steel foreign bodies were distinguishable, but they were not sufficiently dissimilar to suggest any great or fundamental differences in the chemistry of the coating material.

In Figure 3 are the IR spectra of the equilibrium surface deposits on the plastic Lippes loop and on the metal spring device after five years *in utero*. It is remarkable how similar these spectra are considering the difference in the IUD materials and that they had been exposed to the biological milieu of different women who lived in different parts of the country and had different habits of personal hygiene, sexual activity, and diet. Each spectrum can be most directly interpreted as representing a substantially pure or configurationally monodisperse glycoprotein on the basis of the major spectral bands present. Our identification of the material as a true glycoprotein is supported by the observation that, upon extensive rinsing or leaching of the film in distilled water or saline solutions, the relative ratios of the IR absorption bands that typify the polyamide backbone and the saccharide or carbohydrate groupings could

Table I.   **Contact Angles of Wetting Liquids on Surfaces of Unused and Used IUDs**

			Slowly Advancing Contact Angles, degrees		
Wetting Liquids			*Lippes Loop after Implantation*		*Coating from Steel Spring IUD[c]*
Liquid	*Surface Tension, dynes/cm at 20°C*	*Lippes Loop[a] as Implanted*	*Lippes Loop after Implantation 1963–1968[b]*	*Coating from Lippes Loop[c]*	*Coating from Steel Spring IUD[c]*
Water	72.8	99	68	80	75
Glycerol	63.4	88	65	67	71
Formamide	58.2	84	36	62	69
Thiodiglycol	54.0	70	46	54	51
Methylene iodide	50.8	46	67	51	44
Sym-tetrabromoethane	47.5	36	63	46	45
1-Bromonaphthalene	44.6	27	54	38	38
0-Dibromobenzene	42.0	19	44	34	31
1-Methylnaphthalene	38.7	12	41	28	25
Dicyclohexyl	33.0	0	19	22	0
n-Hexadecane	27.7	0	13	12	0
n-Decane	23.9	0	0	7	0

[a] Low density polyethylene.
[b] Coated with adsorbed film.
[c] Spread in distilled H_2O on germanium prism.

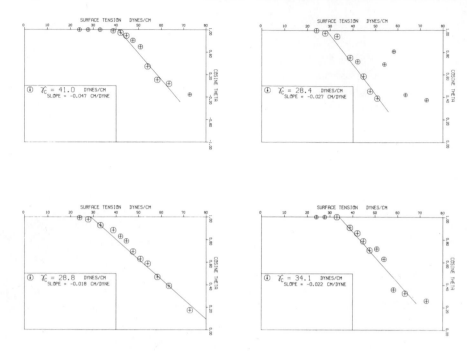

Figure 2.   Plots of contact angle data

Top left: unused Lippes loop D; top right: Lippes loop D after 5 yrs of intrauterine service; and bottom: glycoproteinaceous coatings extracted with distilled water from plastic (left) and steel (right) intrauterine foreign bodies that were in place 5 yrs. Points with small symbols were not included in straight-line fit.

not be changed. In particular, the oligosaccharide contribution could not be eliminated by differential extraction into aqueous solvents as does happen with pure polysaccharides and proteoglycans. Microscopic inspection of the surfaces of these devices revealed no evidence of erythrocyte accumulation. Rather, only a dried amorphous mass, typical of simply dried, adsorbed films of protein [even as spontaneously adsorbed from blood on intravascular prosthetic devices (8)] was found. X-ray diffraction studies of scrapings from these and similarly studied intrauterine foreign bodies showed evidence of an additional component which must have been present in only minor proportions on the basis of the IR spectra; it was tentatively identified as cholesterol stearate.

In Figure 4 are presented other typical IR spectra that characterize the equilibrium surface deposits on intrauterine foreign bodies of different materials and configurations with different periods of intrauterine exposure. Time of exposure varied from 4.5 yrs (the upper left-hand spectrum of the coating on a stainless-steel ring) to only 1.5 yrs (the lower right-hand spectrum of the coating on a Lippes loop); all exposures occurred between 1964 and 1969. It is clear from perusal of these 12 additional spectra that—independent of the wearer of the IUD, its composition and configuration, and the period of intrauterine residence—all surface coatings are essentially the same, *i.e.* predominantly glycoproteins. We conclude that the initial and sometimes permanent modification of nonphysiologic materials placed within the uterine cavity is provided by spontaneously adsorbed layers of proteinaceous material which may be involved in interfering with reproduction.

Discussion

Mechanism of Fertility Regulation. IUDs are successful means of contraception; they are particularly suitable for large-scale use when motivation is minimum and continuous surveillance of the population is difficult. For most couples, the use of an IUD in family planning is

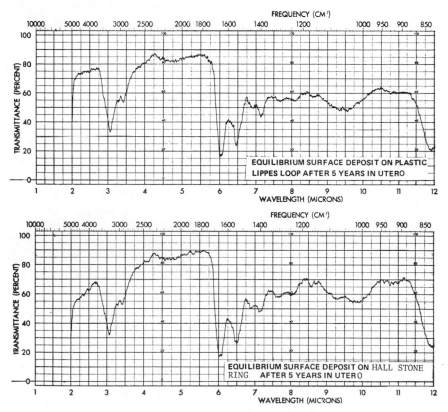

Figure 3.   Internal reflection IR spectra characterizing the equilibrium surface layers coating plastic (top) and steel (bottom) intrauterine foreign bodies, each after 5 yrs in utero

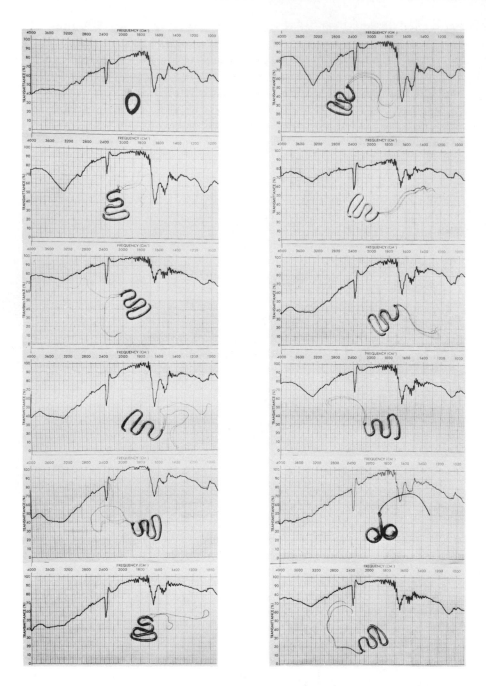

Figure 4. Internal reflection IR spectra illustrating the similarity of the coating acquired in utero *on the surfaces of various intrauterine foreign bodies that were in place for various lengths of time*

considered effective, safe, and acceptable. Its use is dissociated from the time and circumstances of sexual activities, a desirable quality which characterizes few contraceptive methods. In reflection of the fact that intrauterine foreign bodies are economical and that only paramedical skills are required for programs of mass application, the number of women in the world currently wearing a modern version of the intrauterine contraceptive device has increased to more than six million (*1*). Oral contraceptive pills and intrauterine foreign bodies are the most effective contraceptives currently available. Although oral contraceptives are more effective in preventing pregnancies, IUDs have a higher continuation rate (*9*).

What is the mechanism by which inert or relatively inert foreign bodies exert their antifertility function? One possible explanation is the induction of a mild antibody–rejection response to the presence of an indwelling, reasonably large surface-area, foreign protein—the configurationally distorted glycoproteinaceous layer that is spontaneously acquired by the IUD immediately after insertion from uterine secretions.

We speculate that altered glycoproteins forming at the surface of an IUD inhibit reproduction by evoking an inflammatory response to the long-term, indwelling nonphysiologic material (otherwise inert enough to minimize pain, bleeding, abortive, and other complications). This response includes mobilization of macrophages which may be sensitized to sperm antigens. It is possible that macrophages so sensitized may even reach the oviduct. Supporting this speculation are the observations in humans that uterine foreign bodies do cause infiltration of the endometrium by macrophages (*10*) whereas in rats leukocytes are found in the endometrium (*11*). In 50% of women, foreign particles (India ink) placed in the uterus can be found in the oviduct (*12*). Although it has not been demonstrated that capacitation of human sperm is necessary for human fertilization, adherence of sperm to the zona pellucida of the ovum is generally species specific (*13*). Could alteration of uterine or oviductal proteins interfere with sperm adherence to the zona pellucida? This might be caused by removal or addition of modified, adsorbable glycoproteins. On the basis of observations in rabbits, earlier workers suggested that inhibition of sperm capacitation might possibly be an antifertility effect of IUDs (*14*).

It is not known whether the predominant effect in the human female is interference with sperm physiology, fertilization, ovum transport, or nidation. Greater quantities of the glycoprotein which accumulates on all IUD surfaces should be isolated for use in studies to determine specifically which reproductive processes are inhibited.

Some alternative theories seem less probable. Tubal patency tests have established that IUDs do not cause mechanical obstruction or spasm of the tubes (15). Premature expulsion of the ovum is not the mechanism of the antifertility action of intrauterine foreign bodies in most mammalian species and certainly not in primates (16). The incidence of uterine carcinoma is not altered in human recipients of intrauterine foreign bodies (17, 18).

Surface Properties of IUDs. Most IUDs currently in use cause only minimum changes in the uterine endometrium (19). Excessive menstrual bleeding, intermenstrual spotting, and bleeding between menstrual periods have been noted as sources of annoyance, and they constitute a major reason for removal of IUDs. Such bleeding is probably secondary to local hyperemia, edema, pressure necrosis, and sometimes endometritis (20). After IUDs are inserted into the uterus, some patients complain of pain and low backache, but pain is a minor reason for discontinuing intrauterine contraception.

It was reported that the degree of adverse reaction differs with the type and quantity of material in the IUD (21), and this subject should concern specialists in the surface chemistry of biological systems. It is important to ascertain whether their surface properties play any significant role in IUD acceptability or efficacy because there have been numerous observations that specific materials are better accepted in a variety of biological environments than are most other materials (22). It may be stated tentatively that yes, surface properties are significant since very high surface-energy metal devices irritate the uterus more than only relatively high surface-free-energy poly(vinyl chloride) and nylon. The most inert, lower surface-free-energy polyethylenes apparently irritate the least, approaching the passivity of natural rubber, but none of these materials equals the non-irritating performance of medical grade silicone rubber which has a critical surface tension in the biocompatible range of earlier criteria (8, 22). When unexpected pregnancies did occur in wearers of IUDs, spontaneous abortion occurred more frequently when metal devices were present than when the devices were of other material.

The challenge remains to identify more precisely the specific surface chemical arrays on a foreign material which, when it is inserted into the uterine cavity, might trigger reactions that produce these adverse effects as well as inhibit reproductive activities (23).

Effect of Adsorbed Protein on New IUDs. A new, promising improvement in contraceptive technology involves the installation of intrauterine foreign bodies loaded with certain steroids (24) which can diffuse through the wall of these devices in microquantities at slow and constant rates. Also, the addition of copper to the IUD surface greatly improves contraceptive effectiveness (25, 26). Since these devices introduce new materials to the intrauterine environment, they create the need for more detailed understanding of the effect of that enviroment on the surface properties of the devices and *vice versa*. For example, diffusion rates through medical grade silicone rubber were 8–10 times higher when human plasma surrounded the capsules rather than normal saline solution (*27, 28*).

Our work suggests that IUD effects may be mediated through a rather specific layer of adsorbed proteinaceous material. With the large-scale introduction of steroid-dispensing subcutaneous and/or intrauterine devices on the horizon, it is of critical importance to evaluate as well the effect of adsorbed protein on the metering of such potent biological agents to the remainder of the body. The glycoproteins which adhere to copper IUDs have not been analyzed. This is planned for the near future. If the presence of adsorbed glycoproteinaceous films on the IUD surface decelerates or accelerates the rate of permeation of steroids or copper ions into the uterine environment, the incidences of pregnancy and side effects as well as serious complications may be increased or decreased. Only further investigation will provide this information.

Acknowledgments

The authors appreciate the assistance of J. Wallace and V. DePalma of the Department of Biophysics, Roswell Park Memorial Institute, Buffalo, N. Y., for their assistance with x-ray diffraction analyses, and of E. Gasiecki and S. Perlmutter of Calspan Corp. for assistance with IR spectroscopic analyses.

Literature Cited

1. Segal, S. J., Tietze, C., "Contraceptive Technology: Current and Prospective Methods," Rep. on Population/Family Planning **1**, Population Council, 245 Park Ave., New York, 1971.
2. Ryder, N. B., "Time Series of Pill and IUD Use: United States, 1961–1970," *Stud. Fam. Plan.* (1972) **3**, 233.
3. El Sahwi, S., Moyer, D. L., "Antifertility Effects of the Intrauterine Foreign Body," *Contraception* (1970) **2**, 1.
4. Coleman, D. L., King, R. N., Andrade, J. D., "The Foreign Body Reaction: A Chronic Inflammatory Response," *J. Biomed. Mater. Res.* (1974) **8**, 199–211.
5. Stern, I. J., Kapsalis, A. A., DeLuca, B. L., Pieczynski, W., "Immunogenic Effects of Foreign Materials on Plasma Proteins," *Nature* (1972) **238**, 151.
6. Zisman, W. A., "Relation of the Equilibrium Contact Angle to Liquid and Solid Constitution," ADVAN. CHEM. SER. (1964) **43**, 1.
7. Baier, R. E., "Applied Chemistry at Protein Interfaces," ADVAN. CHEM. SER. (1975) **145**, 1.
8. Baier, R. E., DePalma, V. A., in "Management of Arterial Occlusive Disease," W. A. Dale, Ed., chap. 9, Year Book Medical, Chicago, 1971.
9. Lippes, J., Feldman, J. G., "A Five-Year Comparison of the Continuation Rates Between Women Using Loop D and Oral Contraceptives," *Contraception* (1971) **3**, 313.

10. Sagiroglu, N., Sagiroglu, E., "Biologic Mode of Action of the Lippes Loop in Intrauterine Contraception," *Amer. J. Obstet. Gynecol.* (1970) **106**, 506.
11. Parr, E., "Intrauterine Foreign Bodies: A Toxoid Effect of Leukocyte Extracts on Rat Morulae *In Vitro*," *Biol. Reprod.* (1969) **1**, 1.
12. DeBoer, C. H., "Transport of Particulate Matter Through the Human Female Genital Tract," *J. Reprod. Fert.* (1972) **28**, 295–297.
13. Hartmann, J. F., Gwatkin, R. B. L., "Alteration of Sites on the Mammalian Sperm Surface Following Capacitation," *Nature* (1971) **234**, 479.
14. Hussein, M., Ledger, W. J., "Preimplantation Effect of an Intrauterine Device in the Rabbit," *Amer. J. Obstet. Gynecol.* (1969) **103**, 221.
15. Siegler, A. M., Hellman, L. M., "The Effect of the Intrauterine Contraceptive Coils on the Oviduct," *Obstet. Gynecol.* (1964) **23**, 173.
16. Marston, J. H., Kelly, W. A., Eckstein, P., "Effect of an Intrauterine Device on Uterine Motility in the Rhesus Monkey," *J. Reprod. Fert.* (1969) **19**, 321.
17. Ishihama, A., "Clinical Studies on Intra-Uterine Rings, Especially the Present State of Contraception in Japan," *Yokohama Med. J.* (1959) **10**, 89.
18. Oppenheimer, W., "Prevention of Pregnancy by the Graefenberg Ring Method," *Amer. J. Obstet. Gynecol.* (1959) **78**, 446.
19. Corfman, P. A., Segal, S. J., "Biologic Effects of Intrauterine Devices," *Amer. J. Obstet. Gynecol.* (1968) **100**, 448.
20. Buchman, M. I., "A Study of the Intrauterine Contraceptive Device with and without an Extracervical Appendage or Tail," *Fert. Steril.* (1970) **21**, 348.
21. Sakurabayashi, M., "Studies on Intrauterine Contraceptive Devices (III) Applications of High Polymers in Cavity of Uterus and a New Radiopaque Phycon-X Ring," *Acta Obst. Gynaecol. Japan* (1969) **16**, 56.
22. Baier, R. E., in "Adhesion in Biological Systems," R. S. Manly, Ed., chap. 2, Academic, New York, 1970.
23. Arehart-Treichel, J., "Birth Control in the Brave New World," *Sci. News* (1973) **103**, 93.
24. Scommegna, A., *et al.*, "Intrauterine Administration of Progesterone by a Slow Releasing Device," *Fert. Steril.* (1970) **21**, 201.
25. Zipper, J. A., Tatum, H. J., Medel, M., Pastene, L., Rivera, M., "Metallic Copper as an Intrauterine Contraceptive Adjunct to the 'T' Device," *Amer. J. Obstet. Gynecol.* (1969) **105**, 1274–1278.
26. Tatum, H. J., Zipper, J. A., "The 'T' Intrauterine Contraceptive Device and Recent Advances in Hormonal Anticontraceptional Therapy," Sixth Northeast Obstet. Gynecol. Congr., Bahia, Brazil, October, 1968, pp. 78–85.
27. Mullison, E. G., "Silastic Silicone Rubber with Steroids used in Contraception," *Bull. Dow Corning Center Aid Med. Res.* (1970) **12**, 5.
28. Chang, C., Kincl, F., "Sustained Release Hormonal Preparation: Biologic Effectiveness of Steroid Hormones," *Fert. Steril.* (1970) **21**, 134.

RECEIVED June 19, 1974.

Marine Conditioning Films

GEORGE I. LOEB and REX A. NEIHOF

Naval Research Laboratory, Washington, D.C.

Changes on the surfaces of well-defined solids exposed to sea water from diverse sources have been observed by microelectrophoresis, ellipsometry, and contact angle measurements. These observations indicate that an adsorbed film forms rapidly on a variety of surfaces. Other experiments with artificial and photo-oxidized sea waters show that the film is organic. It lowers the surface energy of platinum and imparts a moderately negative electrical charge to most surfaces. Dissolved polymeric materials of biological origin have properties required for this interaction. Fluorescence measurements give a direct indication of some participation by humic substances in film formation. Adsorption of dissolved organic materials thus constitutes a very early step in fouling, a step that may affect the sequence of events leading to macrofouling.

Marine organisms select their habitats on the basis of many factors (*1*). Among these, the nature of the surface of the material on which an organism settles or through which it burrows has a pronounced effect. The surfaces of materials immersed in sea water may be altered by primary films described by Henrici (*2*), ZoBell and Allen (*3*), and others. The primary film consists of a community of microorganisms, their slimy exudate and components adsorbed from the water. Many researchers report that preliminary exposure of culture vessels to sea water, which allows the formation of a primary film, favors the subsequent growth of marine organisms in them. Most studies of immersed surfaces have not distinguished between the effects of adsorption of dissolved substances from the natural water and colonization by microorganisms. However, membrane-filtered sea water does not render clean sand attractive to certain burrowing marine worms without replacing the residue left on the filter (*4*).

Adsorption of dissolved material without simultaneous adsorption of microorganisms is sufficient to change the suitability of a surface for subsequent biological settlement. This is recognized by many organ tissue culturists who condition culture vessels in serum (5). Marine biologists also have exposed surfaces to cell-free extracts of adult barnacles to enhance settling of larvae (6); extracts of different algal species adsorbed on immersed surfaces render the surfaces selectively attractive to the species of *Spirobis* normally found on that particular type of algae (7, 8). Quantities of material equivalent to those found in adsorbed films are sufficient to modify the attractiveness of surfaces for attaching organisms.

These cases are clear evidence that extracts of species closely allied to the life cycle of an attaching organism may have pronounced effects on its attachment; mechanisms based on molecular recognition similar to enzyme–substrate or antigen–antibody complementarity may be envisioned. However, it is conceivable that dissolved components in the sea may alter surface properties of immersed surfaces in more general ways that would affect attachment and colonization. Thus, adhesion of organisms to a surface might be influenced by hydrophilic or hydrophobic characteristics of the surface, the surface charge and charge density, and the chemical functional groups available for reaction. Zisman (9) and Baier, Shafrin, and Zisman (10) have reviewed the surface chemistry and physics of biological adhesion emphasizing wetting phenomena; Marshall, Stout, and Mitchell (11) have presented arguments and evidence that emphasize the possible importance of electrical charge and dipole interactions in attracting marine bacteria to surfaces in the reversible phase of their interaction. These properties of surfaces may be drastically altered by adsorption of dissolved substances from the surrounding medium (12, 13).

We have investigated the adsorption of soluble components from sea water and its effects on the wettability and charge characteristics of several well characterized solid surfaces in order to determine whether changes do indeed occur from this cause and whether their magnitude is likely to be sufficient to affect significantly later processes on the surface. We have called this adsorption molecular fouling (14).

Experimental

Water Samples and Their Treatment. Samples of sea water from a variety of sources were used. Collection methods evolved with experience. The most satisfactory method uses a vacuum chamber into which a Teflon bottle is placed. A Teflon tube dipping into the sea draws the water sample into the chamber and bottle. Earlier samples were taken by a well-rinsed polyethylene bucket, or by suction through well-rinsed polyethylene or polypropylene tubes and stored in polyethylene or polypropylene bottles that were rinsed with sea water at the time of the collection.

Glass and Teflon ware was acid-cleaned and rinsed; Teflon ware was subsequently steamed. Other plastic ware was detergent-washed, rinsed with water followed by methanol, and exhaustively rinsed with distilled water (*15, 16*).

Sea water samples were collected from (a) Chesapeake Bay at the Patuxent Naval Air Station at high tide with salinity of the samples varying between 9‰ and 16‰ and pH between 7.9 and 8.2; (b) the Atlantic Ocean about 30 km east of Ocean City, Md., salinity 31‰, pH 8.2; (c) the Caribbean Sea off Coco Solo, Panama Canal Zone, salinity 32.8‰; and (d) the Gulf of Mexico, 350 km WNW of Key West, Fla., salinity 35.4‰, pH 8.1. The particulate material found in natural sea water samples was concentrated and separated from the bulk of the water sample by centrifugation at 18,000 g.

Artificial sea water made up according to the formulation of Lyman and Fleming (*17*) but containing only the seven major ions and called seven-ion sea water was used as a control after dilution to match the salinity of the sample of interest for a particular experiment. Dissolved organic matter was removed when required from natural and seven-ion sea waters by UV irradiation according to a modification (*14, 16*) of the photo-oxidation procedure of Armstrong *et al.* (*18*). Dissolved organic carbon analyses were performed according to Menzel and Vaccaro (*19*).

Measurements. The following parameters were used to characterize the surfaces investigated: electrical surface charge by microelectrophoresis; wettability by contact angle measurements; and kinetics of adsorbed film formation by ellipsometry. Details of the microelectrophoresis procedures are given by Neihof and Loeb (*14, 16*), and contact angle measurements are described by Zisman (*20*). Ellipsometry, described by Rothen (*21*), involves the analysis of the elliptical polarization of a beam of polarized light after reflection from the surface of interest. The data yields the thickness of a film if the refractive index is known. In favorable cases, thickness and refractive index may both be evaluated (*22, 23*). Technical difficulties have delayed our application of this technique to surfaces other than platinum. Measurements were made using a Rudolph Research instrument with a He–Ne laser source with a 70° angle of incidence at 20°C.

Fluorescence analyses of water samples were made with a Perkin-Elmer MPF2A spectrophoto-fluorimeter using 1-cm cells. Settings were fast scan, 10 nm bandpass, and sensitivity 4. Excitation wavelength for humic acid type spectra was 330 nm. An analysis for amino groups was made using the fluorescamine reagent as described by Udenfriend *et al.* (*24*) except that a more consistent response was obtained in our instrument by reading the fluorescence emission at 480 nm rather than 475 nm; excitation was at 390 nm. Reagent blanks in all cases were obtained by substituting photo-oxidized sea water for the sample. Microscopic examination of the water samples at various times ruled out the possibility that significant bacterial growth took place during the experimental procedures.

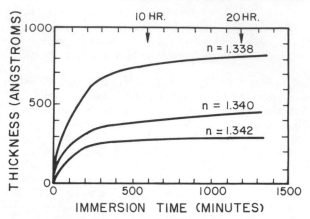

Figure 1.   Formation of an adsorbed film from Chesa-peake Bay water on polished platinum as followed by ellipsometry. Calculated thicknesses of a film with dif-ferent assumed values of refractive index (n) are plotted against immersion time.

Test Surfaces. Electrophoretic technique requires particles small enough to remain in suspension in the sea water samples; the size range used here was 0.5–5 μm. The particles used were a strongly positively charged quaternary ammonium ion exchange resin (Dowex 21K), pow-dered Pyrex glass, powdered dextran gel (Sephadex G-25, Pharmacia), a chlorinated wax (Chlorex, Dover Chem. Co.), germanium powder, and quartz powder. Details of preparative and cleaning procedures used have been described previously (*15, 16*).

Platinum plates with 1 cm² area per side were metallographically polished to a mirror finish, rinsed with distilled water, and flamed just before exposure to sea water for either ellipsometric or contact angle measurements. The platinum plates were wetted with photo-oxidized sea water before immersing into experimental sea water, so that passage of the surface through the air–sea water interface would not cause any film present there to be transferred to the plate. For the same reason plates removed from the experimental sea water were immediately im-mersed while still visibly wet in photo-oxidized sea water and then rinsed.

Adsorption on plates used for contact angle measurements was allowed to proceed from a volume of about 40 ml for one week at 3°–4°C. The plates were then rinsed three times with photo-oxidized sea water and five times with triple distilled water to remove residual salts before drying.

Results

Do Adsorbed Films Form on Immersed Surfaces? To answer this question, clean surfaces were immersed in centrifuged sea water from a number of sources and examined by several techniques. The surfaces of natural particulate matter were also investigated.

Ellipsometry in principle yields both the refractive index and thickness of transparent films (*21*) but large errors may arise in both determinations. Accepting the uncertainty of the refractive index of the film, we may still follow the kinetics of film formation on a reflective surface with the *proviso* that the calculated value of film thickness will require a correction factor when the true film refractive index is determined. This procedure was followed here. All measurements were made on platinum surfaces during continuous submersion in sea water; details of the procedure will be presented elsewhere. Figure 1 shows the results of a typical experiment in which a platinum plate of 2 cm² area is immersed in 80 ml of Chesapeake Bay water. The curves show thickness of film against time; a different value of refractive index for the film was assumed for each curve. Although precise values cannot be given, it is apparent that a film forms very rapidly during the first few minutes, continues to grow at an appreciable rate for a period of hours, and is still building somewhat after 20 hrs. Reasonable values of assumed refractive index lead to film thicknesses similar to those found on adsorption of known macromolecules (*23*).

Further evidence of film formation has been developed by study of the wettability of polished platinum plates after immersion for one week at 3°C in Chesapeake Bay and Atlantic sea waters followed by rinsing with distilled water and drying. The contact angle of a liquid on a surface becomes smaller with increasing tendency of the liquid to spread on the surface. The contact angles of pure water and methylene iodide on these samples are 27° and 28° for Atlantic water, and 45° and 33° for Bay water (Table I). When compared with values of less than 10° for either liquid on clean platinum, or platinum which has been rinsed in organic media and dried, the presence of a film is apparent.

Surfaces other than platinum have been examined by microelectrophoresis. The electrophoretic mobilities (at 25°C) of carefully cleaned

Table I.   Contact Angles for Film Samples

Sample	Contact Angle with	
	Methylene Iodide, °	*Water,* °
Clean Pt	< 10	0
Pt after Immersion in		
Atlantic water	28	27
Bay water	33	45
Known Sample of:		
Nylon 2	30	49
poly(methyl glutamate)	30	49
polyethylene	52	94

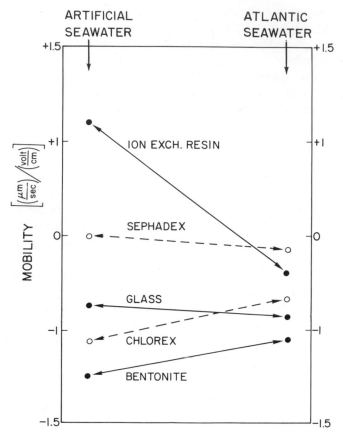

Figure 2.   Electrophoretic mobilities of particles in seven-ion (left) and natural sea water (right). Tie lines connect values for the same kind of particle. Data from Ref. 15.

solid particles of a wide range of surface charge densities after immersion in an artificial inorganic salt solution (seven-ion sea water) and in natural sea water are shown in Figure 2. The particles exhibit a range of mobilities in seven-ion sea water, both positive and negative, but after immersion in natural sea water the mobilities fall in a much more restricted, moderately negative range, indicating that the surfaces have become more nearly alike. The coefficient of variation for measurments on a given type of particle is 5–10% of the mean mobility.

The particulate material found in natural waters was also observed to have a net negative charge when examined by microelectrophoresis. This was true of all particles studied in all natural sea waters; the histogram for particles from Atlantic water (Figure 3A) illustrates the range of mobilities observed. When the salinity is lowered to 1% of the initial value, the mean mobility increases, and more deviation from the mean mobility seems to be evident (Figure 3B). Analysis shows, however, that the coefficient of variation remains about the same for both salinities at $33 \pm 4\%$. The increase in mobility of a given sign as ionic strength is decreased, as observed here, is typical of colloidal systems and can be explained in terms of an expanding ionic double layer outside the surface of shear surrounding the particle (25). Figure 4 shows the mean mobilities of particles from Atlantic sea water as a function of salinity and of particles from Caribbean and Chesapeake Bay immersed in their supernatants. The Caribbean particulate sample is not in good agreement with the Atlantic particles (perhaps because of changes during the longer transit time to the laboratory or because of significant inherent differences in the nature of the samples), but particles from the Bay do agree well with the curve for the Atlantic sample. The natural particles were identified by microscopic examination as mainly bacteria, small algae, and detritus.

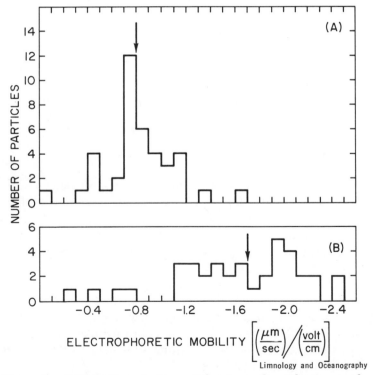

Figure 3. Distribution of 40 particles from the Atlantic sample measured (A) in Atlantic sea water, salinity 31.5‰ and (B) in diluted Atlantic sea water, salinity 0.32‰. Arrows designate mean values (15).

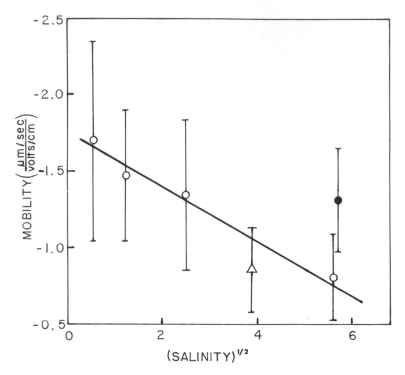

Figure 4. Variation of electrophoretic mobilities of natural particles with salinity of the medium. Standard deviations are indicated. Data from Ref. 15.

———○———, *Diluted Atlantic sea water;* ———●———, *Caribbean sea water;* ———△———, *Chesapeake Bay water*

The similarity of charge of all the natural particulates is consistent with the postulate that a film of electronegative material forms on particles in the sea but would also be consistent with an inherently negatively charged particulate population. However, the negative mobilities exhibited by introduced test particles, especially those positively charged in artificial sea water, favor the film hypothesis.

The coefficient of variation of electrophoretic mobilities in the case of natural particles (33%) was greater than that found for immersed test particles all of a single kind in natural water (< 10%). However, the range of mobilities spanned by the entire population of many kinds of test particles in natural water is quite similar to that found with the natural particles shown in Figure 3.

This convergence of mobilities is reminiscent of early work reviewed by Abramson *et al.* (25) in which particles of different materials and widely divergent electrophoretic mobilities were shown to exhibit similar mobilities after equilibration with protein solutions. More recently Bull (*12*) studied the electrophoretic behavior of glass and of cation and anion exchange resins before and after they were allowed to interact with protein solutions. Although the mobilities of the different particles in simple salt solutions varied widely from highly positive to highly negative, exposure to protein solutions caused the mobilities of all of the particles to approach each other. This behavior may be explained by a layer of adsorbed macromolecules coating the surface which then dominates the electrokinetic properties. We may be dealing with this phenomenon here. It is possible of course that the various surfaces may be reacting with different components in the sea water or to different extents with varying surface coverage.

The mobilities of the cleaned particles in seven-ion water are consistent with their known chemical compositions. The ionogenic materials, glass, and anion exchange resin have negative and positive charges, respectively, reflecting the nature of the principal ionized groups on their surface. These oppositely charged materials should not adsorb materials of the same charge if the mechanism of adsorption is primarily electrostatic. However, both these substances become more negative in sea water, implying that the negative charge is a result of at least partly non-electrostatic mechanisms and that some negatively charged material is more surface active than any of positive charge which may be present. The interactions of nonionogenic surfaces are expected to involve the weaker van der Waals forces, hydrogen bonds, or dipole interactions, and hydrophobic forces as discussed previously (*15*).

Is the Film Organic? The discussion so far has clearly shown that immersed surfaces are modified by sea water but establishes very little about the nature of the film. To establish whether the changes observed arise from organic material in sea water, a control sea water free of organic matter is necessary. Artificial sea water is not entirely satisfactory for this purpose because traces of surface active organic matter are known to be present in even the best reagent grade salts (*26, 27*). It is also conceivable that differences in minor inorganic constituents of artificial as compared with natural sea water might be sufficient to cause differences on immersed surfaces. An attractive alternative to artificial sea water lies in the use of the UV photo-oxidation procedure of Armstrong *et al.* (*18*) which eliminates most of the dissolved organic matter in sea water without apparent change in inorganic composition. We have used this irradiation technique to treat both natural waters and artificial sea water so that the influences of dissolved organic materials on the measurements can be more explicitly determined.

If an ellipsometer reading is made on a platinum surface immersed in triple-distilled water and the water is replaced by photo-oxidized sea

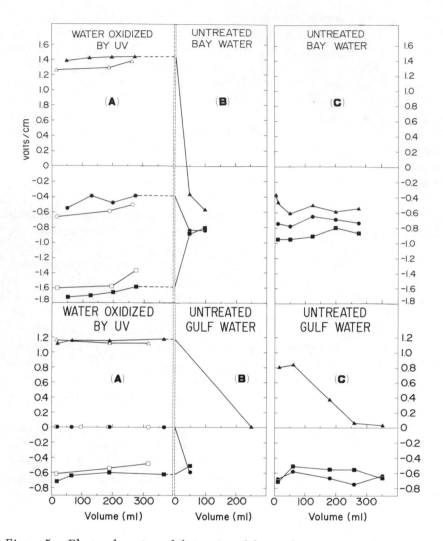

Figure 5.   Electrophoretic mobilities of model particles in untreated and photo-oxidized sea water from Cheasapeake Bay (top) and Gulf of Mexico (bottom) and in photo-oxidized, seven-ion water of comparable salinities.

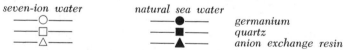

seven-ion water   natural sea water
———○———   ———●———   *germanium*
———□———   ———■———   *quartz*
———△———   ———▲———   *anion exchange resin*

water, a small instantaneous change in reading occurs which is proportional to salinity within the error. However, no further change with time occurs, and reverting to distilled water again brings the reading back to its original value. This is true for both the artificial and natural photo-oxidized sea waters. If the photo-oxidized sea water is replaced by centri-fuged natural sea water of the same salinity, there is no instantaneous shift, but a progressive change is observed resulting in the data which were discussed above and shown in Figure 1. Thus the ellipsometric results give no evidence of a stable film formation in photo-oxidized seven-ion water or in photo-oxidized Chesapeake Bay or Atlantic sea waters, but such a film does appear to form in untreated natural sea waters.

Electrophoretic analysis of a number of different kinds of test par-ticles, ranging from highly positive to highly negative when in artificial seven-ion sea water, retained their characteristic mobilities in photo-oxidized sea water, as shown in Figure 5, Curves A. The abscissa indicates the volume of sea water to which a batch of particles (initially 3×10^7/ml) were exposed in increments after the first treatment. These particles were then treated with centrifuged but not photo-oxidized natural waters with the results shown in Curves B of Figure 5. Finally, another portion of clean particles was introduced directly into centrifuged natural water with the result shown in Curves C of Figure 5.

No significant difference was found between photo-oxidized, artificial seven-ion sea water and natural photo-oxidized sea water of the same salinity. When exposed to the centrifuged natural waters, either directly from photo-oxidized water or as clean dry particles, the mobilities fall in a more restricted range similar to that found with naturally occurring particles. Thus, we again find that the difference between film-forming and non-film-forming water lies in the organic material decomposed by irradiation, and the seven-ion sea water is sufficiently characteristic of the inorganic components of natural sea water to yield similar mobilities when organics are removed.

As before, the observed mobility of a particle in photo-oxidized waters is consistent with its composition. The anion exchange resin is positive and the quartz negative since they bear quaternary ammonium groups and silanol groups, respectively. Germanium is slightly negative in low-salinity, photo-oxidized Bay water which is consistent with ioniza-tion of surface germanic acid groups, but its mobility is relatively low and is reduced to zero in the higher salinity photo-oxidized Gulf water. The higher mobilities in Bay water can be attributed to the lower salinity, as discussed in connection with natural particulates.

There is a very significant difference in the volume of Gulf water required to bring about a constant mobility of anion exchange resin as compared with Bay water: 250 ml as compared with 25 ml in Bay water. The volumes are inversely correlated with dissolved organic matter content: 0.75 mg carbon/l for Gulf water compared to 2.3 mg carbon/l for Bay water. In addition, the resin even in Bay water does not quite reach its constant value when treated with only 5 ml whereas both other types of particles reach their constant values in either water sample with 5 ml contact. Greater volumes are required for the resin equilibration because of its porosity and higher density of ion exchange groups compared with the smooth non-porous quartz and germanium. The failure of ion exchange resin mobilities to approach the mobilities of quartz and germanium as closely in Gulf as in Bay water may be because of incomplete coverage in the lower organic content water, but other causes, such as differences in the nature of the dissolved organic matter in the waters, cannot be ruled out.

Properties of the Film. There are many possible organic substances which may lead to electronegative films on adsorption; it is possible that no single class of substances is fully responsible. A number of experimental approaches have been made, and although our picture is still somewhat unclear, several features are now recognizable.

MOLECULAR SIZE OF THE FILM FORMING MATERIAL. Using a dialysis membrane nominally of pore size corresponding to a protein molecular weight of 12000 daltons, Atlantic sea water was dialyzed against seven-ion sea water and seven-ion sea water against Atlantic sea water. The results of measurement of electrophoretic mobility of ion exchange resin equilibrated with each sample indicate that both the retained fraction in dialysis and the fraction passed by the membrane are implicated in the charge modification of the immersed particles (15). Further work along these lines is in progress.

Adsorbed films of polymers are normally tenaciously retained on surfaces, even after extensive rinsing with solvent because the large number of regions on each polymer molecule which may bind cooperatively drastically reduces the probability that all of the regions will desorb simultaneously (28). Assessments of the difficulty of removing adsorbed constituents were made by washing particles previously equilibrated with centrifuged sea water in photo-oxidized sea water. After washing in eight 25-ml portions over a day, the mobility of the ion exchange resin was unchanged, and the mobilities of quartz and germanium remained much closer to the values obtained in natural waters than to those in photo-oxidized waters. A strong retention, characteristic of the multiple cooperative binding regions found in macromolecules, is thus indicated.

WETTABILITY. A comparison of contact angles for methylene iodide and pure water on natural films with the contact angles of the same liquids on known materials is shown in Table I. By the contact angle criterion, the natural films obviously resemble the two polar (but non-ionic) polyamide materials much more closely than the high energy platinum or the non-polar polyethylene. There is a particularly close similarity between the film from Bay water and the polyamide surfaces. The Atlantic films yield substantially lower contact angles with water; the small lowering of the methylene iodide angle from that of Bay water may be equivocal. This combination of contact angles indicates a larger polar or dipole contribution to the surface energy of the Atlantic films than of the Bay water films while dispersion contributions are similar.

ADSORPTION OF FLUORESCENT COMPONENTS OF SEA WATER. Further investigation concerning the nature of the film substance has also been undertaken using fluorescence spectroscopy. This technique is quite sensitive and is applicable to aqueous solutions as well as to adsorbed films in contact with aqueous media (*29*). One may observe the intrinsic fluorescence of a material of interest and, in addition, reagents and pro-

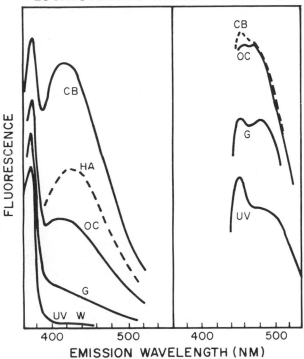

Figure 6.   Fluorescence spectra of sea waters.

Left side: intrinsic fluorescence of waters excited at 330 nm. Right side: fluorescence of water after treatment with fluorescamine regent, excited at 390 nm. Peaks at 372 nm and 450 nm in the left and right sets of curves, respectively, are from Raman emission.

CB, Chesapeake Bay water; OC, Atlantic sea water; G, Gulf of Mexico sea water; UV, photo-oxidized sea water; HA, photo-oxidized sea water extract of humic acid.

Table II.   Change in Fluorescence of Sea Water with
Adsorption on Cabosil at 40 μg/ml

	Fluorescence before Contact with Cabosil[a]		Change (%) after Contact	
Sea Water Sample	420 nm	Amine Signal	420 nm	Amine Signal
Chesapeake Bay	100	100	$-6\%\pm1$	$-2\%\pm2$
Atlantic	39	91	$-7\%\pm0.5$	$-10\%\pm2$
Gulf of Mexico	15	41	$-4\%\pm2$	$+4\%\pm2$

[a] Normalized so that 100 is intensity of Bay water sample.

cedures have been developed which yield fluorescent signals when specific functional groups are present. The procedure developed by Udenfriend *et al.* (*24*) involves fluorescamine, and it permits detection of amino groups.

The intrinsic blue fluorescence of sea water has been known for many years (*30, 31*) and has been detected and reported from many areas (*32, 33*). Our fluorescence spectra are presented in Figure 6. As shown, the fluorescence spectrum of water from the Chesapeake Bay is very similar to that of a sample of humic acid (Calbiochem Corp.) extract in photo-oxidized Bay water. Both the humic acid extract and the sea water sample show a linear decrease in fluorescence intensity with concentration as the solutions are diluted with photo-oxidized sea water. When assayed with the fluorescamine reagent the fluorescent intensity of the amine-generated signal is also linear in concentration with both the humic acid, sample extract, and the sea water. To determine whether the humic acid, or other amine-containing substance may be adsorbed to an immersed surface, fumed silica particles (Cabosil, grade M5, Cabot Corp.) were introduced to the extent of 40 mg/ml in samples of Atlantic, Gulf of Mexico, and Chesapeake Bay sea waters. The pH change resulting from this was compensated for by adding 1M NaOH. The slurries were allowed to stand overnight at 30°C then centrifuged, and the fluorescent signals were attributable to the humic acid-type spectrum (intensity at 420 nm under excitation at 336 nm). The amine reagent spectrum (corrected for the reagent blank response of photo-oxidized water sample) were compared. The results are shown in Table II. (All natural waters after photo-oxidation yield only the spectrum of distilled water).

In all cases, the fractional decrement of the fluorescent humic acid-type signal at 420 nm is at least as great, within the error, as the change in amine signal. Thus, since the amine signal intensity and the intrinsic fluorescence of the humic acid extract at 420 nm are both linear in concentration, there would not be a quantitatively comparable loss from the supernatant of non-humic, amine-containing material (such as lysyl residues of a protein chain or amino sugar-containing carbohydrates) adsorbing preferentially with respect to the humic fraction in sea water. Small amounts of other amino-containing dissolved substances might become preferentially adsorbed without significantly depleting the amino signal in the supernatant, however, and acetylated amino sugars would not be detected by this procedure.

Humic materials are the product of a complex and ill understood sequence of reactions involving biological compounds excreted from living organisms or derived from decomposition of organisms after death. These decomposition products are then acted on by microorganisms,

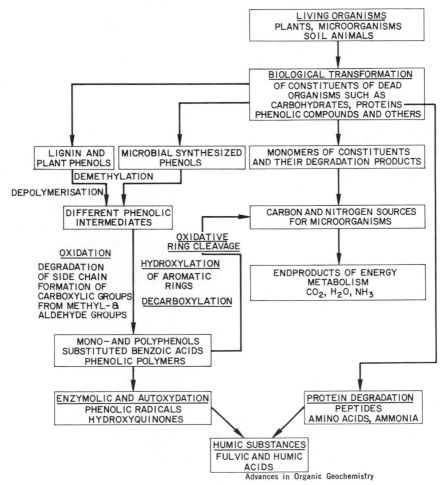

Advances in Organic Geochemistry

Figure 7.   Scheme for synthesis of humic substances (34)

affected by environmental factors, and transformed into the humic and fulvic acids. A general scheme by Flaig (*34*) is presented as Figure 7.

The nitrogen content of humic acids is between 1% and 5% and of this approximately 50% is in heterocyclic moieties. The remaining portion, after acid hydrolysis, is distributed among amino sugars, ammonia nitrogen, and α-amino nitrogen, with up to 5% of the total nitrogen described as originating in peptide linkages (*35*). The manner in which amino acids are incorporated into the humus structure is not clear. Insofar as they are present, the humus substances are related to proteins to some extent, and they thus meet the requirements for inclusion in this discussion.

Although the complex character of humic substances makes analysis of their structure difficult, and their characterization is incomplete, there is general agreement that they contain caroboxyl groups and are anionic and that their formation is associated with blue fluorescence (*31*). The significance of the fluorescence is not clear. Kalle (*31*), on the basis of his own work and that of Jerlov in the Baltic area, has concluded that since the ratio of fluorescence to optical density of sea water increases as salinity increases, oceanic water contains organic material of a more fluorescent nature than terrestrial waters. Rashid and King (*36*) and Rashid and Prakash (*37*) also have found differences between marine and terrestrial humus. Karabeshev and Zangalis (*38*) made measurements in the Baltic area and concluded that all their data could be normalized so as to yield a standard luminescence curve, and direct proportionality between the fluorescent intensity and optical density was found.

Our data indicate a clear difference in the organic content of Atlantic and Chesapeake Bay sea waters. The ratio of intrinsic humus fluorescence at 420 nm to the amine-generated fluorescence with fluorescamine is quite different in the two samples. The different contact angles produced by films adsorbed from these two samples are indicative of different materials adsorbed. The data shown in Table II indicate that the ratio of depletion of humus fluorescence to depletion of amine signal is apparently higher in Bay water than in Atlantic waters and, although somewhat equivocal, a difference in reaction of the dissolved organics with surfaces is implied.

Acknowledgments

We thank E. Shafrin for the contact angle measurements and R. Lamontagne and G. Bugg for determinations of dissolved organic carbon.

Literature Cited

1. Meadows, P. S., Campbell, J. I., *Advan. Mar. Biol.* (1972) **10**, 271.
2. Henrici, A. T., *J. Bacteriol.* (1933) **25**, 277.
3. ZoBell, C. E., Allen, E. C., *J. Bacteriol.* (1935) **29**, 239.
4. Meadows, P. S., Williams, G. B., *Nature* (London) (1963) **198**, 610.
5. Rappaport, C., Poole, J. P., Rappaport, H. P., *Exp. Cell Res.* (1960) **20**, 465.
6. Crisp, D. J., Meadows, P. S., *Proc. Roy. Soc. (London)* (1963) **B158**, 364.
7. Gee, J. M., *Anim. Behav.* (1965) **181**, 1.
8. Williams, G. B., *J. Mar. Biol. Assn. U.K.* (1964) **44**, 397.
9. Zisman, W. A., *J. Paint Technol.* (1972) **44**, 42.
10. Baier, R. E., Shafrin, E. G., Zisman, W. A., *Science* (1968) **162**, 1360.
11. Marshall, K. C., Stout, R., Mitchell, R., *J. Gen. Microbiol.* (1971) **68**, 377.
12. Bull, H. B., *Arch. Biochem. Biophys.* (1962) **98**, 427.
13. Bull, H. B., "An Introduction to Physical Biochemistry," pp 269, 321, F. A. Davis, Philadelphia, 1964.
14. Neihof, R. A., Loeb, G. I., "Proc. Third International Congress on Marine Corrosion and Fouling" (1973) 710.
15. Neihof, R. A., Loeb, G. I., *Limnol. Oceanogr.* (1972) **17**, 7.
16. Neihof, R. A., Loeb, G. I., *J. Mar. Res.* (1974) **32**, 5.
17. Lyman, J., Fleming, R. H., *J. Mar. Res.* (1940) **3**, 134.
18. Armstrong, F. A. J., Williams, P. M., Strickland, J. D. H., *Nature* (London) (1966) **211**, 481.
19. Menzel, D. W., Vaccaro, R. F., *Limnol. Oceanogr.* (1964) **9**, 138.
20. Zisman, W. A., ADVAN. CHEM. SER. (1964) **43**, 1.
21. Rothen, A., "Physical Techniques in Biological Research," Vol. II, p. 155, G. Oster, A. W. Pollister, Eds., Academic, New York (1956).
22. McCrackin, F. L., NBS Technical Note 479 "A Fortran Program for Analysis of Ellipsometer Measurements" (1969).
23. Smith, L. E., Stromberg, R. R., Fenstermaker, C. A., Grant, W. H., "Abstracts of Papers," 166th National Meeting, ACS, Aug. 1973, COLL 132.
24. Udenfriend, S., Stein, S., Bohlen, P., Dairman, W., Leimgruber, W., Weigle, M., *Science* (1972) **178**, 871.
25. Abramson, H. A., Moyer, L. S., Gorin, M. H., "Electrophoresis of Proteins and the Chemistry of Cell Surfaces," Reinhold, New York, 1942.
26. Jarvis, N. L., Scheiman, M. A., *J. Phys. Chem.* (1968) **72**, 74.
27. Wallace, G. T., Loeb, G. I., Wilson, D. F., *J. Geophys. Res.* (1972) **77**, 5293.
28. Silberberg, A., *J. Phys. Chem.* (1962) **66**, 1884.
29. Harrick, N. J., Loeb, G. I., *Anal. Chem.* (1973) **45**, 687.
30. Kalle, K., *Deutschen Hydrographischen Zeitsch.* (1949) **2**, 117.
31. Kalle, K., *Oceanogr. Mar. Biol. Ann. Rev.* (1966) **4**, 91.
32. Hornig, A. W., Eastwood, D., "A Study of Marine Luminescence Signatures," NASA CR 114578 (Part I) and 114579 (Part II) (1973).
33. Traganza, E. D., *Bull. Mar. Sci.* (1969) **19**, 897.
34. Flaig, W., *Advan. in Org. Geochem.* (1971) **33**, 50.
35. Bremmer, J. N., *Agronomy* (1965) **9**, 1238.
36. Rashid, M. A., King, L. H., *Geochim. Cosmochim. Acta.* (1970) **34**, 193.
37. Rashid, M. A., Prakash, A., *J. Fish Res. Bd. Canada* (1972) **29**, 55.
38. Karabeshev, G. S., Zangalis, K. P., Isv., *Atmospheric Oceanic Physics* (1971) **7**, 1013.

RECEIVED January 17, 1974.

Proteins at Gas/Liquid Interfaces: Introduction

ROBERT E. BAIER

The previous sections have referenced important protein interfaces of skin, teeth, blood, horn, hoof, nails, tissue, and individual cells, allowing a reasonable claim for having completely considered the animal body. An important interface dominated by protein matter which has not been considered in the foregoing sections is the gas/liquid interface at the major refracting surface of the eye where the thin fluid film (the tear film) over the cornea is exposed to air (1). Here, we present only the extremes of the vast domain where proteins at gas–liquid interfaces are of primary and practical importance since the middle ground has been covered in numerous dedicated volumes on the subject of biological membranes and their models.

Ben Malcolm of the University of Edinburgh reviews the special interactions of the side chains of proteins, and of their synthetic polypeptide analogs, in films formed at air/water interfaces in small laboratory devices. This review calls special attention to the molecular configurations present and their interactions with water which have been repeatedly remarked on in the foregoing manuscripts and extends the excellent summary already provided by Malcolm in "Progress in Surface and Membrane Science" (2). Concluding this volume is a well-documented review of the importance of protein-containing interfacial layers on a truly global scale. Duncan Blanchard demonstrates convincingly that water-to-air transfer of material, and probably air-to-water transfer as well, is mediated by a microscopically thin organic layer. The gateway for such two-way traffic is that organic skin at the quiescent ocean/atmosphere interface or, more importantly, at the dynamic gas/liquid interfaces continuously being produced and eliminated by rising and breaking bubbles. Recognizing the enormous oceanographic/meteorologic implications of the existence, behavior, and chemical properties of such organic films on the stability of the ocean-atmosphere couple, numerous potential research avenues are suggested. An immediate need is for a reliable estimate of the geometric area and bubble-breaking activity in those coastal zones where surficial flow and wind skimming combine to concentrate the organic, protein-dominated components of the sea in films which can be microtomed (by the breaking bubbles) into the breeze and delivered to inland populations. Better analytical techniques are also required for these interfacial layers of organic matter to supplement the published findings which have so heavily emphasized solvent extracts of the natural films collected. Further, experiments must be designed to assess the specificity of elemental binding or concentration in the natural films so that the observed ionic and elemental fractionations which occur across the air/sea boundary can be understood, predicted, and controlled.

Literature Cited

1. F. J. Holly, "Surface Chemistry of Tear Film Component Analogs," *J. Coll. Interface Sci.* (1974) **49**, 221-231.
2. B. R. Malcolm, "The Structure and Properties of Monolayers of Synthetic Polypeptides at the Air-Water Interface," *Prog. Surf. Membrane Sci.* (1973) **7**, 183.

17

Hydrophobic Side Chain Interactions in Synthetic Polypeptides and Proteins at the Air–Water Interface

B. R. MALCOLM

Department of Molecular Biology, University of Edinburgh,
Edinburgh, Great Britain, EH9 3JR

The shapes of the pressure-area isotherms of monolayers of synthetic polypeptides in the α-helical conformation depend on the nature of the side chain interactions. Poly(γ-n-decyl-L-glutamate), poly(L-leucine), poly(L-norleucine), and poly(L-methionine) show differences related to side chain flexibility and dipolar interactions. Comparison of the isotherms of monolayers of the enantiomorphic and racemic forms of polymers [poly(alanine), poly(γ-benzyl-glutamate), poly(β-benzyl-aspartate), poly(ε-benzyloxycarbonyllysine)] similarly show features related to side chain properties. The results support the view that when a monolayer consists of α-helices, the shape of the isotherm depends on the difference between the energies of interaction of parallel and antiparallel molecules. These conclusions are discussed in relation to proteins.

The classical methods of surface chemistry led to the important conclusion that the hydrophobic side chains of globular proteins were predominantly in the interior of the molecule (*1*). Crystallographic studies have now fully confirmed this view and given a detailed picture. They show that in many cases internal pockets in protein molecules are filled with non-polar side chains in close contact with one another while polar, and more particularly ionizable side chains, are distributed on the surface. However, many non-polar side chains are found on the surface, and important ionizable side chains such as histidines may be buried, *e.g.*, as part of the active center of an enzyme (*2*). But the methods of surface chemistry do not give a reliable and detailed picture of the state

338

of a protein at an interface; the conformations proposed are speculative and limited by the available knowledge of protein conformations. No method for the study of an interfacial structure exists that is comparable in power with x-ray crystallography for studying protein crystals.

The loss of solubility of a protein at an interface and the observed area per residue suggested that the molecule unfolded and adopted the β conformation. This structure with intermolecular hydrogen bonds adequately accounted for the area observed and is still postulated by some workers (*3*). Cheesman and Davies (*4*) suggested other extended conformations with the orientation of the side chains largely determined by their hydrophobic or polar character. Their proposals, mainly based on early work on synthetic polypeptides, do not conform to present stereochemical criteria nor do they take account of the possibility that the α-helix or related helical conformations might be present.

The complexity and diversity of structures in the native proteins eluded any attempt to produce some simple conformation that accounted for their interfacial properties. The study of synthetic polypeptides with non-polar side chains has provided good evidence to support the view that the α-helix can be stable at the air–water interface (*5*), and it is therefore possible that the interfacial denaturation of proteins is mainly a loss of the tertiary structure (*6, 7, 8*). Since for a typical protein an α-helix takes up about the same area per residue as the β conformation, it can be accommodated as easily. Moreover, like the β conformation but unlike a more randomly coiled structure, it is linear and therefore compatible with a plane surface without loss of configurational entropy (*5*). In this respect a plane surface may favor an ordered over a more random structure. The loss of solubility of the spread protein can then be attributed to intermolecular association between hydrophobic side chains exposed as a result of the action of the interface on the polar exterior of the molecules.

One of the few methods by which these ideas can be tested is IR spectroscopic examination of proteins at or removed from interfaces since the frequencies of the amide bands are sensitive to the conformations present. Such experiments show that in general the β conformation is not predominant, and the proportions of the various conformations that might be present are very similar to those of the native molecule (*7, 9*), although in β-lactoglobulin the amount of the β conformation depends on the surface pressure (*10*).

Synthetic polypeptides are valuable in understanding more fully both the properties of structures such as the α-helix at interfaces and the part played by different types of hydrophobic side chains. While in some respects they are unlike any natural protein, their relative simplicity enables us to highlight features of proteins that would otherwise be difficult to isolate and study in detail; experimental and theoretical techniques can be applied more readily, and the results can be interpreted more precisely than for any protein. Moreover they are available with a considerable diversity of side chains so that patterns of behavior in related polymers can be seen.

When α-helices form in synthetic polypeptides at the air–water interface, their rigid rod-like nature promotes side-by-side association of the molecules into highly ordered arrays or micelles, just as liquid crystalline structures form in solution at sufficiently high concentration (11). When such a monolayer is compressed on a Langmuir trough, the pressure rises when the surface area has reached a value expected for close-packed α-helices. At a pressure which appears a characteristic for the polymer, a transition is observed which is either an almost flat plateau in the pressure–area curve or simply an inflexion, first noted by Crisp (13), if the side chain is short (12). An inflexion also occurs if the side chain is inflexible. Normally the pressure rises again as the area is decreased, and in some instances further transitions are observed (14).

A simple theory has been proposed to account for this behavior (15, 16). It is supposed that at the plateau, the monolayer collapses under the action of the interfacial forces and the surface pressure (W), so that

$$\gamma_{lv} - \gamma_{pv} - \gamma_{pl} - W = 0 \tag{1}$$

where γ_{lv}, γ_{pv}, and γ_{pl} are the surface free energies of the liquid–vapor, polymer–vapor, and polymer–liquid interfaces, respectively. This equation is similar to the equation for the spreading of one liquid on another if W is regarded as a negative spreading coefficient. If the angle of contact (θ) for water is measured on bulk specimens separately, the relation can

$$\gamma_{pv} = \gamma_{pl} + \gamma_{lv} \cos\theta \tag{2}$$

be used in conjunction with Equation 1 to calculate values for γ_{pv} and γ_{pl}. The values obtained appear consistent (5). γ_{pl} increases with increasing hydrophobicity, and where data is available, γ_{pv} appears close to the critical surface tension for the polymer. At least to a first approximation the theory gives a correct physical interpretation of the nature of the transition. It does not explain why the transition shows only as an inflexion and not a plateau, if the side chain is short or inflexible. The probable arrangement of the molecules offers a simple explanation (5). It is very likely that when the monolayer is formed, the molecules align next to each other randomly parallel and antiparallel. This is supported by x-ray analysis where available for the solid state, when no strong forces sufficient to promote a more regular arrangement of the molecules exist. It is probable that the same situation prevails commonly in monolayers. Thus, when side chains are short or inflexible, the energies of interaction between parallel and between antiparallel molecules will be different, and the monolayer will behave essentially as if it were a mixture of two components. If the side chains are long and flexible, their interactions with those on adjacent molecules will be essentially independent of the direction of the backbone and of the same energy so that the plateau is flat.

To test the basic idea, some experiments are possible. When the side chains are long and flexible, there will be little difference between the energy of interaction between one molecule and another either of the same type or of the opposite enantiomorphic form, irrespective of the backbone directions. For example the properties of poly(ε-benzyloxy-carbonyl-L-lysine) can be compared with a 1:1 mixture with the D-lysine enantiomorph. When side chains are short and inflexible, as in poly(L-alanine), significant differences should exist between its monolayer properties and that of a 1:1 racemic mixture, in this case of poly(L-alanine) and poly(D-alanine).

An instructive experiment is to look at three polymers with rather similar side chains: poly(L-leucine), poly(L-norleucine), and poly(L-methionine) to compare their surface properties. Finally as an example of extreme side chain flexibility we consider poly(γ-n-decyl-L-glutamate), which while quite unlike any protein (more like fat bacon), its surface properties are of some interest and perhaps help us to understand the properties of more protein-like molecules.

The experimental strategy is to look at not only the monolayer properties, recording surface area, pressure, and potential but to also study the structure of the fully collapsed monolayer using polarized IR spectroscopy and electron diffraction. Such indirect methods can give valuable information of the conformation of the monolayer by correlating the results with experiments using isotope exchange and by directly relating the monolayer properties to changes in the IR spectra (5, 17). IR spectroscopy is particularly useful when the crystallinity of a specimen for study by electron diffraction is low—for example, in proteins. Electron diffraction, besides giving detailed conformational information, has the advantage that it requires only sufficient material for viewing in the electron microscope. The orientation of the molecules as removed from the water surface can be deduced (in favorable cases), and intermolecular side chain interactions are revealed more clearly than by x-ray diffraction. This last feature, first noted explicitly by Vainshtein and Tatarinova (18), is valuable in interpreting the properties of polymer mixtures in relation to their monolayer properties.

These powerful indirect methods are particularly suitable as an aid in studying surface chemistry of high molecular weight polypeptides as

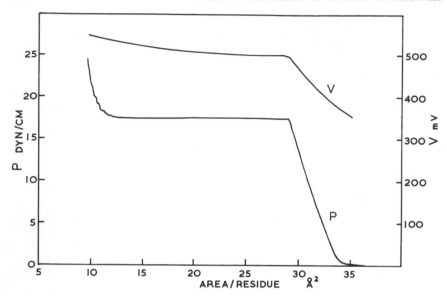

*Figure 1.   Surface pressure (P) and surface potential (V) vs. area for a mono-
layer of poly(γ-n-decyl-L-glutamate) on distilled water, pH 5.7, 20°C*

opposed to molecules of low molecular weight such as lipids. Because
of hysteresis effects, the conformation of these polymers is frequently an
expression of their past history and manner of preparation. Correlating
the information with the monolayer properties and the extensive knowledge now available on the factors that determine the conformation of
these molecules in the solid state and in solution enables the monolayer
conformation to be determined with reasonable certainty.

Experimental Methods

Except where otherwise stated all polymers were spread at 20°C
from solution in chloroform containing 10% dichloracetic acid, on the
surface of water in a Langmuir trough made of polytetrafluorethylene
(PTFE). Compression of the monolayer was a continuous 0.25 A²/resi-
due/min (barrier speed, 2 mm/min). Surface potentials were measured
routinely with an americium 241 air ionizing electrode connected to a
Vibron Electrometer model 33B2 (Electronic Instruments Ltd.) and a
recorder. Surface pressures were measured continuously with a flexure
device (19) consisting essentially of a waxed phosphor bronze strip ¼ in.
wide, 0.003 in. thick with its lower edge in the interface. The deflection of
the strip was recorded with an optical automatic following device connected to a recorder.

Studies of the Fully Collapsed Monolayer. Collapsed monolayers of
all polymers were examined by IR spectroscopy and in most cases by
electron diffraction using a Phillips EM300 electron microscope set for
selected area diffraction. Viewing of the specimens in the microscope

was kept to the absolute minimum necessary to select a uniform area (to minimize radiation damage) and the diffraction pattern from an area 40μ in diameter was recorded with a working voltage of 80 kV. The exposure time was usually 64 sec using Ilford plates type EM4. In other respects the experimental procedures followed those in earlier work (*12, 14, 17*).

Polymers. Poly(L-leucine) and poly(β-benzyl-D-aspartate) were kindly provided by E. M. Bradbury, Portsmouth Polytechnic. Poly(L-alanine), poly(D-alanine), and poly(L-methionine), MW 40,000, were obtained from Sigma Chemical Co. A further sample of poly(L-methionine), MW 36,000, (Van Slyke method) was obtained from Mann Research Labs. Poly(β-benzyl-L-aspartate), MW 210,000, poly(ϵ-benzyloxycarbonyl-L-lysine), MW 400,000, poly(ϵ-benzyloxycarbonyl-D-lysine), MW 900,000, poly(γ-*n*-decyl-L-glutamate), MW 860,000, and poly(γ-benzyl-D-glutamate), MW 235,000, were all obtained from Pilot Chemicals. The remaining polymers were the same as those used previously (*12, 14*).

Experimental Results and Discussion

Poly(γ-*n*-decyl-L-glutamate). Monolayers of this polymer were spread from solution in chloroform in which it is freely soluble. The surface pressure–area isotherm (Figure 1) has several notable features. In contrast to most polymers previously studied, there is very little tail at low pressures, and there is an almost linear rise to the commencement of the plateau just below 30 A^2/residue. There is very little indication of hysteresis, shown by a small hump at the beginning of the plateau with certain other polymers, and the plateau extends almost completely level down to 12.5 A^2. The pressure then rises, and final collapse occurs at a pressure well below that of most other polymers.

While collapsed films of this polymer can be lifted off the surface on electron microscope grids, viewed under the light microscope they are seen to break under the action of surface forces within a few minutes. Electron diffraction observations are evidently not feasible, but good polarized IR spectra are obtainable (Figure 2). The parallel dichroism of the Amide A band (3300 cm^{-1}) and the Amide I band (1660 cm^{-1}), and the perpendicular dichroism of the Amide II band (1555 cm^{-1}) is strong evidence that the collapsed monolayer is in the α-helical conformation with the molecules aligned on the water surface more or less parallel to the barrier. There is not sufficient dichroism in the bands associated with the *n*-decyl side chain for it to be orientated predominantly either parallel or perpendicular to the backbone. Since the side chains are very flexible it is probable that during collapse of the monolayer the side chains fold to form a more compact non-dichroic structure.

CONTACT ANGLE OBSERVATIONS. Thin films of polymer were made by spreading a solution in chloroform on a microscope slide and allowing

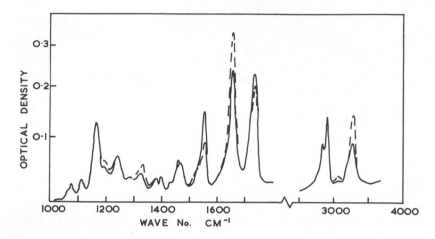

Figure 2.   Polarized IR absorption spectrum of a collapsed monolayer of poly(γ-n-decyl-L-glutamate) air dried on a silver chloride plate.
Broken line, electric vector parallel to the barrier used to collapse the monolayer;
full line, electric vector perpendicular.

the solvent to evaporate. The meniscus at the polymer–water interface was then observed using the dipping plate technique. A value of about 100° was obtained for the angle of contact (20°C) provided the surface had not been immersed. If the plate was lowered into the water and raised again, an angle as low as 60° could be observed. The plate could then be rotated to 100° where a flat meniscus could again be obtained. Within these limits the meniscus never appeared far from flat so that precise determination of the limits was not possible. At 5°C the range over which a flat or nearly flat meniscus could be observed was 70° to 103°, and raising the temperature to 32°C gave a range about 100° to 45°.

DISCUSSION. The general behavior of the monolayer is very similar to monolayers of poly(γ-methyl-L-glutamate) and poly(γ-ethyl-L-glutamate) (*12, 14*) for which there is strong evidence of the existence of the α-helix in the monolayer state (*5*). But while the observed area per residue in this instance cannot be compared with that expected from diffraction studies, it is possible to extrapolate from poly(γ-ethyl-L-glutamate) which has an area of 19.6 A²/residue, if the side chains between molecules are assumed to stick out at right angles and interpenetrate as far. The molecules will then be separated by an additional eight methylene groups or approximately 10 A. With an axial increment of 1.5 A per residue for an α-helix, the area will be 19.6 + 15 = 34.6 A²/residue which agrees well with the observed area of 34.5 ± 0.5 A²/residue. An additional feature of similarity between these polymers is the behavior of the surface potential (Figure 1). The correspondence in shape of the surface potential–area curve with the pressure–area isotherm and with similar behavior in the other glutamate esters suggests a similar explanation for its origin. An isolated α-helix has no net dipole moment perpendicular to its axis, and it has therefore been proposed that the potential arises mainly from interaction of water dipoles with the peptide group and side chain dipoles (*16*). However the potential at the start of the transition at 29.5 A^2/residue is approximately 500 mV compared with 540 mV at 17.5 A^2/residue for poly(γ-methyl-L-glutamate) and 635 mV at 18 A^2/residue for poly(γ-ethyl-L-glutamate). Therefore considered in relation to the number of residues per unit area, the potential observed in poly(γ-*n*-decyl-L-glutamate) is significantly higher. This result is unexpected but consistent with the unexplained observation that the surface potential of the ethyl ester is higher than the methyl ester. It suggests that the hydrocarbon side groups are also promoting some type of interaction with the water that gives rise to a dipole with a component perpendicular to the interface.

While this polymer has not so far been as fully investigated as some, it fits into the general pattern (*5*) and appears to have the α-helical conformation in the monolayer state. But normally the pressure rises at about half the monolayer area, and it has been suggested that over the length of the plateau to this point molecules are being forced out of the surface forming a second layer (*12, 14*). In some instances there is also clear evidence for the formation of further layers at higher pressures (*19*). Shuler and Zisman however did not find an additional rise at about half the monolayer area in monolayers of poly(γ-methyl-L-glutamate) in the presence of capillary waves or any change in the wave damping coefficient (*20*). While accepting the view that the monolayer consisted of α-helices, they concluded that in earlier experiments (*14, 21*) a true equilibrium was not reached and that their results did not support the view that a bimolecular layer formed at the onset of the plateau.

But apart from the need to provide a satisfactory explanation for the observed decrease in area, the case for an orderly formation of a second layer of molecules rests on the following considerations (*5*):

The implicit order in the monolayer arising from the side by side packing of rigid α-helices;

The flatness of the plateau when the side chains are flexible. This contrasts with the type of collapse envisaged by Yin and Wu (*22*) in a flexible polymer where the isotherm beyond the point of collapse has in general either a positive or negative slope;

The observation of further transitions in some cases associated with the consecutive formation of up to five layers of molecules. These occur at higher pressures than the transition indicative of the formation of the second layer of molecules in a region where Equation 1 is not applicable and where hysteresis effects are evident. The underlying molecular process is nevertheless clearly of a similar nature;

Fully collapsed films normally show alignment of the molecules and, in some cases, three dimensional structures with preferred orientation with respect to the water surface;

The surface potential remains almost constant over the transition, consistent with non-polar molecules forming a second layer.

In the case of the polymer considered here, which has almost fluid side chains, the behavior is close to simple theory with the plateau extending to smaller areas until the pressure ultimately rises and the film becomes unstable and collapses. Similarly at low pressures, assuming the molecules form condensed ordered micelles, the absence of a marked tail to the pressure–area curve can be understood as arising from the very easy plastic deformation of the micelles to fit together when compression starts. The monolayer properties of this polymer approximate well the ideal case of simple theory, primarily as a consequence of the flexibility of the side chains.

The curious behavior of the contact angle is clearly related to the flexible nature of the side chains. It is probable that an angle of contact around 100° is a consequence of the n-decyl side groups being folded over the backbone so that the surface properties are similar to poly(ethylene). The lower angle of about 60° probably arises from the side chains in the interface seeking to spread over the surface of the water. It may be that two stable states for the side chains are possible depending on the orientation of the polymer in the liquid. Further studies would be of interest. Hysteresis effects in contact angle observations have been reported previously in polypeptides (12), and a similar though much less marked behavior may be responsible. The observations recorded here are clearly related to and their interpretation consistent with the behavior of the monolayer.

Poly(L-leucine), Poly(L-norleucine), and Poly(L-methionine). MONOLAYER PROPERTIES. For solubility reasons it was necessary to spread poly(L-leucine) from chloroform containing 20% trifluoracetic acid. This solvent causes drops of liquid to creep up the outside of the pipet, and the observed area per residue of 16 A² may be slightly low. Use of a glass pipet treated with a silicone minimized the problem. Both specimens of poly(L-methionine) behaved in a similar manner. Poly(L-norleucine) has been investigated previously (12, 16). The results presented here are in general agreement, but the work has been repeated under the same experimental conditions as for the other polymers to enable a precise comparison (Figure 3).

While the surface potential behaves in a very similar manner on compression of all the monolayers and the values recorded are close, there are quite marked differences in the surface pressure–area curves. A feature of all three and also of all the other polymers in this study is a progressive but regular departure from smoothness of the curves above the plateau or transition. This is not an instrumental artefact nor is it easily detectable with point-by-point recording, but its precise form undoubtedly depends on the experimental conditions.

STRUCTURE OF THE COLLAPSED MONOLAYERS. IR spectra of specimens prepared from air dried collapsed monolayers were typical of specimens in the α-helical conformation with no indication of any β conformation. Electron diffraction patterns gave a similar result. The patterns for poly-(L-leucine) and poly(L-norleucine) are similar to poly(L-norvaline) (*12*) with low crystallinity. A strong equatorial reflection at 10.94 ± 0.10 A is observed in poly(L-leucine). If we assume as previously (*5*) that this is the 100 reflection from a hexagonal cell, the calculated area per residue in the monolayer is 17.3 A², assuming the molecular separation is the same as in the collapsed film. This figure is in agreement with the observed area of 16 A² in view of the difficulties encountered in spreading the monolayer.

The diffraction pattern of poly(L-methionine), Figure 4a, is much more crystalline than that of the other two polymers and in several respects resembles that of collapsed monolayers of poly(γ-methyl-L-glutamate). The correspondence is shown clearly by the reciprocal lattice diagrams (Figure 5). For an extended account of the basic theory *see* Stokes (*23*) and Elliott (*24*).

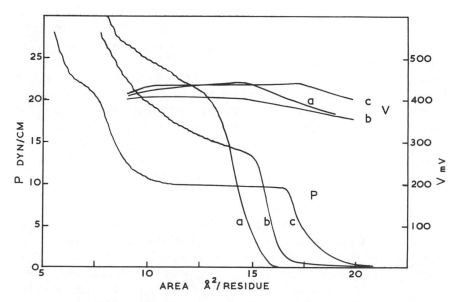

Figure 3.   Surface pressure (P) and surface potential (V) vs. area for (a) poly(L-leucine), (b) poly(L-methione), (c) poly(L-norleucine) on distilled water

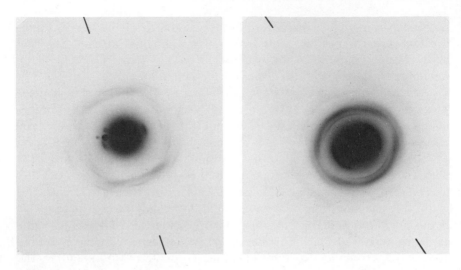

Figure 4.   Electron diffraction patterns from collapsed monolayers. (a) Poly(L-methionine). The outermost reflection at 1.49 A is not visible on this photograph; the sharp meridional reflection is 006. (b) 1:1 poly(L-alanine) + poly(D-alanine).

In (b) because of folding in the latter stages of formation of the specimen, the arcs are extended. The outermost reflection is at 1.489 A. Note the sharpness of the meridional arc at 5.36 A and also the strength of the equatorial 100 reflection close to the center in comparison with that in (a). The lines mark the direction of the meridian.

In the case of high voltage electron diffraction, as opposed to x-ray diffraction, the observed distances on the photograph are directly proportional to the reciprocals of the perpendicular distances between the lattice planes responsible for the reflections. The reflections marked on a reciprocal lattice diagram correspond directly to the diffraction photograph. Not all points on the reciprocal lattice are associated with reflections; a regular α-helix with 18 residues in five turns (18/5 helix) should produce no meridional reflection of lower order than 00,18 corresponding to the 1.5 A residue repeat distance along the molecular axis. The presence of an 006 reflection in poly(L-methionine) indicates distortion of the side chains arising from the hexagonal packing of the molecules in the lattice. This gives rise to a sixfold periodicity and has been observed before in poly(γ-methyl-L-glutamate) (*18, 24*).

The close correspondence in size between the hexagonal unit cells of poly(γ-methyl-L-glutamate) and poly(L-methionine) implies a similarity in the intensity expected in the observed strong reflections. In particular those on the fifth layer line associated with the pitch of the helix (or on the corresponding layer line of a higher order helix) indexed as 105 and 115 are normally very strong (*18, 24*) in poly(γ-methyl-L-glutamate). But the 105 reflection is much weaker than the 115 reflection in monolayer specimens of either polymer. The corresponding 100 equatorial reflection, almost absent in poly(γ-methyl-L-glutamate), is observed in poly(L-methionine) but it is not so intense as in a fiber specimen. The weakness of the 100 and 105 reflections from collapsed monolayers probably arises because the (100) planes develop predominantly and preferentially parallel to the surface. When mounted with the specimen perpendicular to the beam, these planes are unable to reflect (5). In contrast, in bulk fiber specimens, some of the planes are correctly oriented. The 100 planes are the most closely packed and their formation parallel to the surface would be expected since this promotes maximum hydrophobic bonding (5).

A further very marked feature of the diffraction photographs is a streak along the fifth layer line (as opposed to a crystalline arc), similar to that observed in poly(L-alanine) (25) and shown there to be a consequence of random parallel–antiparallel packing of the molecules in the

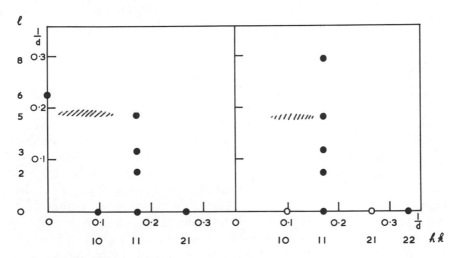

Figure 5.   Reciprocal lattice rotation diagrams are for (left) poly(L-methionine) and (right) poly(γ-methyl-L-glutamate) from collapsed monolayers using electron diffraction. The reciprocals of the spacings d, (A) correspond to their positions on the diffraction photographs [compare the diagram for poly(L-methionine) with the top right hand quadrant of Figure 6a].

Full circles denote observed reflections, open circles absent reflections that would be observed if all orientations about the fiber axis were present (fiber type orientation). While the 100 reflection is observed in poly(L-methionine), it is weaker than would be expected if the orientation were fiber type, and it appears that the (100) planes develop preferentially parallel to the surface and (110) planes perpendicular (5). The diagram for poly(γ-methyl-L-glutamate) is constructed from earlier data (12) with a very similar unit cell; a = 11.8 A, c = 27.0 A, compared with a = 11.46 A, c = 26.8 A for poly(L-methionine). The cell size and the positions of the reflections are consistent with the presence of an α-helix containing 18 residues in five turns. The hatched areas denote the position of layer line streaks. The meridional 00,18 reflections at about 1.5 A are not shown.

crystal. A similar interpretation is reasonable in the polymers considered here, and it shows clearly that the interactions between molecules are not sufficiently strong for them to segregate into a more orderly structure. This is further indirect support for the assumption of a random parallel–antiparallel packing in the monolayer state.

From the electron diffraction observations the calculated area per residue in a monolayer of poly(L-methionine) is 17.2 A^2; the area observed is 16.5 $\pm$ 0.5 A^2.

CONTACT ANGLE MEASUREMENTS. The angle of contact of 93° for poly(L-norleucine) has been reported previously. Problems of preparing uniform adherent films of poly(L-methionine) and hysteresis effects combined to make measurements of θ very imprecise; it is estimated at 55° $\pm$ 10°. No measurements at all were possible with poly(L-leucine) for similar reasons.

DISCUSSION. Considering first the pair of polymers poly(L-norleucine) and poly(L-leucine), the marked differences in the pressure–area isotherms (Figure 3) contrast with the closeness in constitution. The results are good support for the view that the more flexible side chain of poly(L-norleucine) will cause the energy of interaction of adjacent molecules to be less sensitive to their relative directions, compared with poly(L-leucine), so that the plateau is flatter.

Comparing poly(L-norleucine) with poly(L-methionine) it is unlikely that the flexibility of the side chain is greatly different in the two cases. The crystal structures of the amino acids are very similar, with the side chains forming double layers with the terminal groups opposed (26, 27). In the monolayer state however, the geometry of the α-helix is such that the side chains of adjacent helices can interpenetrate to form good hydrophobic contacts. If the side chains were opposed, as in the amino acid crystals, the areas per residue would be much higher than are observed. Moreover there is direct evidence for side chain overlap in the collapsed films of poly(L-methionine) from the electron diffraction observations. In this situation the dipolar nature of the sulfide group is probably important [dimethyl sulfide has a dipole moment of 1.45 D (28)] and sufficient to cause the energy of interaction of parallel and antiparallel molecules to differ, thereby causing the plateau to depart from flatness. In the absence of a dipole, as in poly(L-norleucine), no energy difference might be expected, and the plateau is flat.

From Equations 1 and 2 the calculated values of γ_{pl} for poly(L-methionine) and poly(L-norleucine) are 8.5 erg cm^{-2} (taking $W = 14$ dyne cm^{-1}) and 28. erg cm^{-2} (5), respectively. The relative magnitudes of these figures may be compared with measurements of the hydrophobicity of methionine and leucine side chains (relative to glycine) from solubility measurements by Nozaki and Tanford (29) who find energies of 1300 and 2600 cal/mole, respectively. Their measurements are strictly the free energies of transfer of the side chains from ethanol to water. Using the assumption (*16*) that one third of the side chains on a helix interact with the underlying water, the figures for γ_{pl} can be converted to molar energies. The values so obtained are 635 cal/mole and 2780 cal/mole for poly(L-methionine) and poly(L-norleucine), respectively, taking monolayer areas from the spacings calculated from the electron diffraction data (though the observed surface areas do not significantly differ). Now clearly the accuracy of the monolayer data is not high, particularly on account of the difficulties experienced with the contact angle measurements for poly(L-methionine) where a higher value of θ would give better agreement. Moreover the free energies obtained by Nozaki and Tanford are obtained in a somewhat different system; the agreement is therefore perhaps as good as should be expected.

The value obtained for γ_{pv} for poly(L-methionine), 50 erg cm^{-2}, is significantly higher than the corresponding value of 34 erg cm^{-2} for poly(L-norleucine). Since for a sufficiently hydrophobic polymer γ_{pv} is closely related to the work of cohesion, this would contribute to the markedly higher crystallinity of poly(L-methionine).

The similarity of the surface potential of poly(L-methionine) to that of the other two polymers is surprising. With the latter, interaction of water dipoles with the peptide groups is probably mainly responsible (*16*), and little difference between the two might be expected. But with poly(L-methionine) additional components of the potential might arise from departure from helical symmetry of the side chain dipoles, so giving rise to a net, though small, dipole moment perpendicular to the surface and also from side chain dipole–water interactions. This is the type of situation envisaged in poly(γ-methyl-L-glutamate) for example (*16*), and it must be concluded that in this instance the net effect is slight.

Monolayers of Racemic Mixtures: Poly(alanine), Poly(γ-benzyl Glutamate), Poly(β-benzyl Aspartate), Poly(Benzyloxycarbonyl Lysine). EXPERIMENTAL RESULTS. The pressure–area and surface potential results for the two enantiomorphic forms of a given polymer were virtually identical except for poly(benzyl aspartate) where the plateau of poly(β-benzyl-D-aspartate) was about 2 dynes/cm higher than that of the L enantiomorph (Figure 6). This may result from the incomplete benzylation of the D-aspartate.

Electron diffraction patterns have been obtained from collapsed monolayers of poly(alanine) and poly(γ-benzyl glutamate). The enantiomorphic forms of the other two polymers give patterns with very poor crystallinity, and their racemic mixtures have not therefore yet been investigated. The principal features of the diffraction pattern of poly-

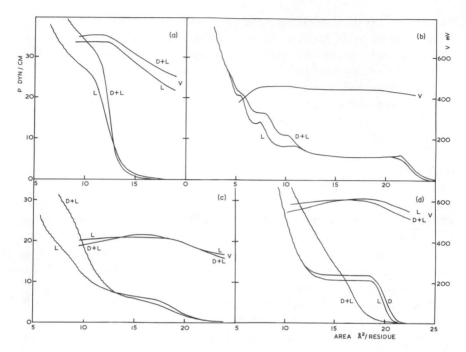

Figure 6. Surface pressure (P) and surface potential (V) vs. area for mono-layers on distilled water, pH 5.6–5.7. (a) poly-alanine, (b) poly(ε-benzyloxy-carbonyl lysine), (c) poly(γ-benzyl-glutamate), (d) poly(β-benzyl-aspartate).

In general the curves for the two enantiomorphic forms are virtually identical and one curve only, L, is shown; the exception is (d) where both are shown. The curves of the 1:1 mixtures from the same solutions spread under identical conditions are denoted by D + L.

(alanine) (Figure 4b) are similar to those of poly(L-alanine) (*12*) with a strong equatorial reflection at 7.40 ± 0.02 A. There is however a meridional reflection at 1.489 ± 0.002 A [compared with 1.495 A in poly-(L-alanine)] and a strong sharp reflection at 5.36 ± 0.02 A [not evident in poly(L-alanine)] which appears to be on the meridian. Otherwise the pattern is rather diffuse.

The diffraction pattern of poly(γ-benzyl glutamate) is very diffuse, the only sharp reflection being at 5.26 A on or close to the meridian. A reflection at about 1.50 A on the meridian is observed, but in other respects the diffraction pattern shows none of the detail found in the x-ray diffraction patterns of poly(γ-benzyl glutamate) fibers (*30*) (where it should be noted the specimens had been heated to promote crystal-linity). Polarized IR spectra were obtained from collapsed monolayers which showed them to be in the α-helical conformation in all cases. Poly(β-benzyl aspartate) proved particularly interesting in that the spec-trum of the collapsed monolayer was similar to that of poly(β-benzyl-L-aspartate) in the left-handed helical conformation but with a small amount of the right-handed form also present. This is in contrast to the spectrum of poly(β-benzyl-L-aspartate) when prepared from collapsed monolayers where the spectrum is clearly that of the right-handed α-helix (17).

DISCUSSION. The shape of the transition in the pressure–area curve of poly(L-alanine) can be accounted for as a consequence of the difference in the energy of interaction between parallel and between antiparallel arrays in the monolayer. Calculations by Parry and Suzuki (31) show that a small difference in the energies is to be expected, *ca.* 1.0 kcal/mole residue. At the air–water interface the energy is probably less on account of the presence of intervening water molecules. A calculation of the shape of the transition would clearly need to take into consideration the various random arrangements with respect to chain direction in the monolayer and their interactions as molecules formed a second layer. Parry and Suzuki find that when various assemblies, both parallel and antiparallel, of more than two molecules are considered, the energy difference per molecule lies within the limits of the two molecule case. A similar situation is probable in considering the range of energies in the molecules involved in the transition. The observed rise in the pressure in the case of poly(L-alanine) is not less than about 5 dyne-cm^{-1} which at the monolayer area of about 14 Å^2/residue corresponds to an energy of about 100 calories. The energy difference for a pair of isolated molecules is probably higher for the reasons given, and therefore this figure is not inconsistent with the calculated value of about 1.0 kcal which itself may be subject to considerable error.

Further support for this general explanation comes from the markedly raised transition in the racemic mixture. Model building shows that molecules of poly(L-alanine) pack well together (25), but the packing of the racemic mixture is much less satisfactory. This is reflected in the fact that if fibers are prepared from the racemic mixture, the two components segregate, and x-ray studies show no evidence of a true mixture (32). This shows that under conditions used normally to prepare fibers the interaction of the racemic molecules is weaker than between molecules of the same enantiomorphic form. In the monolayer state, segregation of the two forms is not able to take place since the process of spreading the molecules is not one favorable to equilibration of the system. The molecules are constrained to pack in the monolayer without being able to move over each other to find a more satisfactory partner. This is evident also from the pressure–area curve, for if the molecules were able to segregate, it would be similar to that of either component. Now if the molecules in the mixture attract each other less strongly than those of the same type, γ_{pv} will be less, and consequently more work will have to be done on the monolayer to promote the transition, *i.e.*, W will be higher which is observed.

Support for this general picture comes from the electron diffraction observations on the collapsed monolayer. If the sharp reflection at 5.36 A is a true meridional reflection (better orientation is necessary to be certain), it can be indexed as 005 and the 1.489 A meridional reflection as 00,18 for a hexagonal cell with c = 26.8 A. An 005 reflection would not be produced by a perfect helix, but it can probably be accounted for assuming a distorted α-helix, the distortions being caused by the unsatisfactory packing of the two forms. The reduced axial increment per residue, 1.489 A compared with 1.495 A in the enantiomorphic form, shows that the packing causes the helix to shorten slightly.

Compared with the enantiomorphic form, racemic poly(γ-benzyl glutamate) has a transition at a slightly lower pressure in contrast to the situation above. But in this case, both in the solid and liquid crystalline states, the racemic polymer produces mixed crystals (33), and following from the line of argument used for poly(alanine), a lower transition is therefore expected. The difference in height in this instance is much less, and probably a consequence of the longer and more flexible side chains. In fact in analyzing the structures of the racemic form, Squire and Elliott (33) suppose that a given molecule will have sufficient flexibility of the side chain to pack equally well irrespective of the direction of the backbone. If this were strictly true, and on the basis of the ideas put forward here, the transition in the enantiomorphic polymer should be perfectly flat which is not so. An additional factor in comparing the monolayer properties with ordered mixed crystals is to note that the diffraction pattern of the collapsed racemic monolayer has much lower crystallinity. This suggests that again the monolayer is not organized in such a manner that during collapse the maximum intermolecular interaction (which would lower the transition further) is possible.

Comparing the pressure–area curves of poly(β-benzyl-L-aspartate) with poly(γ-benzyl-L-glutamate) (Figures 6c and 6d), the difference in height of the transitions has already been accounted for as arising essentially from the additional methylene group in the glutamate polymer causing γ_{pl} to be higher (16). But the flatter plateau of the aspartate polymer cannot be accounted for on the basis that in all cases the energy of interaction of parallel and antiparallel molecules will be more nearly equal the more flexible the side chain. Since this view is essentially an aid to qualitative reasoning, and perhaps particularly where strongly interacting benzyl groups are concerned, an additional methylene group in the chain may promote some specific benzyl interaction which is sensitive to the chain direction. Such explanation is reasonable when it is noted that fibers of the α-helical form of poly(β-benzyl-L-aspartate) (34) show much lower crystallinity than poly(γ-benzyl-L-glutamate) (*30*). In the former case the side chains may well interact less precisely so the energy may not be at all sensitive to the chain direction. Indeed if specimens are heated to promote crystallinity, the side chains then interpenetrate and interact with sufficient strength to change the structure from the α-helix to the ω-helix (*34*). Since this involves distortion of the amide group from planarity, it is clearly a strong interaction. The packing of the benzyl groups in the monolayer state of poly(γ-benzyl-L-glutamate) might resemble that of the ω form of the benzyl aspartate which does not appear to be its monolayer conformation. For these reasons simple comparisons between the two polymers considering only side chain flexibility are probably unfruitful.

Considering now the very marked difference between the pressure–area curve of poly-(β-benzyl-L-aspartate) and the racemic mixture (Figure 6d), it is probably true to say in the light of the above discussion that some particular favorable interaction occurs between the benzyl groups in the racemic mixture. The marked decrease in the area per residue of 2 Å² and the almost complete disappearance of the plateau suggest that the spread polymer is only just stable in the monolayer state and that the transition proceeds almost spontaneously, *i.e.* W is close to zero. If there is a strong interaction between the two enantiomorphic forms markedly greater than between molecules of the same kind, it is to be expected that in the monolayer the transition will appear as an inflection rather than a flat plateau. The conformation of the monolayer is as yet uncertain. In the solid state and in solution, poly(β-benzyl-L-aspartate) normally is left rather than right-handed α-helix. But in the monolayer state the right handed helix appears to be present, and this is probably a consequence of the polar water molecules weakening the side chain–backbone interactions that otherwise cause the helix to be left handed (*17*). The observation in the present work that the collapsed dried monolayer of the racemic mixture has a spectrum indicative of the left-handed α-helix for L residues may simply be a consequence of a conformational change when the polymer is removed from the water surface since the right handed helix of the L-aspartate is then unstable and can be converted to the left handed form by swelling it with dichloracetic acid vapor (*35*). The similarity of the magnitude of the surface potential of the mixture and the enantiomorphic form gives no indication that the monolayer conformations are different. A change in helix sense would alter the orientation of the side chain ester groups with respect to the backbone, and this in turn would modify side chain–water interactions so that the resulting surface potential could be different.

Finally the behavior of racemic poly(ε-benzyloxycarbonyl lysine) in the monolayer state is particularly straightforward; the area per residue and the height and shape of the plateau is not significantly different from that of the enantiomorph. Clearly the four methylene groups close to the backbone confer sufficient flexibility on the side chains for the backbone to have little effect. At pressures above the plateau, the inflections indicative of the formation of further layers of molecules (14) become less pronounced. This is a region of almost perfect plastic deformation, and it must be concluded that these further transitions are exceptionally sensitive to the molecular structure. This polymer is similar to poly(γ-methyl glutamate) where again the racemic mixture has a pressure–area isotherm with a flat plateau at the same height as the enantiomorphic form (5).

General Observations

This work is essentially an extension of the ideas first put forward by Crisp who not only noted the correspondence between the shape of the transition and the length of the side chain (13) but also developed the two dimensional phase rule (36). This rule has been applied qualitatively and with caution to these polymers to account for the shape of the transition. The only additional idea involved is the realization that an organized polymer monolayer of polar molecules may behave as if it consisted of two components if the energies of interaction of parallel and antiparallel molecules are significantly different. In addition when considering mixtures of two enantiomorphic forms, the monolayer behaves as if it consisted of only one component when the energies of interaction between the molecules (in this case involving four types of contact) are the same. In those polymers that show only an inflection rather than a flat plateau in the pressure–area isotherm, a similar inflection at a higher or lower pressure in the racemic mixture occurs depending on whether the two forms attract each other less or more strongly than molecules of the same type.

Clearly these ideas can be extended to mixtures of more dissimilar molecules. For example, it is possible to see differences between the pressure–area curves of poly(L-alanine) mixed with poly(D-α-amino-n-butyric acid) and the corresponding mixture containing poly(D-alanine). With such mixtures there are six different types of molecular contact involved, and the interpretation is therefore inevitably more complex.

This work provides more evidence that at least in some cases the collapse of the monolayer represents quite a remarkable molecular rearrangement. The diffraction pattern of poly(L-methionine) shows that not only do the molecules collapse in a crystalline array but there remains a measure of preferred orientation of the crystals with respect to the water surface. This has been observed previously with poly(γ-methyl-L-glutamate), and it might be observable with sufficient care in other polymers. It fits in with the interpretation of the additional transitions in poly(ε-benzyloxycarbonyl-L-lysine) which evidently arise from the consecutive formation of several layers of molecules one at a time. In comparing the crystallinity of the diffraction pictures from collapsed monolayers with those of specimens prepared as fibers, it should be realized that the fiber specimens have usually been heated to promote maximum crystallinity, as well as being stretched in some way to orientate the molecules. That well orientated and crystalline specimens can be prepared simply by collapsing a monolayer and allowing it to dry at room temperature is probably a reflection of the highly ordered initial state of the monolayer that promotes the free movement of molecules, both at the initial transition and during further compression, as the collapsed film increases in thickness. How far the specimens build up as a molecular process before gross folding of the monolayer takes place is not known, but in the instances referred to here it seems that a number of layers must form before folding starts.

The slight irregularities observed in the pressure–area curves above the transition pressure are an interesting general feature. They become progressively more marked as the pressure rises and are only clearly evident with continuous automatic recording of the pressure. Particularly where the transition is not a well-defined flat plateau, they are a good indication of the onset of the transition. They probably arise from varying rates of collapse within the film. It should be realized that with these high molecular weight polymers hysteresis effects are common. It is possible, for example, to observe relaxation effects if compression of the film is stopped (5, 20). Irregularities might be seen once the pressure to produce the transition has been exceeded where simple theory suggests the film should be unstable. But this does not preclude experimental observations above the transition pressure being used to interpret the real behavior of the system, as in poly(ε-benzyloxycarbonyl-L-lysine), provided equilibrium is not implied in any theoretical interpretation.

Conclusion

The work described here strengthens the idea that when the α-helix is present in a polypeptide monolayer, the transition in the pressure–area curve arises from the collapse of the monolayer under the action of surface forces and that at least in some cases this is in the nature of a phase transition and proceeds in an orderly manner. The transition is in fact rather analogous to the development of the tertiary structure in a protein under the action of hydrophobic forces. We can therefore now have additional confidence in the use of these methods to obtain inter-facial energies and to gain insight into the role of hydrophobic forces in proteins both in maintenance of the tertiary structure of the native molecule and in understanding how this is modified by the presence of an interface. Indeed the methods are some of the few ways by which an experimental attack can be made in this area. Better data, particularly of measurements of the contact angles, is desirable as well as a more rigorous discussion of the theoretical implications.

Incidental to this work is more evidence that the α-helix exists at the air–water interface. While some have appeared reluctant to accept this view, no good theoretical reason exists why it should not be stable. Where the nature of the side chain might provoke other conformations [as in poly(L-valine)] or the molecular weight is low or monolayers are spread from poor solvents miscible with water, other conformations are detectable, and the monolayer properties are significantly different (5).

The proportion of α-helix in native proteins is variable and some-times not very great. It is a mistake to over emphasize its role in inter-facial structures, but where it is present its radial distribution of side chains means that its orientation in the interface will be governed by its over-all hydrophobicity. The presence of hydrophobic groups directed into the water is then possible as well as others contributing to cohesion between adjacent molecules as in the monolayers considered here. Those directed into the water may function as sites for the binding of other molecules in the aqueous phase. This is also a possibility for the β con-formation but not for extended chain conformations where under pressure the hydrophobic side chains are directed away from the surface and the hydrophilic ones into the water. While this latter model has been accepted by surface chemists (37), the conformation appears unlikely both from a biological and a stereochemical standpoint. Indeed except where there is a regular alternation of hydrophobic and hydrophilic side chains, the conformation is probably not one acceptable within the usual criteria for polypeptide structures (5).

Polymers of the naturally occurring amino acids alanine, leucine, and methionine all show interactions which depend on the relative directions of the backbones. In contrast, poly(L-norleucine) shows less specific interactions; clearly for the hydrophobic regions of proteins to function in a precise manner the natural amino acids are most suitable.

NOTE ADDED IN PROOF. Alternative explanations have recently been put forward to account for the height and shape of the transition in certain synthetic polypeptide monolayers (38). These do no however account for many of the results here presented or for the general pattern of behavior observed in monolayers when the α-helical conformation is present (5).

Acknowledgments

I am indebted to Sir John Randall for the use of the electron microscope and to A. C. Everid for his skilled assistance in obtaining the diffraction photographs.

Literature Cited

1. Langmuir, I., Cold Spring Harb. *Symp. Quant. Biol.* (1938) **6**, 171.
2. North, A. C. T., Phillips, D. C., *Prog. Biophys. Mol. Biol.* (1969) **19**, Part I, 5.
3. Biridi, K. S., *J. Colloid Interface Sci.* (1973) **43**, 545.
4. Cheesman, D. F., Davies, J. T., *Advan. Protein Chem.* (1954) **9**, 439.
5. Malcolm, B. R., *Progr. Surface Membrane Sci.* (1973) **7**, 183.
6. Malcolm, B. R., *Nature* (1962) **195**, 901.
7. Malcolm, B. R., *Soc. Chem. Ind. London* (1965) **19**, 102.
8. Miller, I. R., *Progr. Surface Membrane Sci.* (1971) **4**, 299.
9. Loeb, G. I., *J. Colloid Interface Sci.* (1969) **31**, 572.
10. *Ibid.* (1968) **26**, 236.
11. Robinson, C., *Trans. Faraday Soc.* (1956) **52**, 571.
12. Malcolm, B. R., *Proc. Roy. Soc., Ser. A* (1968) **305**, 363.
13. Crisp, D. J., in "Surface Phenomena in Chemistry and Biology," J. F. Danielli, K. G. A. Pankhurst, A. C. Riddiford, Eds., pp 23-54, Pergamon, Oxford, 1958.
14. Malcolm, B. R., *Biochem. J.* (1968) **110**, 733.
15. Malcolm, B. R., *Polymer* (1966) **7**, 595.
16. Malcolm, B. R., *J. Polym. Sci. Part C* (1971) **34**, 87.
17. Malcolm, B. R., *Biopolymers* (1970) **9**, 911.
18. Vainshtein, B. K., Tatarinova, L. I., *Soviet Physics, Doklady* (1962) **6**, 663.
19. Malcolm, B. R., Davies, R. E., *J. Sci. Inst.* (1965) **42**, 359.
20. Shuler, R. L., Zisman, W. A., *Macromolecules* (1972) **5**, 487.
21. Loeb, G. I., Baier, R. E., *J. Colloid Interface Sci.* (1968) **27**, 38.
22. Yin, T. P., Wu, S., *J. Polym. Sci., Part C* (1971) **34**, 265.
23. Stokes, A. R., *Progr. Biophys. Biophys. Chem.* (1955) **5**, 140.
24. Elliott, A., in "Poly-α-Amino Acids," G. D. Fosman, Ed., pp 1-67, Dekker, New York, 1967.
25. Elliott, A., Malcolm, B. R., *Proc. Roy. Soc. Ser. A* (1959) **249**, 30.
26. Mathieson, A. McL., *Acta. Cryst.* (1952) **5**, 332.
27. *Ibid.* (1953) **6**, 399.
28. Lumbroso, H., Dumas, G., *Bull. soc. chim. France* (1955) 651.
29. Nozaki, Y., Tanford, C., *J. Biol. Chem.* (1971) **246**, 2211.
30. Elliott, A., Fraser, R. D. B., MacRae, T. P., *J. Mol. Biol.* (1965) **11**, 821.
31. Parry, D. A. D, Suzuki, E., *Biopolymers* (1969) **7**, 189, 199.
32. Mitsui, Y., Iitaka, Y., Tsuboi, M., *J. Mol. Biol.* (1967) **24**, 15.
33. Squire, J. M., Elliott, A., *J. Mol. Biol.* (1972) **65**, 291.
34. Bradbury, E. M., Brown, L., Downie, A. R., Elliott, A., Fraser, R. D. B., Hanby, W. E., *J. Mol. Biol.* (1962) **5**, 230.
35. Malcolm, B. R., *Nature (London)* (1968) **219**, 929.
36. Crisp, D. J., *Research (London) Suppl.* (1949) 17.
37. Adamson, A. W., "Physical Chemistry of Surfaces," 2nd ed., p 173, Interscience, New York, 1967.
38. Biridi, K. S., Fasman, G. D., *J. Polym. Sci., Part C* (1973) **42**, 1099.

RECEIVED December 4, 1973. This work was supported by the Science Research Council.

18

Bubble Scavenging and the Water-to-Air Transfer of Organic Material in the Sea

DUNCAN C. BLANCHARD

Atmospheric Sciences Research Center, State University of New York, Albany, N.Y. 12222

Surface-active organic material (SAOM) in the sea tends to concentrate at the surface. It is brought there by diffusion, by Langmuir circulations, and by the surfaces of air bubbles rising through the water. These bubbles, produced primarily by breaking waves, not only carry SAOM to the surface but upon breaking eject it into the air. This process may account for the airborne droplets of sea water in the marine atmosphere which have SAOM concentrations several thousand times that found in the sea. The SAOM on the droplets has given marine meteorologists a tracer which enables them to understand better the role of these droplets in rain formation.

Although adsorptive bubble separation has been used commercially for more than half a century [principally in froth flotation to separate minerals from ores (1)], oceanographers and marine meteorologists have only become aware of the importance of natural sea bubble processes in the past 20 years. Marine biologists consider the bubbles as a possible mechanism to convert dissolved organic material to particulate organic material (2); meteorologists consider the role of the bubbles both in the production of a sea-salt aerosol and the ejection of organic material into the atmosphere (3, 4).

Organic Material in the Sea

Composition of Sea Water. Sea water constitutes about 98% of all the water on the face of the earth and contains all of the naturally occurring elements known. It is about 96.5% water and 3.5% salt, most of which is in dissolved form. The salt content, or the salinity of sea water, is defined as the total amount of solid material in grams contained in 1 kg of sea water when all the carbonate has been converted to oxide, bromine and iodine have been replaced by chlorine, and all organic matter has been completely oxidized (5). Depending on location, the salinity will be about 32–38‰ (thousand parts/million), 35‰ being the average.

Although the salinity may vary by 10% from the average, ionic ratios remain constant within narrow limits for all oceanic waters. This constancy is such that oceanographers can physically describe sea water by simply giving its temperature, pressure, and salinity. As shown in Table I, 99.95% of the salinity of sea water is contributed by eight ions and the largely undissociated boric acid. Two salts, NaCl and $MgSO_4$, account for 97% of the salinity of sea water. As Horne states it (6), "... roughly speaking, sea water is an aqueous 0.5M NaCl solution, 0.05M in $MgSO_4$, and containing a pinch or trace of just about everything imaginable."

Concentration and Forms of Organic Material. Concentrations of organic material in sea water are orders of magnitude less than the salinity, but the variations around the mean are at least an order of magnitude more. Although salinity does not vary by more than 10% from 35‰, the organic content of sea water is only on the order of 1 part per million (about 1 mg/l), and variations of well over 100% are common. These variations in a small yet important constituent have made studies difficult and tedious. Only within the past 10 years have detailed studies been made on the distribution of organic material in the sea. Before that reliable instrumentation to determine organic concentrations, much of which is based on the oxidation of the organic material and the

Table I.   Major Salt Constituents of Seawater[a] (115)

Component	Concentration, ‰	% of Total Salt
Cl^-	18.980	55.04
Na^+	10.543	30.61
SO_4^{2-}	2.465	7.68
Mg^{2+}	1.272	3.69
Ca^{2+}	0.400	1.16
K^+	0.380	1.10
HCO_3^{-} [b]	0.140	0.41
Br^- [c]	0.065	0.19
H_3BO_3	0.024	0.07
Total	34.455	99.95

[a] Values in g per kg (‰) based on chlorinity of 19‰.
[b] Varies to give equivalent CO_3^{2-} depending on pH. Value given is essentially true for pH 7.50 at 20°C.
[c] Corresponds to a salinity of 34.325‰.

various means of detecting the resultant carbon dioxide (7), had not been developed.

Organic material exists in one or more forms (dissolved, planktonic, etc.) at all depths in the sea, but its mass concentration is greatest in the euphotic zone which is the upper 100 m of the sea where photosynthesis can occur. This region is where most of the bubbles are produced and where they interact with the organic material. In this euphotic zone, organic material is found in five different forms. More than 98% of the mass of organics is non-living in either dissolved or particulate form, with the former about 100 times the latter (8, 9). Phytoplankton, zooplankton, and fish constitute less than 2% of the organic material in the sea.

Phytoplankton. The cycle of organic material begins with the phytoplankton, unicellular and colonial plant cells that range in size from about 1 μ to 1 mm. Organic material is produced in the cells, primarily by photosynthesis and the utilization of inorganic nutrients. The most important nutrients and their respective concentrations are: soluble inorganic phosphates (0.1–3.5 μg atoms/l), nitrates (0.1–43), and the nitrites (0.1–3.5). The concentration of phytoplankton cells is highly variable and, depending on the time of year and location, may vary from less than 10^3 to more than 10^8 cells/l. The number of species may range from less than 10 to greater than 250 (10). The annual net organic carbon production by phytoplankton for all the oceans has been estimated at about 2×10^{10} metric tons (11). The weight of the organic material would be twice this value since it is customary to multiply by two to convert the weight of organic carbon to that of organic material. As shown in Table II, most of the material is produced on the open ocean, although the production rate per unit column of water (carbon/m²/yr) can be higher in the coastal zone and in upwelling areas.

Zooplankton and Fish. A portion of the organic carbon produced by the phytoplankton is used by the zooplankton. The zooplankton constitute an amazingly diverse group of animals that differ widely in size, shape, and concentration. Their size ranges from less than a millimeter to more than a meter. The numbers of zooplankton in the euphotic zone of the Sargasso Sea, for example, are of the order of 200–400/m³, constituting a mass of 2–4 mg/m³ (12) or about 0.002–0.004 ppm. This is far less than the concentration of the non-living dissolved and particulate organic material. Fish constitute only 0.002% of the total amount of organic material in the euphotic zone and for our purposes can be neglected.

Dissolved and Particulate Material. It was earlier believed that the dissolved and particulate organic material was produced primarily by the death and breakdown of the phytoplankton and zooplankton. It is now clear that the sequence of events is more complicated. There is no doubt that both dissolved and particulate organic material exists at all depths of the sea. Although its concentration is less than that in the euphotic zone, the dissolved organic material can be found uniformly distributed at depths greater than 200 m (*8, 13, 14, 15*). This uniformity in the vertical is not found in the horizontal direction, indicating that the organic material varies with water mass. The organic material in the deep waters represents from 30 to 150 times the average annual production of organic material in the sea (*16*). Since most of the newly produced organic material in the euphotic zone is rapidly used or decomposed, leaving only perhaps 1% as a contribution to the dissolved organics in the deep waters, it is possible that most of the deep-water material ". . . may represent the gradual accumulation of several hundreds or thousands of years . . ." (*9*). The long-term survival of organic material in the deep waters may occur because the bacteria, which break down the organic material in the euphotic zone, cannot do so as easily in the deep water (*17*).

The production of dissolved material in the euphotic zone occurs not only after the death of phytoplankton organisms but also by the diffusion of organic molecules from the organisms during normal growth (*18, 19, 20, 21*). "Almost every phytoplankton organism studied has been shown to diffuse small organic molecules into its medium, and almost any organic molecule of biological interest has been shown to be given off by some organisms" (*22*). Zooplankton release enough amino acids in one month to equal the amount in solution (*18*), and the rate of release of dissolved material by growing phytoplankton is 10–30% of that being fixed by photosynthesis (*19, 21*).

The composition of the dissolved organic material has been the subject of numerous papers (*23*). Jeffrey (*24*) found that lipids (chloroform extractable organic compounds) make up 10–20% of the dissolved organic material in semi-tropical waters and suggested it may be from 40 to 55% in Antarctic waters. Probably the bulk of the dissolved organic material is composed of proteins and protein-derived metabolites (*25, 26*). The free amino acids, fatty acids, sugars, and phenols represent less than 10% of the organic material in sea water. The identification of individual proteins and peptides has proved difficult, and data have been published only on the amino acid spectrum of the proteinaceous material (*25*). A majority of the amino acids occur as compounds with molecular weights between 400 and 10,000 (*25*).

Concentration of Organic Material at the Surface of the Sea

Although Benjamin Franklin (*27*) was the first to show that the slicks commonly observed on the surface of the sea were composed of thin layers or monolayers of surface-active organic material, it had been well known that these slicks were produced by biological activity in the sea. European fishermen of the 18th century located schools of fish by first looking for high concentrations of slicks on the surface of the sea (28). The scientific community did not give serious attention to the correlation of slicks to organisms within the sea (29, 30) until 250 years later.

Surface-Active Material and Surface Pressure. If a slick or any surface-active film is to spread and increase in area it must exert a horizontal surface pressure. The relation between surface pressure and surface tension is (both measured in dynes/cm):

$$P_f = \gamma_w - \gamma_f$$

where $\gamma_w =$ surface tension of the water, $\gamma_f =$ surface tension of the film, $P_f =$ the surface pressure exerted by the film. A postive surface pressure means that the surface tension of the film is less than that of the water, and if the surface pressure increases for any reason, the surface tension of the film must decrease. On a clean, quiet, unconfined body of water a film can spread until it becomes a monolayer under zero surface pressure, at which point γ_w equals γ_f. Nature seldom provides such ideal conditions, and all surface-active material on natural bodies of water do not spread to form uncompressed monolayers. Both the water and the air are usually in motion and provide horizontal surface stresses which produce surface pressure in the films.

Surface-active organic material undoubtedly exists on the surface of nearly all fresh water bodies, as was found on all those studied in England (31). The surface pressures produced by this material were from 1 to greater than 30 dynes/cm. By appearance, the films were assumed to be composed of natural protein compounds. When the winds were sufficiently strong a surface stress was generated which drove the films downwind to the lee side of the lake where they collapsed. Microscopic examination of this protein material revealed many small organisms imbedded in the material, presumably there to eat it. Indeed, Cheesman (32) reports that a snail uses his foot as a miniature Langmuir trough and is able to compress and collapse surface-active films to obtain clumps of organic material that are easily eaten.

Visible slicks are often seen within 100 km of land. Possibly these are caused by man-made pollution (oil from ships, sewage, *etc.*) and/or by seaweeds and higher plankton concentrations near the continent (Table II). On the open sea, slicks are less common (30). This does not mean that organic material is absent from the surface but simply that the surface-active molecules are not present in sufficient quantity to produce a surface pressure that will cause damping of capillary waves (*33*) which smooth the sea and expose the slick-like or glassy appearance of a monolayer under pressure.

Sea Surface Composition. Sampling of the upper 150 μm (*34*) of both the Atlantic and the Pacific and in both slick-covered and non-slick areas has revealed surface-active materials (*35, 36*). The greatest quantities were found in the biologically-rich areas, but even the inactive waters contained some material. The primary surface-active, chloroform-soluble organic components were free fatty acids, fatty esters, fatty alcohols, and hydrocarbons. Although these species were both in and out of slick areas and in samples taken a few meters beneath the surface, there were indications that a high proportion of the higher molecular weight and less water-soluble fatty acids and alcohols were in the slick areas. Both Garrett (*35*) and Jarvis *et al.* (*36*) suggest that the surface pressure in the slick areas forces the more water-soluble and less surface-active species from the surface into the underlying water.

The microlayer at the surface of the sea was found by Harvey (*37*) to contain bacteria, dinoflagellates, and other plankton in much higher concentrations than in the water beneath. Other workers have found both organic and inorganic nitrogen and phosphorus concentrated in the surface microlayer (*38*), bacteria enrichment (*39*), enrichments of DDT of up to 10^5 times over that of sub-surface water (*40*), and enrichments of heavy metals in the surface films of perhaps 10^4 (*41*). Although the heavy metal and DDT enrichment was found within 50 km of the U.S. Eastern Shore, it illustrates how material injected or produced in the main body of the sea can become concentrated at the surface. It is possible that some of this surface concentration, especially near land, is fallout of material from the atmosphere.

The cycle of surface-active material in the sea is shown schematically in Figure 1. Man-made sources are unimportant. The annual input of

Table II.   Division of the Ocean into Provinces According to Their Level of Primary Organic Production[a]

Province	Percentage of Ocean	Area (km²)	Mean productivity (grams of carbon/ m²/yr)	Total productivity (10⁹ tons of carbon/yr)
Open ocean	90	326×10^6	50	16.3
Coastal zone[b]	9.9	36×10^6	100	3.6
Upwelling areas	0.1	3.6×10^5	300	0.1
Total				20.0

[a] Data from Ref. 11.
[b] Includes offshore areas of high productivity.

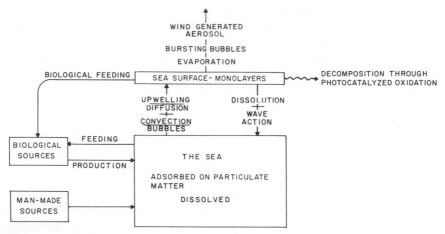

Figure 1. Cycle of surface-active material in the sea. From Garrett (114)
with permission of the author.

oil into the seas by man (river discharges, spills from ships, etc.) plus
the atmospheric fallout of petroleum products is several orders of magnitude less than the organic material produced annually in the euphotic
zone (42, 43). Even were it to concentrate entirely at the surface it would
produce a film only 58 A thick, about twice the thickness of a monolayer
(42). In a day or two, such a film would be removed from the sea by
bacteriological degradation, mixed downward into the sea and diluted,
or transported into the atmosphere.

Langmuir Circulations. Most of the organic material in a surface
monolayer reaches the surface by molecular diffusion, aided greatly by
turbulence or organized convective motions (33). These organized motions, called Langmuir circulations, appear to be very important in the
mixing of water in the upper few tens of meters in the sea. First noted
by Langmuir on the sea, but studied in detail on a fresh-water lake (44),
they are water motions in the form of alternate left and right helical
vortexes (Figure 2) that have their axes approximately parallel to the
direction of the wind. These vortexes produce alternate lines of convergence and divergence on the surface which are also lined up with
the wind. Organic-rich water is carried to the surface in regions of upwelling (surface divergence) where molecular diffusion enables the
surface-active molecules to reach the air–water interface. This surface
water plus the molecules is carried into the region of convergence where
the water sinks leaving an accumulation of surface-active material which
is compressed into the visible surface slicks or windrows that are aligned
with the wind (2, 45, 46). These long lines of compressed surface-active
films are 20–50 m apart on the sea (47), and, although not proved, the

Langmuir circulations that produce the lines are thought to extend that deep into the sea. The little-understood coupling between the sea and the air can produce circulations that transport organic material to the surface and compress it into visible slicks that form lines parallel to the wind. A detailed review of the dynamics of Langmuir circulations can be found in a paper by Faller (*48*).

Formation and Concentration of Air Bubbles in the Sea. Air bubbles in the sea, the final mechanism listed in Figure 2, aid in transporting organic material and may convert dissolved organic material to particulate organic material. For years it had been thought that the zooplankton, most of which are filter feeders and especially adapted for feeding on particulate organic material, fed primarily on phytoplankton and the particulate material formed from dead phytoplankton. There was no known mechanism by which the zooplankton could utilize the vast store of dissolved organic matter in the sea until 1963 when bubbles were reported to provide such a mechanism (*2*).

Bubbles may be produced by several mechanisms, among them biological, water temperature changes, precipitation, and whitecaps. As a result of the inverse relation between temperature and the solubility of air in water, immense quantities of air are given off by the sea during periods of warming. For example, during March through October in the Gulf of Maine, about 3×10^5 cm^3 of oxygen leave each square meter of

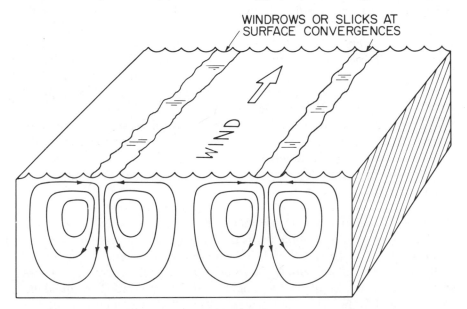

Figure 2.   Production of slicks by Langmuir circulations in the surface waters of the sea

the surface of the sea (49). If this were in the form of air bubbles of 50 μm diameter, the rather staggering number of 2.5×10^7 bubbles/m²/sec would reach the surface of the sea! This is not observed. Most of this air crosses the interface by gaseous diffusion. The air supersaturation needed for bubble formation in the sea probably occurs only under special situations not significant on the global scale (3).

Precipitation both in the form of rain and snow can produce bubbles on the sea (3). Bubbles from moderate rain intensities have diameters of less than 200 μm, and those produced by snow are less than 100 μm. However, since these bubbles are produced only in the upper few centimeters of the sea and only during the precipitation, they do not constitute the major source of bubbles in the sea. The precipitation of the continental aerosol into the sea has been suggested as a source of bubbles (50), but since this aerosol is composed of particles primarily less than 100 μm diameter, and since water drops of 100 μm diameter produce no bubbles when they fall into the sea (3), it is unlikely that the continental aerosol produces a significant quantity of bubbles.

The major source of bubbles is the whitecaps or breaking waves (3) which form whenever the wind speed exceeds 3–4 m/sec (51, 52). When a wave breaks large quantities of entrapped air are carried into the sea. This air produces bubbles that range in size from less than 100 μm to several millimeters diameter (3). The bubble-size distribution is heavily weighted toward the small end, varying inversely with the D^4 or the D^5 power of diameter D. Because of this strong inverse relationship, a majority of the bubbles are less than 200 μm diameter. About 10^8/m³ have been found in the upper meter of the sea a few seconds after a whitecap has formed (3). The only direct measurements of bubble-size distribution in the sea were done in waters near the shore and in whitecaps that were relatively small by open-ocean standards.

The areal distribution of bubbles probably increases in direct proportion to the whitecap coverage, and above the critical wind speed of 3–4 m/sec the latter is a function of the windspeed (51, 52, 53). Knowing this function, the climatic winds can be integrated over any given area of the sea to obtain the average coverage of whitecaps. This has been done (51) and is about 3.5% over the world ocean. Assuming this percent and that the upper meter of the sea directly beneath the whitecap area contains bubble densities of about 10^8/m³, bubble densities of at least 3.5×10^4/m³ could be maintained throughout the entire euphotic zone if mixing processes could operate sufficiently fast. However, this does not happen. Most of the bubbles produced by breaking waves undoubtedly rise to the surface within 30–60 secs (bubbles of 100 and 200 μm diameter have rise speeds of 0.5 and 1.5 cm/sec, respectively). However, through turbulence and the downwelling in the convergent regions of the Langmuir circulations where downward speeds of 3–6 cm/sec (2) have been observed and the positively buoyant *Sargassum* are held beneath the surface (54), small air bubbles could easily be carried many meters beneath the surface of the sea. No data on the vertical gradient of bubbles in the sea are available to confirm this.

Bubbles and the Formation of Particulate Organic Material. With such high concentrations of bubbles in the surface layers of the sea, interactions between the bubbles and the dissolved organic material are significant. Although bubble-organic interaction has been discussed (3, 56) and much basic work existed on the adsorption of organic material at air–water interfaces (55) as well as the long practice of using bubble separation techniques (1), it was not until 1962 that detailed data were obtained. In that year Baylor *et al.* (57) found that inorganic phosphate (PO_4), one of the nutrients necessary for the growth of the phytoplankton, became attached to air bubbles in sea water and was transported to the surface. By collecting the spray from the bursting bubbles they were able to remove over 99% of the PO_4 from the water. When the experiment was repeated in artificial sea water (no organic material) or in PO_4-tagged distilled water (about 1 μg A/l), no PO_4 was removed by bubbling. This indicated that some of the dissolved organic material in natural sea water was required for PO_4 removal. Perhaps surface-active anion binders became attached to the PO_4 and to the bubble, both materials then being carried to the surface and ejected into the atmosphere. Measurements made in the upper ten meters of the vertical PO_4 gradient at sea showed that the gradient increased with wind speed in a manner suggesting the associated increase in bubble production was responsible for PO_4 transport to the surface. In the following year Sutcliffe *et al.* (2) reported that the bubble spray was not only rich in PO_4 but also in an unidentified surface-active material.

These workers (2, 57) also found a great deal of particulate organic material in the spray. This particulate material was visually indistinguishable from some of the particulates found in the sea. The particle sizes were not reported but probably ranged from about a micron to a few tens of microns in diameter. Their organic nature was indicated by their solubility in cyclopentane and the fluorescence of the resulting solution. Organic material found in sea water (58) behaves in a similar manner. Brine shrimp were found to thrive on the particles (59). The production of the particles by bubbling action was verified by carrying out the bubbling in water which had been passed through filters of 0.45 μm pore size. The mechanisms of particle production is not clear. The surface-active, organic-PO_4 films may be carried to the surface by the bubbles and compressed there beyond the collapse pressure to form colloidal micellae or long fibers which could coalesce to generate the organic particles. The bubble spray is then generated from this material. Later work suggests UV radiation may also be involved in the film-to-particle conversion (60).

If organic particles are generated from the collapse of surface films, then particulate organic material should be found concentrated beneath the slick areas (convergent zones) of the Langmuir circulations. Light scattering evidence in these regions (2) appears to confirm this. However, no attempt was made to estimate the percent increase in particles here as opposed to adjacent non-slick covered areas of the sea. These findings have been reproduced (61).

Particle production by bubbling is confirmed (62, 63) although one worker (17) suggested the importance of bacteria in the process. Menzel's (64) work in 1966, using what appeared to be adequate controls, could not find any organic particle production by bubbling. A more recent study appearing in 1969 (65) analyzes prior work and concludes that earlier disagreements arose in part from differences in the size of filters used and further that the presence of particles inhibits particle production by bubbling. Additional particles will form only if the steady-state concentration is removed. No explanation for the particle inhibition exists.

Particulate Organic Material from Bubbles Going into Solution. Although work on this subject has ceased, early papers (2, 57) suggest that some of the small bubbles produced by breaking waves might go into solution completely before reaching the surface, thus releasing absorbed organic material in the form of colloidal micelles which, upon aggregation, could form particles of a size usable by the filter-feeding zooplankton. There is no doubt that small bubbles in sea water, especially those less than 100 μm diameter, go into solution rather quickly, even in water that is 100% saturated with air. In properly designed laboratory experiments this can be observed with the unaided eye. The reason is that all bubbles have an internal pressure that is higher than that in the water just outside the bubble by $2\gamma/R$ where γ is the surface tension and R the bubble radius. Because of this surface curvature effect which increases as bubble size decreases, all bubbles eventually go into solution even in air-saturated water. Bubbles smaller than 300 μm diameter will be forced into solution even in water that is up to 102% saturated, and bubbles less than 20 μm will dissolve in water that is up to 115% saturated (3). This pressure effect produced a rapid rate of decrease and subsequent disappearance of the small bubbles produced when raindrops and snowflakes fell into sea water (3).

It is likely that a significant portion of the bubbles produced by breaking waves go into solution before they reach the surface. Probably at least 20% of the bubbles beneath a breaking wave are less than 100 μm. Bubbles of 100 μm and 50 μm at a depth of 1 m in sea water 100% saturated with air will go into solution in 250 and 100 sec, respectively (3). Since these bubbles rise at speeds of only 0.5 and 0.13 cm/sec, the smaller ones will never reach the surface. Since bubble rising speed decreases with size, calculation would no doubt show that the 100 μm bubbles would also go into solution. In addition, bubbles larger than 200 μm which are carried several meters beneath the sea by mechanisms already discussed or are caught in Langmuir circulation cells (2, 66) would most likely go into solution.

Evidence suggests that the adsorption of organic material onto the surface of a bubble rising through sea water occurs at such a rate as to reach a steady-state value within 20 to 40 sec. Force–area isotherms for numerous samples of natural surface-active material found on the sea indicated that just at the point that a detectable film pressure was noted, the film area was of the order of 0.2 m^2/mg (36). Assuming that a 100 μm bubble attains this coverage, the volume of organic material in the film is about 40 μm^3. If this material were compressed as the bubble went into solution, it could produce a particle 4 μm in diameter. Whether the particle is produced or not depends on two factors. First, as the material is compressed by the dissolving bubble, the surface tension decreases, thus lowering the internal pressure. Secondly, many of the shorter chain, less strongly adsorbed surface-active molecules will be displaced from the bubble surface as it goes into solution (35, 36). Nevertheless, the size distribution of particulate material at a depth of 1 m in the sea peaks in the diameter range 3–8 μm when whitecaps (and therefore bubbles) were observed (61).

An order-of-magnitude calculation of the flux of bubbles entering the sea and dissolving gives 7×10^5 bubbles/m^2/min. The average time of a whitecap is about 1 min; some 3.5% of the sea is covered with whitecaps, and about 20% of the 10^8 bubbles per cubic meter go into solution. In detailed studies Gordon (67, 68) has found particle concentrations of from 3×10^7 to 2.5×10^8/m^3 in the surface waters of the North Atlantic. A majority of the particles were 20 μm or less.

A complete summary of the work done on the perplexing problem of the origin of particulate material in the sea can be found in Riley's review (69). His review of the work on the bubbling makes it clear that mechanisms other than bubbling can produce particulate material. For example, particles appear to develop more or less spontaneously by aggregation of smaller entities (70) and by the adsorption of dissolved organic material onto calcium carbonate particles (71).

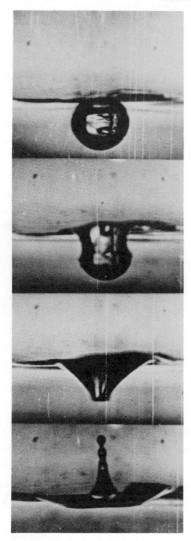
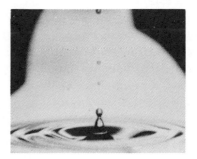

Figure 3. (a) Composite view of high-speed motion pictures illustrating some of the stages in the formation of jet and jet drops upon the collapse of a 1.7 mm diameter bubble. The time interval between top and bottom frames is about 2.3 msec. The angle of view is horizontal through a glass wall. The surface irregularities are due to a meniscus. (b) Oblique view of the jet from a 1 mm diameter bubble.

The Water-to-Air Transfer of Organic Material

The surface of the sea may rise and fall as waves pass, and capillary waves induced by wind gusts can cover the surface; but until there is some physical disruption of the surface there is no known way in which particles or drops of any kind can be ejected into the atmosphere. Surface disruption may occur in several ways (rain, snow, etc.) but the major way appears to be breaking waves or whitecaps. When whitecaps form, and even prior to their formation, fine spray drops are mechanically torn from wave crests by the wind. It is unlikely that these drops constitute the major mechanism of drop ejection from the sea (72). They are much larger than those commonly found in the marine atmosphere, and they undoubtedly fall back to the sea before travelling many meters. The major mechanism appears to be the breaking at the surface of air bubbles produced by the whitecaps.

Dynamics of Bubble Breaking. When an air bubble breaks the bubble cavity quickly collapses producing a jet of water which is ejected upward at very high speed. To satisfy momentum conservation, a downward-moving jet (usually not visible) is also produced (73). The jet becomes unstable after rising a distance of about one bubble diameter and breaks up into one to five drops, depending on bubble size, which continue on upward. These drops are called the jet drops. The bubble–jet-drop mechanism was first proposed in 1937 (74) to account for the sea-salt aerosol found in marine atmospheres (75), but it was not until 1953 that a high speed camera was used to confirm the details of this mechanism (76). Figure 3 illustrates the formation of the jet and the jet drops.

For a given bubble size the ejection height and size of the drops is remarkably constant, not varying by more than a few percent. As shown in Figure 4, compiled with data from several sources (3, 51, 77, 78), the ejection height of the top jet drop from bubbles breaking in sea water increases with bubble size, reaching a maximum of about 18 cm for a 2 mm bubble and then decreasing to near zero for bubbles larger than about 6 mm. The difference between the sea water curve and the one for distilled water for bubbles in excess of 1 mm is not significant. Apparently it is related to differences in delay times when a bubble reaches the surface and breaks (51). Although the relation between bubble and jet drop diameter is well-known (51), a good rule of thumb is that the jet drop diameter is one-tenth that of the bubble. The top drop ejection height, at least for bubbles of less than 2 mm, is about 100 times the bubble diameter. For large bubbles there is only one drop ejected from the jet while for smaller bubbles several drops appear. The jet drop phenomenon extends down to the very smallest bubbles; the production of jet drops as small as 2 μm diameter have been observed from bubbles smaller than 50 μm (79). The data of Figure 4 do not hold for a bubble coated with surface-active organic material.

The energy source for the ejection of the jet drops can be understood from calculating the speeds of ejection required to attain the observed

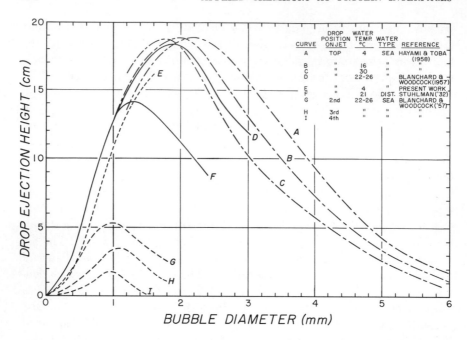

Figure 4. Jet drop ejection height as a function of bubble diameter, water temperature and salinity

ejection heights (51). These calculations, which took into account the high drag forces associated with large Reynold's numbers, showed that the ejection speeds increased with decreasing bubble size and were extraordinarily high. For example, the top jet drop from a 2 mm bubble rises about 18 cm after being ejected from the bubble with a speed of 350 cm/sec while the jet drop from a 70 μm bubble in water at 4°C rises only 0.17 cm and has an initial speed of 8×10^3 cm/sec. Without the frictional drag of the air, the latter drop would rise to a height of 335 m, about 2×10^5 times higher than is observed.

The kinetic energy of the jet drops is calculated from the ejection speeds and is proportional to the square of the bubble diameter. The source of the kinetic energy is not the gravitational potential energy, which is an order of magnitude less than the drop kinetic energy, nor is it the potential energy of the slightly compressed gas within a bubble at rest. The only source of energy that can account for the jet drop kinetic energy is the surface free energy of the bubble. This varies with the square of the bubble size, as does the kinetic energy, and in magnitude is from five to 10 times the kinetic energy (51). When an air bubble breaks, some of the surface free energy is dissipated in circular capillary waves that move outward horizontally from the bubble. The rest is in a wave that moves down along the surface of the bubble. It is this energy that produces the jet and jet drops. A detailed analysis of these waves has been made by McIntyre (73). A sketch that shows both these waves (80) is given in Figure 5. This sketch, based in part on the high-speed movies shown in Figure 3, shows 10 time sequences of the profiles of a collapsing bubble. Sequence 1 shows the bubble just after the bubble film (that thin film of water which separates the air in the bubble from that above) has broken, and sequence 10 shows the jet fully formed and ready to break into drops.

Jet drops are not the only drops produced by a bursting bubble. There are film drops also. Film drops are produced from the bubble film during the time interval from breaking to time Sequence 2 in Figure 5. They are much smaller than jet drops. Most are less than a micrometer but some are as large as 20 or 30 μm. Unlike jet drops, whose numbers vary between one and five and decrease with bubble size, the number of film drops per bubble increases rapidly with size (51, 81). Bubbles of less than 0.3 mm do not appear to produce any film drops at all, but a 2 mm bubble produces a maximum of about 100, and a 6 mm bubble a maximum of about 1000. For bubbles breaking individually, the presence of a surface-active material at the bulk air–water interface decreases the film drop production (51, 82). However, if the bubble concentrations are high enough to produce clustering at the surface, the presence of the material causes up to a threefold increase in film drop production (83).

Simple Experiments on Sea-to-Air Organic Transfer. Both jet and film drops, mixed upward into the atmosphere by turbulence, account

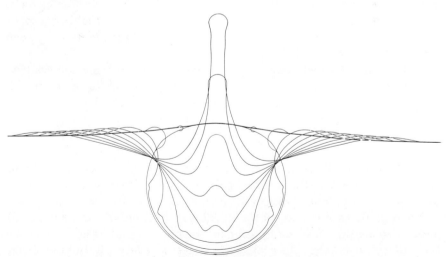

Figure 5.   Ten stages in the time sequence of collapse of a 1.7 mm diameter bubble.   Profiles are ∼1/6000 sec apart.   Data from Ref. 80.

for the distribution of sea-salt nuclei (also called sea-salt particles, the sea salt left when water evaporates from drops of sea water) that is found in the lowest 2 to 3 km of the marine atmosphere (75). The rate of injection of these drops into the atmosphere is of the order of 10^{10} metric tons of sea salt per year (51). The concentration of surface-active organic material on these drops is considerably higher than that in the sea, indicating that the bubble or the mechanism of bubble breaking is capable of concentrating organic material.

By bubbling air through sea water surface-active material can be stripped from the surface and carried into the atmosphere by jet and film drops (2, 51, 83). To study jet drops in this process, bubbles of known size were allowed to break, one at a time, beneath an organic duplex film (called indicator oil by Langmuir and Schaefer (84) because the film was thick enough to give interference colors and thus indicate its presence), and the film carried by the top jet drop when it fell into a clean water surface (51) was observed. This experiment, illustrated in Figure 6, showed that the jet drop from bubbles larger than 1 mm carried some of the duplex film while the drops from smaller bubbles did not. These films were under zero pressure so a hypothesis was formulated that related film travel down the insides of the bubble to the speed at which jet drops formed. Inasmuch as organic films on the sea can be under pressure, the experiments were repeated with an oleic acid film under pressure. Experimental difficulties obscured the interpretation of this experiment, but the surface pressure seemed responsible for more organic material on the jet drops. Little is known of the manner in which jet drops accomplish the sea-to-air organic transfer. Present techniques can detect organic material in extremely small samples and permit a variation of the Figure 6 experiment in a rotating tank (85) with a device which can generate more than 1000 bubbles per min of any desired size.

No reported experiments detect surface-active organic material on film drops, but several report an enrichment of various elements in the film drops (86, 87, 88). New research in this area has been presented (89).

Adsorption of Organic Material on Bubbles Rising through the Sea. Although some of the organic material on jet and film drops originates in monolayers on the surface of the sea, most of the material carried into the air is adsorbed on the surface of the bubble while it is rising through the water. Adsorption of dissolved surface-active organic material (1, 90, 91) and particulate material (85, 92, 93, 94) to a bubble surface is well known but not in sea-to-air transfer of organic material. Figure 7 shows the process that takes place on the surface of every air bubble rising through the sea (90). Molecules of dissolved organic material diffuse to the surface of the bubbles where the downward drag of the water

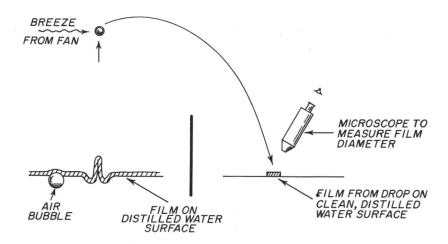

*Figure 6.   A method of detection of the removal of surface-active films by
jet drops*

sweeps them around to the lower or downstream surface. Thus a compressed monolayer will build up on the bottom of the bubble, the amount
of compression depending in part on bubble rise speed, the time it is in
the water, and the concentration of dissolved organic material.

As the concentration of surface-active material on a rising bubble
increases with time, the surface tension and thus the surface free energy
decreases. Since the latter is the source of energy for jet drop ejection,
the ejection heights of the drops decrease with increasing bubble age.
This is observed both with bubbles rising through aqueous suspensions
of bacteria (85) and through samples of lake, river, and sea water. The
decrease in ejection height is generally paralleled by a decrease in drop
size. An example of the ejection height decrease for bubbles in sea water
collected in Long Island Sound (95) is shown in Figure 8. Note the rapid
decrease in the top jet drop ejection height during the first 20 sec of bubble
aging and little change after that time. The second jet drop shows little
effect of the bubble aging. No data are reported on the time rate of increase in surface-active organic material on the bubble or on the top jet
drop. The probable magnitude is suggested by the fact that the numbers
of bacteria attached to a rising bubble can increase 10^3 times (and in
some cases 10^4) during the first 20 sec of bubble life (85).

The steady adsorption of dissolved organic material onto a rising
bubble, with the resulting increase in surface pressure, should progressively force the more water soluble and less surface-active species out of
the bubble surface (35). Thus, for a given sized bubble, the proportion
of various species varies with bubble age. This is reflected in the com-

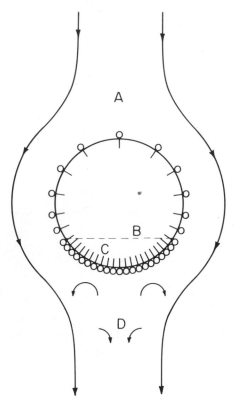

Figure 7.   Adsorption of dissolved surface-active material onto the surface of an air bubble rising through sea water. (A) The upstream region; (B) rough dividing line between compressed and non-compressed monolayer; (C) compressed monolayer; (D) bubble wake. Data from Ref. 90.

position of the material carried by the jet drops. Not only is the total amount of organic material carried by a jet drop increased with bubble age but so is the relative composition. The complexity of the nature and amount of surface-active organic material transferred from sea to air and the interactions between adsorbed organic material on bubbles and the dynamics of jet drop ejection has not been investigated adequately. The relation between drop ejection height and bubble diameter shown in Figure 4, obtained for bubbles less than 1 sec old, cannot be applied to the oceans or to any natural bodies of water where bubble ages of more than 20 sec are common. In such case the behavior shown in Figure 8 is probably more typical though perhaps exaggerated. These data were obtained with Long Island Sound water which may have had a higher dissolved organic content than water on the open sea.

Amount of Organic Material Ejected into the Atmosphere. A rough estimate can be made of both the enrichment of organic material on jet drops and film drops and of the total amount of surface-active organic material ejected from sea to air per year. Analysis of collections of sea-salt particles and sea water drops in the atmosphere near Hawaii (*4, 96*)

showed that the amount of surface-active material on the aerosol was
about 50% that of the sea salt. Other work (*97*) in the same area suggested it was closer to 10%, but a probable error in the calculations made
it too low by a factor of two (*98*). Thus, the weight of surface-active
organic material on the marine sea-salt aerosol is from 20 to 50% that
of the sea salt, constituting an enrichment of 7000–17,000 times what is
found in the sea since sea water is 3.5% sea salt and contains about 1 ppm
organic material. The enrichment is attributable to the bubble mechanisms just discussed.

Barger and Garrett (*97*) have shown that the organic material found
on the aerosol did indeed come from the sea with the jet and film drops
and not from any continental source. They found that the aerosol contained a mixture of surface-active compounds and nonpolar hydrocarbons,
and specifically identified five fatty acids (C_{14}–C_{18}) to be in the same
relative proportions as have been found in sea surface slicks. Further,
the film pressure *vs.* area curves for the surface-active material on the
aerosol were of the liquid-expanded type, quite similar in shape to those
reported for sea surface samples (*36*).

The total amount of surface-active material ejected into the atmosphere per year is 20 to 50% of the estimated 10^{10} metric tons of sea salt

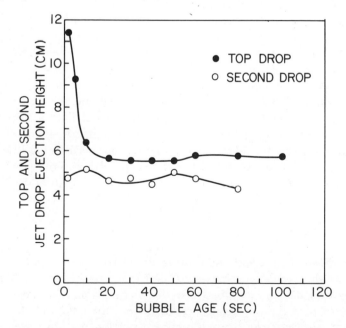

*Figure 8.   Decrease in ejection height of jet drops as a
function of bubble age. After Lee (95) with permission
of the author.*

thought to be involved in the annual cycle (51). This amounts to 2 to 5×10^9 tons. Possibly it is more than this since the organics-to-salt ratio in the atmosphere near Hawaii is likely to be less than that for the world ocean average. Biological productivity is low in Hawaiian waters. Probably most of this material returns directly to the sea since only about 10% of the particulate material ejected from the sea falls out on the continents (99).

In terms of the total productivity of the oceans (Table II) the amount of airborne surface-active material is only 5 to 12% of 2×10^{10} tons of carbon per year (about 4×10^{10} tons of organic material).

The water-to-air transfer of organic material by bubbles occurs not only in the sea but also in bodies of fresh water. Baier (100), in a study of oil films produced on Lake Chautauqua, New York, by human activity, concluded that natural bubble mechanisms were largely responsible for cleaning the lake surface by ejecting the films into the atmosphere. If future work confirms this finding, bubbles may be a mechanism for converting water pollution to air pollution (101).

Other Methods for Organic Removal from the Surface of the Sea. Three other mechanisms are shown in Figure 2 for the removal of organic material from the surface of the sea. The first is biological consumption. High concentrations of zooplankton and bacteria (39) are found in the surface microlayer of the sea indicating the use of the film as food. How significant this is for film removal is not known. The second mechanism, photochemical oxidation by UV irradiation, acts on chemically-unsaturated components of the surface films to break them into smaller, more soluble fragments which are easily displaced from the surface (102). It has also been suggested that UV irradiation can be the first step in a series of reactions with lipid films to produce particulate material (60). The third mechanism, the breaking up of the surface film and mixing back into the sea by turbulence and Langmuir circulations, is little understood. The relative significance of these mechanisms is not known except that it varies with wind, temperature lapse rates both in and above the sea, and biological activity in the sea.

Organics from the Sea and the Formation of Rain

At a given cloud depth, rain forms more easily in clouds over the sea than in clouds over the continent (103). The reason lies in the differences in the aerosol distributions between the marine and continental atmospheres (104). Although aerosol particles less than 0.1 μm in diameter may not come from the sea, a significant portion of those greater than 0.1 μm and essentially all larger than 1 μm originate at the surface of the sea (3, 51, 75, 105). Particles coming from the sea carry organic material which can influence rain-forming and other weather mechanisms in the atmosphere over the sea.

Organic Films and Evaporation. One such mechanism is the rate of evaporation of surface-active film-coated water drops in the atmosphere. It is often suggested that the films can significantly retard the evaporation, but it has been pointed out (*97*) that this is based on laboratory work where monomolecular films of highly adlineated, straight-chain molecules were used. Some of the surface-active material found on the sea and in the atmosphere contains permanently bent (chemically unsaturated) hydrocarbon sections which prevents close packing in compressed films. Films containing these molecules have holes and are thus unable to prevent evaporation (*106*). The validity of this idea has been shown convincingly in experiments where the droplets in water fogs were coated with various surface-active materials. Cetyl alcohol, composed of linear molecules, appreciably retarded the evaporation of the fog while oleyl alcohol, composed of non-linear molecules, did not (*107*). Thus, the surface-active material on jet and film drops have little effect on drop evaporation. Evaporation leaves a high ratio of organic material to sea salt.

Organic Surface-Active Material and the Formation of Rain. Organic-laden drops from the sea provide a clue to the mechanism for the formation of marine rain. Rain forms in marine clouds when a sufficient number of giant (those larger than 1 μm) sea-salt particles are carried into clouds which have the proper updrafts and vertical thickness. Presumably the giant particles provide the nuclei for the formation of raindrops, the raindrops forming by a coalescence process as the nuclei fall down through the cloud of smaller cloud drops (*108, 109*). Recently, evidence has been presented which suggests that it is not the giant salt particles that provide the nuclei for rain formation, but the so-called large sea-salt particles, those in the range 0.1–1 μm (*110*).

The evidence for this mechanism is provided by the surface-active organic material on the sea-salt particles. The iodine-to-chlorine ratio of the sea-salt particles is 100–1000 times that of sea water and varies inversely with particle size (*111, 112*). As the particles rise from the sea as jet or film drops, a fractionation process occurs as they are ejected into the atmosphere resulting in a relative increase in iodine. This iodine is organically bound in surface-active material and is thus ejected into the atmosphere with the organic material (*96*). From a knowledge of the concentration of iodine in organic material in the sea and the amount of organic material on the airborne salt particles, one can deduce iodine-to-chlorine ratios in the range actually observed (*96*).

The inverse dependence of these ratios with particle size is also consistent with the organic film hypothesis. Since the iodine is bound in the film which is on the surface of the drop, it varies directly with surface area, R^2. The chlorine in the drop itself varies with R^3. Therefore, the iodine-to-chlorine ratio should vary as R^{-1} which is about what is observed (112). Although it seems likely that these ratios are caused by organically-bound iodine in surface films, an alternative hypothesis has been presented which suggests that the iodine is not on the aerosol when it leaves the sea but diffuses to it in the form of gaseous iodine from the atmosphere (113).

The inverse relation between particle size and I/Cl provided a tracer by which the role of the sea-salt particles in the rain-forming process can be studied. Comparison of the I/Cl ratios in the sea-salt particles in marine air to those in raindrops from clouds formed in this air indicated that only the small end of the sea-salt particle spectrum plays any role in the rain formation (110).

Acknowledgments

All, or sections of, the first draft of this paper were read by William Sutcliffe, Jr., Donald Gordon, Jr., John Wheeler, and Peter Wangersky, of the Bedford Institute in Nova Scotia and by Bruce Parker of the Virginia Polytechnic Institute and State University. I thank them for their helpful constructive criticism.

Literature Cited

1. Fuerstenau, D. W., Healy, T. W., in "Adsorptive Bubble Separation Techniques," pp. 91-131, R. Lemlich, Ed., Academic, New York, 1972.
2. Sutcliffe, W. H., Jr., Baylor, E. R., Menzel, D. W., "Sea Surface Chemistry and Langmuir Circulations," *Deep-Sea Res.* (1963) **10**, 233-243.
3. Blanchard, D. C., Woodcock, A. H., "Bubble Formation and Modification in the Sea and Its Meteorological Significance," *Tellus* (1957) **9**, 145-158.
4. Blanchard, D. C., "Sea-to-Air Transport of Surface Active Material," *Science* (1964) **146**, 396-397.
5. Hood, D. W., in "The Encyclopedia of Oceanography," pp. 792-799, R. W. Fairbridge, Ed., Reinhold Pub. Corp., 1966.
6. Horne, R. A., "Marine Chemistry," see p. 4, Wiley-Interscience, 1969.
7. Gordon, D. C., Jr., Sutcliffe, W. H., Jr., "Marine Chemistry," in press, 1973.
8. Gordon, D. C., Jr., "Distribution of Particulate Organic Carbon and Nitrogen at an Oceanic Station in the Central Pacific," *Deep-Sea Res.* (1971) **18**, 1127-1134.
9. Menzel, D. W., Ryther, J. H., "Organic Matter in Natural Waters," pp. 31-54, D. W. Hood, Ed., Univ. of Alaska, 1970.
10. Smayda, T. J., "The Encyclopedia of Oceanography," pp. 712-716, R. W. Fairbridge, Ed., Reinhold, 1966.
11. Ryther, J. H., "Photosynthesis and Fish Production in the Sea," *Science* (1969) **166**, 72-76.

12. Deevey, G. B., "The Annual Cycle in Quantity and Composition of the Zooplankton of the Sargasso Sea off Bermuda," *Limn. and Oceanog.* (1971) **16**, 219-240.
13. Provasoli, L., "The Sea," pp. 165-219, M. N. Hill, Ed., Interscience, New York, 1963.
14. Menzel, D. W., "Particulate Organic Carbon in the Deep Sea," *Deep-Sea Res.* (1967) **14**, 229-238.
15. Menzel, D. W., "The Role of *In Situ* Decomposition of Organic Matter on the Concentration of Non-conservative Properties in the Sea," *Deep-Sea Res.* (1970) **17**, 751-764.
16. Ryther, J. H., "The Sea," pp. 347-380, M. N. Hill, Ed., Interscience, New York, 1963.
17. Barber, R. T., "Dissolved Organic Carbon from Deep Waters Resists Microbial Oxidation," *Nature* (1968) **220**, 274-275.
18. Webb, K. L., Johannes, R. E., "Studies of the Release of Dissolved Free Amino Acids by Marine Zooplankton," *Limn. and Oceanog.* (1967) **12**, 376-382.
19. Anderson, G. C., Zeutschel, R. P., "Release of Dissolved Organic Matter by Marine Phytoplankton in Coastal and Offshore Areas of the Northeast Pacific Ocean," *Limn. and Oceanog.* (1970) **15**, 402-407.
20. Wilson, W. B., Collier, A., "The Production of Surface-active Material by Marine Phytoplankton Cultures," *J. Mar. Res.* (1972) **30**, 15-26.
21. Choi, C. I., "Primary Production and Release of Dissolved Organic Carbon from Phytoplankton in the Western North Atlantic Ocean," *Deep-Sea Res.* (1972) **19**, 731-735.
22. Wangersky, P., "The Organic Chemistry of Sea Water," *Amer. Scientist* (1965) **53**, 358-374.
23. Hood, D. W., Ed., "Organic Matter in Natural Waters," 625 pp., Univ. of Alaska, 1970.
24. Jeffrey, L. M., "Organic Matter in Natural Waters," pp. 55-76, D. W. Hood, Ed., Univ. of Alaska, 1970.
25. Degens, E. T., "Organic Matter in Natural Waters," pp. 77-106, D. W. Hood, Ed., Univ. of Alaska, 1970.
26. Duursma, E. K., "Chemical Oceanography," pp. 433-475, J. P. Riley and G. Skirrow, Eds., Academic, 1965.
27. Giles, C. H., "Franklin's Teaspoonful of Oil. Studies in the Early History of Surface Chemistry, Part 1," *Chemistry and Industry* (1969) No. 45, 1616-1624.
28. Otto, I. F. W., "Observations on the Property Ascribed to Oil, of Calming the Waves of the Sea," *Phil. Mag.* (1799) **4**, 225-233.
29. Sieburth, J. McN., Conover, J. T., "Slicks Associated with *Trichodesmium* Blooms in the Sargasso Sea," *Nature* (1965) **205**, 830-831.
30. Dietz, R. S., LaFond, E. C., "Natural Slicks on the Ocean," *J. Mar. Res.* (1950) **9**, 69-76.
31. Goldacre, R. J., "Surface Films on Natural Bodies of Water," *J. Animal Ecology* (1949) **18**, 36-39.
32. Cheesman, D. F., "The Snail's Foot as a Langmuir Trough," *Nature* (1956) **178**, 987-988.
33. Garrett, W. D., "Damping of Capillary Waves at the Air-sea Interface by Oceanic Surface-active Material," *J. Mar. Res.* (1967) **25**, 279-291.
34. Garrett, W. D., "Collection of Slick-Forming Materials from the Sea Surface," *Limn. and Oceanog.* (1965) **10** ,602-605.
35. Garrett, W. D., "The Organic Chemical Composition of the Ocean Surface," *Deep-Sea Res.* (1967) **14**, 221-227.
36. Jarvis, N. L., Garrett, W. D., Scheiman, M. A., Timmons, C. O., "Surface Chemical Characterization of Surface-active Material in Sea Water," *Limn. and Oceanog.* (1967) **12**, 88-96.

37. Harvey, G. W., "Microlayer Collection from the Sea Surface: A New Method and Initial Results," *Limn. and Oceanog.* (1966) **11**, 608-613.
38. Williams, P. M., "Sea Surface Chemistry: Organic Carbon and Organic and Inorganic Nitrogen and Phosphorus in Surface Films and Subsurface Waters," *Deep-Sea Res.* (1967) **14**, 791-800.
39. Sieburth, J. McN., "Distribution and Activity of Oceanic Bacteria," *Deep-Sea Res.* (1971) **18**, 1111-1121.
40. Seba, D. B., Corcoran, E. F., "Surface Slicks as Concentrators of Pesticides in the Marine Environment," *Pesticides Monit. J.* (1969) **3**, 190-193.
41. Duce, R. A., Quinn, J. G., Olney, C. E., Piotrowicz, S. R., Ray, B. J., Wade, T. L., "Enrichment of Heavy Metals and Organic Compounds in the Surface Microlayer of Narragansett Bay, Rhode Island," *Science* (1972) **176**, 161-163.
42. Garrett, W. D., "Impact of Petroleum Spills on the Chemical and Physical Properties of the Air/Sea Interface," NRL Report 7372, Naval Research Laboratory, Washington, D.C., 15 pp., 1972.
43. "Man's Impact on the Global Environment," MIT Press, Cambridge, Mass., pp. 266-267, 1970.
44. Langmuir, I., "Surface Motion of Water Induced by Wind," *Science* (1938) **87**, 119-123.
45. Woodcock, A. H., "A Theory of Surface Water Motion Deduced from the Wind-Induced Motion of the Physalia," *J. Mar. Res.* (1944) **5**, 196-205.
46. Scott, J. T., Myer, G. E., Stewart, R., Walther, E. G., "On the Mechanism of Langmuir Circulations and Their Role in Epilimnion Mixing," *Limn. and Oceanog.* (1969) **14**, 493-503.
47. Faller, A. J., Woodcock, A. H., "The Spacing of Windrows of *Sargassum* in the Ocean," *J. Mar. Res.* (1964) **22**, 22-29.
48. Faller, A. J., "Oceanic Turbulence and the Langmuir Circulations," *Ann. Rev. Ecology and Systematics* (1971) **2**, 201-236.
49. Redfield, A. C., "The Exchange of Oxygen Across the Sea Surface," *J. Mar. Res.* (1948) **7**, 347-361.
50. Medwin. H., "*In Situ* Acoustic Measurements of Bubble Populations in Coastal Ocean Waters," *J. Geophys. Res.* (1970) **75**, 599-611.
51. Blanchard, D. C., "The Electrification of the Atmosphere by Particles from Bubbles in the Sea," *Prog. Oceanog.* (1963) **1**, 71-202.
52. Monahan, E. C., "Oceanic Whitecaps," *J. Phys. Oceanog.* (1971) **1**, 139-144.
53. Blanchard, D. C., "Whitecaps at Sea," *J. Atmos. Sci.* (1971) **28** ,645.
54. Woodcock, A. H., "Subsurface Pelagic *Sargassum*," *J. Mar. Res.* (1950) **9**, 77-92.
55. Gaines, G. L., Jr., "Insoluble Monolayers at Liquid–Gas Interfaces," 386 pp., John Wiley, 1966.
56. Fox, F. E., Herzfeld, K. F., "Gas Bubbles with Organic Skins as Cavitation Nuclei," *J. Acoust. Soc. Amer.* (1955) **26**, 984-989.
57. Baylor, E. R., Sutcliffe, W. H., Jr., Hirschfeld, D. S., "Adsorption of Phosphates onto Bubbles," *Deep-Sea Res.* (1962) **9**, 120-124.
58. Hood, D. W., Park, K., Prescott, J. M., "Organic Matter in Sea Water; Amino Acids, Fatty Acids, and Monosaccharides from Hydrolysates," Abstract: *Bull. Geol. Soc. Amer.* (1960) **71** ,1890.
59. Baylor, E. R., Sutcliffe, W. H., Jr., "Dissolved Organic Matter in Sea Water as a Source of Particulate Food," *Limn. and Oceanog.* (1963) **8**, 369-371.
60. Wheeler, J., "Some Effects of Solar Levels of Ultraviolet Radiation on Lipids in Artificial Sea Water," *J. Geophys. Res.* (1972) **77**, 5302-5306.
61. Sutcliffe, W. H., Jr., Sheldon, R. W., Prakash, A., Gordon, D. C., Jr., "Relations between Wind Speed, Langmuir Circulation and Particle Circulation in the Ocean," *Deep-Sea Res.* (1971) **18**, 639-643.

62. Riley, G. A., "Organic Aggregates in Sea Water and the Dynamics of Their Formation and Utilization," *Limn. and Oceanog.* (1963) **8**, 372-381.
63. Riley, G. A., Van Hemert, D., Wangersky, P. J., "Organic Aggregates in Surface and Deep Waters of the Sargasso Sea," *Limn. and Oceanog.* (1965) **10**, 354-363.
64. Menzel, D. W., "Bubbling of Sea Water and the Production of Organic Particles: A Re-evaluation," *Deep-Sea Res.* (1966) **13**, 963-966.
65. Batoosingh, E., Riley, G. A., Keshwar, B., "An Analysis of Experimental Methods for Producing Particulate Organic Matter in Sea Water by Bubbling," *Deep-Sea Res.* (1969) **16**, 213-219.
66. Stommel, H., "Trajectories of Small Bodies Sinking Slowly Through Convection Cells," *J. Mar. Res.* (1949) **8**, 24-29.
67. Gordon, D. C., Jr., "A Microscopic Study of Organic Particles in the North Atlantic Ocean," *Deep-Sea Res.* (1970) **17**, 175-185.
68. Gordon, D. C., Jr., "Some Studies on the Distribution and Composition of Particulate Organic Carbon in the North Atlantic Ocean," *Deep-Sea Res.* (1970) **17**, 233-243.
69. Riley, G. A., "Particulate Organic Matter in Sea Water," *Advan. Mar. Biol.* (1970) **8**, 1-118.
70. Sheldon, R. W., Evelyn, T. P. T., Parsons, T. R., "On the Occurrence and Formation of Small Particles in Sea Water," *Limn. and Oceanog.* (1967) **12**, 367-375.
71. Chave, K. E., "Carbonates: Association with Organic Matter in Surface Sea Water," *Science* (1965) **148**, 1723-1724.
72. Boyce, S. G., "Source of Atmospheric Salts," *Science* (1951) **113**, 620-621.
73. MacIntyre, F., "Flow Patterns in Breaking Bubbles," *J. Geophys. Res.* (1972) **77**, 5211-5228.
74. Jacobs, W. C., "Preliminary Report on a Study of Atmospheric Chlorides," *Mon. Wea. Rev.* (1937) **65**, 147-151.
75. Woodcock, A. H., "Salt Nuclei in Marine Air as a Function of Altitude and Wind Force," *J. Meteor.* (1953) **10**, 362-371.
76. Woodcock, A. H., Kientzler, C. F., Arons, A. B., Blanchard, D. C., "Giant Condensation Nuclei from Bursting Bubbles," *Nature* (1953) **172**, 1144.
77. Stuhlman, O., "The Mechanics of Effervescence," *Physics* (1932) **2**, 457-466.
78. Hayami, S., Toba, Y., "Drop Production by Bursting of Air Bubbles on the Sea Surface (1) Experiments at Still Sea Water Surface," *J. Ocean. Soc. Japan* (1958) **14**, 145-150.
79. Blanchard, D. C., "Bursting of Bubbles at an Air-water Interface," *Nature* (1954) **173**, 1048.
80. MacIntyre, F., "Bubbles: a Boundary-layer 'Microtome' for Micron-thick Samples of a Liquid Surface," *J. Phys. Chem.* (1968) **72**, 589-592.
81. Day, J. A., "Production of Droplets and Salt Nuclei by the Bursting of Air Bubble Films," *Quart. J. Roy. Met. Soc.* (1964) **90**, 72-78.
82. Paterson, M. P., Spillane, K. T., "Surface Films and the Production of Sea-Salt Aerosol," *Quart. J. Roy. Met. Soc.* (1969) **95**, 526-534.
83. Garrett, W. D., "The Influence of Monomolecular Surface Films on the Production of Condensation Nuclei from Bubbled Sea Water," *J. Geophys. Res.* (1968) **73**, 5145-5150.
84. Langmuir, I., Schaefer, V. J., "The Effect of Dissolved Salts on Insoluble Monolayers," *J. Amer. Chem. Soc.* (1937) **59**, 2400-2414.
85. Blanchard, D. C., Syzdek, L. D., "Concentration of Bacteria in Jet Drops from Bursting Bubbles," *J. Geophys. Res.* (1972) **77**, 5087-5099.
86. MacIntyre, F., "Geochemical Fractionation During Mass Transfer from Sea to Air by Breaking Bubbles," *Tellus* (1970) **22**, 451-461.

87. Seto, F. Y. B., Duce, R. A., "A Laboratory Study of Iodine Enrichment on Atmospheric Sea-salt Particles Produced by Bubbles," *J. Geophys. Res.* (1972) **77**, 5339-5349.
88. Glass, S. J., Jr., Matteson, M. J., "Ion Enrichment in Aerosols Dispersed from Bursting Bubbles in Aqueous Salt Solutions," *Tellus* (1973) **25**, 272-280.
89. *J. Geophys. Res.* (1972) **77**, 5059-5349.
90. Lemlich, R., "Adsubble Processes: Foam Fractionation and Bubble Fractionation," *J. Geophys. Res.* (1972) **77**, 5204-5210.
91. Lemlich, R., Ed., "Adsorptive Bubble Separation Techniques," 331 pp., Academic, 1972.
92. Grieves, R. B., in "Adsorptive Bubble Separation Techniques," pp. 191-197, R. Lemlich, Ed., Academic, 1972.
93. Wallace, G. T., Jr., Loeb, G. I., Wilson, D. F., "On the Flotation of Particulates in Sea Water by Rising Bubbles," *J. Geophys. Res.* (1972) **77**, 5293-5301.
94. Carlucci, A. F., Bezdek, H. F., "On the Effectiveness of a Bubble for Scavenging Bacteria from Sea Water," *J. Geophys. Res.* (1972) **77**, 6608-6610.
95. Lee, T. M. S., "An Investigation of the Effect of Bubble Age and Organic Contamination on the Ejection Height and Size of Jet Drops," Master's thesis, State Univ. of N.Y., Albany, 81 pp., 1972.
96. Blanchard, D. C., "Surface Active Organic Material on Airborne Salt Particles," *Proc. Int. Conf. Cloud Physics, Toronto* (1968) 24-29.
97. Barger, W. R., Garrett, W. D., "Surface Active Organic Material in the Marine Atmosphere," *J. Geophys. Res.* (1970) **75**, 4561-4566.
98. Blanchard, D. C., Syzdek, L. D., "Variations in Aitken and Giant Nuclei in Marine Air," *J. Phys. Oceanog.* (1972) **2**, 255-262.
99. Eriksson, E., "The Yearly Circulation of Chloride and Sulfur in Nature; Meteorological, Geochemical and Pedological Implications, Part 1," *Tellus* (1959) **11**, 375-403.
100. Baier, R. E., "Organic Films on Natural Waters: Their Retrieval, Identification, and Modes of Elimination," *J. Geophys. Res.* (1972) **77**, 5062-5075.
101. Blanchard, D. C., "The Borderland of Burning Bubbles," *Sat. Rev.*, January 1, 1972, 60-63.
102. Timmons, C. O., "Stability of Plankton Oil Films to Artificial Sunlight," NRL Report 5774, 8 pp., U.S. Naval Research Laboratory, 1962.
103. Mason, B. J., "The Physics of Clouds," 671 pp., Oxford Univ. Press, 1971.
104. Squires, P., "The Microstructure and Colloidal Stability of Warm Clouds. Part II. The Causes of the Variations in Microstructure," *Tellus* (1958) **10**, 262-271.
105. Woodcock, A. H., "Smaller Salt Particles in Oceanic Air and Bubble Behavior in the Sea," *J. Geophys. Res.* (1972) **77**, 5316-5321.
106. Garrett, W. D., "Retardation of Water Drop Evaporation with Monomolecular Surface Films," *J. Atmos. Sci.* (1971) **28**, 816-819.
107. Kocmond, W. C., Garrett, W. D., Mack, E. J., "Modification of Laboratory Fog with Organic Surface Films," *J. Geophys. Res.* (1972) **77**, 3221-3231.
108. Woodcock, A. H., "Atmospheric Salt Particles and Raindrops," *J. Meteorol.* (1952) **9**, 200-212.
109. Woodcock, A. H., Blanchard, D. C., "Tests of the Salt-nuclei Hypothesis of Rain Formation," *Tellus* (1955) **7**, 437-448.
110. Woodcock, A. H., Duce, R. A., Moyers, J. L., "Salt Particles and Raindrops in Hawaii," *J. Atmos. Sci.* (1971) **28**, 1252-1257.

111. Duce, R. A., Winchester, J. W., Van Nahl, T. W., "Iodine, Bromine, and Chlorine in the Hawaiian Marine Atmosphere," *J. Geophys. Res.* (1965) **70**, 1775-1799.
112. Duce, R. A., Woodcock, A. H., Moyers, J. L., "Variation of Ion Ratios with Size among Particles in Tropical Oceanic Air," *Tellus* (1967) **19**, 369-379.
113. Moyers, J. L., Duce, R. A., "Gaseous and Particulate Iodine in the Marine Atmosphere," *J. Geophys. Res.* (1972) **77**, 5229-5238.
114. Garrett, W. D., "Surface-chemical Modification of the Air/Sea Interface," *Ann. Meteorol.* (1969) **22**, 25-29.
115. "Encyclopedia of Oceanography," Rhodes Fairbridge, Ed., Litton Educational Publishing, 1966.

RECEIVED December 4, 1973. Work was supported by the Atmospheric Sciences Section, National Science Foundation, NSF Grant GA-23413.

INDEX

The text of this book is set in 10 point Caledonia with two points of leading. The chapter numerals are set in 30 point Garamond; the chapter titles are set in 18 point Garamond Bold.

The book is printed offset on Danforth 550 Machine Blue White text, 50-pound. The cover is Joanna Book Binding blue linen.

Jacket design by John Sinnett.
Editing and production by Spencer Lockson.

The book was composed by the Mills-Frizell-Evans Co., Baltimore, Md., printed and bound by The Maple Press Co., York, Pa.